FALLOUT

FALLOUT

Conspiracy, Cover-Up, and the Deceitful Case for the Atom Bomb

PETER WATSON

PUBLICAFFAIRS

New York

PublicAffairs
Hachette Book Group
1290 Avenue of the Americas
New York, NY 10104
www.publicaffairsbooks.com
@Public_Affairs

Printed in the United States of America

Originally published in hardcover and ebook by Simon & Schuster UK Ltd, in September 2018

First Edition: September 2018

Published by PublicAffairs, an imprint of Perseus Books, LLC, a subsidiary of Hachette Book Group, Inc. The PublicAffairs name and logo is a trademark of the Hachette Book Group.

Typeset in the UK by M Rules

Library of Congress Cataloging-in-Publication Data has been applied for.

ISBNs: 978-1-61039-961-6 (hardcover), 978-1-61039-962-3 (ebook)

LSC-C

10 9 8 7 6 5 4 3 2 1

'I realized then [1941] that a bomb was not only possible – it was inevitable.'

– Sir James Chadwick,
the most senior British scientist in the Manhattan Project,
speaking to an interviewer shortly before his death in 1974

'The Allies won [the Second World War] because our German scientists were better than their German scientists.'

– Sir Ian Jacob,
military secretary to Winston Churchill

'Atomic weapons can hardly be used without spelling the end of the world.'

– Joseph Stalin

'When you see something that is technically sweet, you go ahead and do it and you argue about it only after you have had your technical success. That is the way it was with the atomic bomb.'

– J. Robert Oppenheimer,
scientific director of Los Alamos

'Don't bother me with your conscientious scruples. After all, the thing's superb physics.'

– Enrico Fermi,
creator of the first sustainable nuclear chain reaction

'The first atomic bomb was an unnecessary experiment ... [the scientists] had this toy and they wanted to try it out, so they dropped it.'

– Admiral William 'Bull' Halsey,
commander of the Third Fleet

'Chadwick's discovery of the neutron marked the unintentional first step towards man's loss of innocence in the field of nuclear energy.'

– Andrew Brown,
James Chadwick's biographer

'When the righteous sin, they add the force of their virtue to all the evil that they do.'

– Lewis Mumford,
paraphrasing Ezekiel 18:24

Fallout

1. Radioactive particles that are carried into the atmosphere after a nuclear explosion and gradually fall back as dust or in precipitation.
2. The adverse results of a situation or action.

Contents

PART THREE

Lives in Parallel: Niels Bohr and Klaus Fuchs

PART FOUR

The Underestimate of the Russians

Photo insert is located between pages 212 and 213

Preface

Cover-Up: 'When the Righteous Sin'

Albert Einstein's famous realisation that $E=mc^2$, that matter and energy are essentially different aspects of the same phenomenon, may be the most rapidly consequential idea in human history. He first published his theory of nuclear energy in May 1905, refining it – with the help of others – until 1917 in the middle of the First World War. Twenty-eight years later – a single generation in human terms – on 6 August 1945, Hiroshima and Nagasaki were each destroyed by an atomic bomb, bringing an end to the Second World War.

History shows that while ideas *can* be consequential – the Renaissance, the Reformation, the Scientific and Romantic revolutions were real enough – it is not always so easy to identify the exact role they played. What were the intellectual origins of the French Revolution? Why did a Marxist revolution occur in Russia when Marx himself expected it to happen in England? Why did modernism emerge in France first – if, indeed, it did?

But with atomic energy – nuclear energy – the chronology is known with an exactitude that is rare in the history of ideas. Beginning in 1898 with the identification of the electron, soon followed by the discovery of the structure of the atom in 1907, and then the realisation of the existence – and importance – of the neutron in 1932, giving rise to the possibility of a chain reaction, the pieces of this jigsaw were put together in a remarkably swift period, the 'heroic age of physics' as Ernest Rutherford, the director of the Cavendish Laboratory in Cambridge, called it.

As someone who has published several books on the history of ideas, this crowded chronology has always fascinated me. Reading into the subject, however, it soon became clear that the story also involves a highly dramatic – even uniquely dramatic – human dimension. The heroic age of physics during the interwar years comprised an elite community of no more than a few dozen physicists, chemists and mathematicians from a limited number of nationalities – British, German, French, American, Danish, Italian, Russian and Japanese – who all knew each other. They studied at the same small number of European institutions in Berlin, Cambridge, Copenhagen and Göttingen, worked together, attended the same conferences, holidayed together, attended each other's weddings, published their work in the same small number of professional journals, co-operating and competing in an impressive array of scientific advances, that were recognised by the award of numerous Nobel Prizes. A case can be made for saying that the 1920s and 1930s were the most exciting and consequential decades not just in physics but in all of science.[1]

But the 1920s and 1930s were notable for something else, no less consequential and no less dramatic: the rise of Nazism in Germany and Fascism in Italy.

From the discovery of the neutron in the year before Hitler came to power in Berlin, some scientists were aware of the theoretical possibility that the massive energy wrapped up in the nucleus of the atom could be unlocked, but hoped against hope that it would never prove practical. And then, over Christmas/New Year 1938/9, on the very eve of war, four scientists from Germany confirmed that they had split – fissioned – the nucleus of uranium, the heaviest element in the periodic table and the most unstable. The possibility of nuclear weapons came a frightening step nearer.

One German scientist, Werner Heisenberg, arguably the most brilliant of all, who had won the Nobel Prize in 1932, aged thirty-one, was much later to say that had a handful of physicists got together in 1939 and refused to do any more work on nuclear weapons, there was nothing the politicians could have done, and the whole nuclear arms race would have been stillborn.[2]

Instead, that year a very small number of highly trained individuals suddenly found themselves in possession of knowledge and skills that could, at least in theory, decide the outcome of war, should it come to pass, which seemed increasingly likely. What could be more dramatic than that – an idea so powerful that it could determine the outcome of a world war? Einstein himself was distraught.

Six years later Einstein's remarkable insight was fully realised. On Monday 16 July 1945 at 5.29 a.m., the first test of an atomic bomb was successfully carried out at the so-called Trinity site in the Alamogordo desert in New Mexico. Though he was told the noise had been heard in three states, Leslie Groves, the general in charge of the project, insisted the test be kept secret. 'Can you give us an easy job, general,' an aide remarked, 'like hiding the Mississippi River?' The very next day, President Harry Truman sat down for his first and only face-to-face meeting with the Soviet premier Joseph Stalin in a suburb of Berlin near Potsdam.[3]

Three weeks later, on 6 August, also a Monday, Einstein's idea was again deployed when Hiroshima was bombed. Two days after that, on 8 August, the Soviet Union declared war on Japan. The very next morning, the Red Army's tanks rolled across the Manchurian border.

These events were hardly coincidences. In recent years, the view has become established – among historians at least if not yet the general public – that the bombs dropped on Hiroshima and Nagasaki were not needed to end the Second World War but had another purpose.

In some ways it is surprising that this view is not more widespread. One early distinguished sceptic about the bomb's use was General Dwight D. Eisenhower, supreme commander of the Allied Expeditionary Force, who directed Allied operations in Europe against Hitler, and was later president of the United States. At the height of the Cold War, and soon after his famous farewell address as president, in which he drew attention to the risks posed by the 'military-industrial complex', he recalled the moment when the secretary of war, Henry Stimson, told him that the atomic bomb was to be used against Japanese cities:

> During his recitation of the relevant facts, I had been conscious of a feeling of depression and so I voiced to him my grave misgivings, first on the basis of my belief that Japan was already defeated and that dropping the bomb was completely unnecessary, and secondly because I thought that our country should avoid shocking world opinion by the use of a weapon whose employment was, I thought, no longer mandatory as a measure to save American lives. It was my belief that Japan was, at that very moment, seeking some way to surrender with a minimum loss of 'face'.[4]

Admiral William 'Bull' Halsey's Third Fleet was meeting almost no resistance as it bombarded Japanese coastal installations, and Admiral Wagner, in charge of air search-and-patrol, found that in all the millions of square miles on the East Asian seas and coasts 'there was literally not a single target worth the powder to blow it up'. Halsey later, echoing Eisenhower, said: 'The first atomic bomb was an unnecessary experiment ... It was a mistake to ever drop it ... [the scientists] had this toy and they wanted to try it out, so they dropped it ... It killed a lot of Japs, but the Japs had put out a lot of peace feelers through Russia long before.'[5]

Perhaps even more telling, less than a year after the bombings an extensive official study by the US Strategic Bombing Survey published its conclusion that 'Japan would likely have surrendered in 1945 without atomic bombing, without a Soviet declaration of war, and without an American invasion.'[6]

Leo Szilard, a Hungarian Jewish émigré who had escaped the Third Reich by the skin of his teeth and had later been the first to conceive the idea of the nuclear chain reaction, had met with James F. Byrnes, President Truman's personal representative on atomic matters and subsequently secretary of state, at his home in Spartanburg, South Carolina in May 1945. In a memoir, Szilard wrote:

> Mr. Byrnes did not argue that it was necessary to use the bomb against the cities of Japan in order to win the war. He knew at the time, as the rest of the government knew, that Japan was essentially defeated and that we could win the war in another

> six months. At that time Mr. Byrnes was much concerned about the spread of Russian influence in Europe . . . [Mr. Byrnes's view was] that our possessing and demonstrating the bomb would make Russia more manageable in Europe.[7]

This much was obvious to the Russians at the time. To Vyacheslav Molotov, Soviet foreign minister during the whole of the Second World War, the two bombs 'were not aimed at Japan but rather at the Soviet Union. They [the Americans] said, bear in mind you don't have an atomic bomb and we do, and this is what the consequences will be like if you make the wrong move!'[8]

These and other conclusions and observations have led a raft of mainly American scholars to examine the whole process of the decision to use the bomb in more detail as more documents have been declassified and become available. The general consensus is now clear. The decision to use the bomb against Japan, in August 1945, was indeed unnecessary – the Japanese were ready to surrender once an agreed face-saving form of words was found that would let them keep the emperor as a constitutional monarch. (This was not popular in America – a third of the respondents in an opinion poll wanted him executed immediately.) It is also agreed that the *primary* reason for dropping the bomb was to end the war before Russia could come into the Far East as a belligerent and thereby make several territorial demands, *and* as a show of superior force designed to impress the nuclear advantage the Western Allies had over the Soviet Union, so that her behaviour would be more amenable to Western interests in the immediate post-war period.[9]

This later research also shows that the decision to use the bomb was made by a small number of individuals, some of whom tried to cover over their real reasons for acting as they did ('clear evidence of outright lying'), advocating instead a pretence that the bombs were dropped to save American and Japanese lives.[10]

These unedifying manoeuvrings that led to the decision to use the atomic bomb on Japan are not, however, the main focus of this book.

Instead it covers an earlier period of the war when the original reason for building a bomb – because it was thought Hitler's scientists were working towards one – was discovered to have no basis in reality. But this knowledge was not properly assimilated, still less shared by the intelligence services, and was then covered up by some of the same people who misled the world about the bombing of Hiroshima and Nagasaki. Using a close reading of recently released archives in different countries – Britain, the United States, Germany, Denmark, Russia – this book presents a new chronology, or a new narrative, of the construction of the bomb and shows how, if important intelligence on atomic matters had been quickly shared, as it surely should have been, the atomic bomb need never have been built in the first place, nor the world thrust into the threatening and precarious balancing act that we still inhabit. Not everyone felt as James Chadwick did, that once a bomb became possible, it also became inevitable. Errors were made – and lies were told – to bring us a weapon that was not needed.

At the heart of this story are two individuals – Niels Bohr and Klaus Fuchs – who, each in his own very different way, foresaw how the bomb threatened to change the post-war world and sought to do something about it. One failed, the other succeeded.

This book faces squarely the fact that, for some people in this narrative, once it became clear that a bomb *could* be built, they ensured that it *would* be built. Bohr and Fuchs both feared this inevitability, but they also knew that, in wartime, more than at other times, chronology is crucial. In wars, events – life-threatening events – follow on one another very quickly and important decisions with momentous consequences have to be taken rapidly. In such circumstances, as this book shows, even the seemingly inevitable is not necessarily inevitable.

The history of atomic bomb wartime intelligence – which is what this book essentially is – presents us with the unmistakable conclusion that a series of momentous mistakes were made, and lies told, by the French, by the Germans, by the British and by the Americans, with the result that the world stumbled, even blundered, unnecessarily into the nuclear age. A world war was raging, the right hand very often didn't know what the left was doing, individuals were overworked in life-and-death situations, very few had access to all the information they ideally needed. Nonetheless, from the latest evidence we now have, we can conclude that, with different personnel in certain key positions, and if they had *shared* what intelligence they had, they may well have concluded that there was no need to build an atomic bomb and we would all have avoided the knife-edge that we now call peace.

The place of nuclear weapons in our lives continues to be as nerve-racking as ever. More than seventy years after Hiroshima we seem no nearer to controlling the use or even the spread of these fearsome weapons. Worldwide there are now 9,500 nuclear warheads. According to scientists this is sufficient to destroy the planet 100 times over.[11] Our predicament is as absurd as it is dangerous and, with recent developments in Iran and North Korea in mind, it risks becoming even more so.

In setting out this new chronology of the way the atomic bomb came into being, which in certain important respects is at variance with the orthodoxy, I am wary of deriving too-easy lessons from it. The world has moved on.

There is one observation worth making, however, because it underlines the seriousness of the new situation we now face.

The main characters in this story – Presidents Roosevelt and Truman, Vannevar Bush and General Leslie Groves, who between them helped to create and then manage the atom bomb project on the American side, and Prime Minister Winston Churchill, Chancellor of the Exchequer Sir John Anderson and James Chadwick, the discoverer of the neutron, on the British side – were all sophisticated men of the world, mature individuals, highly intelligent, vastly experienced, and extremely well informed, with many practical achievements to their name that improved millions of lives. Compared with our leaders today, they were giants.

Nevertheless, they got the atom bomb wrong. Between them, and for what they convinced themselves were the highest motives, they maladministered us into a world that could have been avoided.

And this brings us back to Heisenberg's argument: had the scientists known what the intelligence agencies and their political masters knew in the run-up to Hiroshima, would they have agreed to take part in the building of the bomb? Readers will draw their own conclusions from the evidence presented here. The atom bomb disaster is an instructive story but it underlines above all the fact that, now as then, the chain reactions between people are even more important than the immense forces of nuclear physics.

PART ONE

Incognito: Klaus Fuchs and Niels Bohr

1

Zigzag

Friday 3 December 1943. Shortly after 7 a.m., just as it was getting light, the 25,000-ton RMS *Andes* nosed into the James River as it flowed into Chesapeake Bay and the western reaches of the Atlantic. Squally weather, with showers coming and going. Bound for Newport News, Virginia, the *Andes* was a fairly new ship, originally intended as the jewel in the crown of the Royal Mail Line, which, before the war, had thirty-one vessels distributing British letters and parcels around the world. She had come into service in 1939, on the eve of the Second World War, and as a result had all her luxurious trappings, including an art deco bar, ripped out before they could be used, converting her into a troop ship. Instead of ferrying 600 well-heeled paying passengers, her bread and butter now was transporting 4,000 troops at a time from the United States to Britain.

On the return journey she was normally almost empty. This time, however, she had a very small but extremely precious – not to say highly secret – cargo: a score and more scientists, physicists, chemists and mathematicians who were being hurried to America to take part in the biggest secret of the war – the development and construction of the atomic bomb. They included the chemist Christopher Frank Kearton, a Cheshire man, the son of a bricklayer, and the mathematician Tony Skyrme, a Londoner, Eton-educated, together with three enemy aliens, Germans no less, who had arrived in Britain as refugees from Hitler's murderous adventures: Rudolf Peierls, Otto Frisch and Klaus Fuchs.

The crossing had taken twice as long as transatlantic crossings normally did. The *Andes* was not accompanied on her voyage by any naval escort; she relied on her speed to get her out of trouble. Even so, she had been forced to take a zigzag course, a more than apt metaphor for the narrative we are to follow.

The scientists had been allowed to use the first-class cabins, though even these had been converted to house eight bunks each. The ship's grand piano had been locked away for the duration, a great disappointment in particular to Otto Frisch, newly naturalised as a Briton only a day before he had set sail from Liverpool, and an accomplished, near concert-level musician. He had had to make do with an old, rickety upright instrument, in what had been the ballroom, chained to a pillar to stop it rolling around in rough weather.[1]

3 December 1943 was an important date in the Second World War, quite apart from the arrival of the *Andes* in America. That day the first news had been released about the Teheran Conference of the 'Big Three' – Franklin Roosevelt, Winston Churchill and Joseph Stalin – which had in fact taken place some days earlier but kept secret for obvious reasons. It was the first time the three leaders had met face to face and important decisions had been taken about the future conduct of the war. Hardly less important, the United States announced that same day that in the previous month, November, the country had turned out no fewer than 8,789 aircraft, one entire plane slightly more often than every five minutes. The war was in full swing.[2]

The transfer of so many British scientists to America was an important development. Although, as we shall see, the British had been the first to realise the possibility of an atomic bomb, by the end of 1943 the United States was in the driving seat. Quite apart from its undoubted greater resources (as the production of those aircraft showed), any project carried out in Britain always risked being bombed.

The *Andes* docked at Newport News later that morning. The city, originally known for its export of coal and what was once the largest shipbuilding yard and dry dock in the world, was now a major naval base, well protected from attack by the coastal configuration. From Newport, the scientists were taken by train to Washington, changing at Richmond. During the stopover at Richmond, Frisch wandered off into the streets nearby. He was a tall, handsome Austrian.

In the Richmond streets, 'I was greeted by a completely incredible spectacle: fruit stalls with pyramids of oranges . . . After England's blackout, and not having seen an orange for a couple of years, that sight was enough to send me into hysterical laughter.'[3]

Rudolf Peierls found the train ramshackle and crowded. A small round-faced man, with a puckish face, heavy-duty spectacles and buck teeth, he had a Russian wife, Genia, whom he had met at a physics conference in Odessa in the early 1930s. She went in search of better accommodation and came back to say that she had found an almost empty car, 'in which there were only two very nice Negroes'.[4] To her disappointment and horror, she was told they were in the South of the United States, where transport was still segregated.

In Washington, they had to wait for several days before they could be briefed by General Leslie Groves. Groves was the military commander of the Manhattan Project, as the attempt to build an atomic weapon was known.

When the briefing with the British scientists finally took place, Groves introduced them to his concept of 'compartmentalisation': in order to maintain total secrecy, each specialist would be allowed to know only what was happening inside that speciality – almost no one would have an overall picture. While this made sense to the military mind – and Groves was in general regarded as an excellent commander – many of the scientists thought that compartmentalisation was impractical, that scientists needed to know the wider picture in order to do their job. It would be a bone of contention throughout the rest of the war. James Chadwick, the most senior British physicist in the Manhattan Project, thought it 'bogus', while Leo Szilard, a Hungarian émigré physicist, thought it delayed the development of the bomb by as much as a year.

Groves was widely admired but not liked. Edward Teller, another Hungarian émigré, thought that 'he could have won almost any unpopularity contest'. And he was an uncompromising Anglophobe, with a ready suspicion of all foreigners. He firmly believed that Americans were more moral people than anyone else and he accepted the presence on United States soil of Frisch, Peierls, Fuchs and the others only reluctantly, compelled by orders from above. This Anglophobia would come to matter.[5]

At that first meeting, Groves informed the scientists that they would be going to one of two places. Some of them, and this included Frisch, would be going to Los Alamos in the New Mexico desert, where the bomb would eventually be assembled. At the time, in the spirit of compartmentalisation, Los Alamos was known only as 'Site Y'. Rudolf Peierls had been there on an earlier visit but Klaus Fuchs and the others, who were being sent to New York, did not find out about the actual Los Alamos site for some months.

Peierls and Fuchs were sent to New York because of their expertise in isotope separation. The isotope separation theoretical work was being run in Manhattan by the Kellex Corporation, a subsidiary of a firm of civil engineers. It was set up especially for the building of the isotope separation plant, which was located somewhere in the South although, because of compartmentalisation, neither Peierls nor Fuchs was told at first where it was.[6]

Peierls and his wife, with Fuchs, stayed at first in the Barbizon Plaza Hotel overlooking Central Park for a couple of weeks, then found an apartment on Riverside Drive. Fuchs found a flat at 128 West 77th Street in a 'walk-up', four-storey 'brownstone', as converted houses are known in Manhattan.

Everyday life in New York was in all ways better than it had been in Britain. Peierls and Fuchs and the others were not badly paid, and they could take full advantage of the more abundant life in America. There was some rationing and a 'brown out' in coastal cities, which restricted street lighting. But it was nothing like in Britain, which was austere and where cities were fully blacked out at night. It was not just food and drink and clothes that were more plentiful: the Broadway theatre was flourishing (*Porgy and Bess*, *The Student Prince*, *Carmen Jones*), there was an active night life and classical music nightly, at which Fuchs was a regular.

There were fifteen British scientists working in New York on the Manhattan Project. Not all of them were scheduled to remain in America for any length of time but Fuchs (like Peierls and Frisch) was and so one of the directors of 'Tube Alloys', the code word for Britain's secret atomic bomb project, asked MI5 for a summary of anything that was known about him, since it would be embarrassing to say the least if he was not what he seemed. MI5 knew that Fuchs had been a communist in the past but replied that he was not now politically active and there had been nothing 'objectionable' about his behaviour in Britain.

Unbeknown to Peierls, however, and to the other physicists labouring on the bomb in New York, Fuchs was working in a way that his colleagues would certainly have regarded as objectionable had they known about it. Since August 1941 he had been a Russian spy.[7]

On Monday 6 December, just three days after the *Andes* had arrived in Newport News, another ship, the RMS *Aquitania*, four-funnelled and painted battleship grey, docked in New York Harbour. She too was fast enough to cross the ocean unescorted. The *Aquitania* had many passengers, including two scientists travelling incognito, whose very presence on board was a closely guarded secret.

The names on their British passports showed Nicolas and James Baker but they were in fact the Danes, Niels Bohr and his son Aage. Along with Albert Einstein, Niels Bohr was one of the two most famous physicists in the world at the time. In terms of the atomic bomb, Bohr was even more important than Einstein, because the latter had not taken much of an active role in the development of what some physicists were privately calling 'the gadget', whereas Bohr played a central role in the understanding of nuclear fission, the way the nucleus of an atom may be split, releasing vast amounts of explosive energy.

Moreover, Bohr was arguably the most distinguished physicist of all time on account of the role his institute in Copenhagen fulfilled in the years between the First and Second World Wars. After Bohr won the Nobel Prize for physics in 1922, for explaining the all-important arrangement of electrons around the nucleus of the atom, which showed how physics governed chemical properties and how the two disciplines are intimately linked, he was given his own institute in Copenhagen, and physicists of almost every nationality (British, American, German, Dutch, Swedish, Austrian, Italian, French, Japanese, Russian) congregated there. He was generous, avuncular and completely devoid of those instincts for rivalry that can so easily sour relations. The success of Copenhagen also had something to do with the fact that Denmark was a small country, where national rivalries could be quietly forgotten.

Bohr was always much more than just a scientist. After 30 January 1933, when Hitler became chancellor of Germany, Bohr had watched subsequent developments with a sense of acute unease. He knew Germany, had been there often, spoke its language, had no illusions and knew the day was not far off when action would be needed to help German scientists.[8] And in fact, not long after Hitler had taken power, Bohr went to Germany, officially on a visit to universities, but secretly checking on the security of scientists and gauging how many were likely to be dismissed under the new racial laws. One of the first people he was able to help was Otto Frisch in Hamburg. Frisch was the young nephew of Bohr's long-standing friend and colleague Lise Meitner and, like her, an Austrian Jew. As Austrians, until the Anschluss in 1938, Frisch and Meitner were protected from the vicious Nazi racial laws.

Bohr took a great interest in Frisch's work, on the energy locked up in sodium atoms, and in the course of their encounter he held Frisch by his vest and whispered that he hoped the young man would join him in Copenhagen. As Frisch wrote to his mother that same night, 'The Good Lord himself has just taken me by the waistcoat button and smiled at me.'[9]

Bohr travelled on through Germany as the Nazis swiftly tightened their grip, letting it be known that Copenhagen would be a refuge for those who needed it. And not just for Jews. Among others, Bohr told Max Planck, who had won the Nobel Prize for his discovery of the quantum and was president of the Kaiser Wilhelm Society in Berlin, Georg Placzek, a Czech authority on the neutron in Leipzig, and Georg von Hevesy, a Hungarian Nobel Prize-winning radiochemist in Freiburg, that they could continue their scientific work in Copenhagen, should they need to.[10]

Even this wasn't the end of it. In September 1938 Bohr's institute held its annual seminar, but that year, with war looming, attendance was very poor. By then Bohr had spoken out against the Nazis more than once, so that very few physicists felt confident enough to join him, even for a few days.

One who did attend, however, was Enrico Fermi, from Italy. Fermi was a Rome-based physicist who had discovered 'Beta decay', a process by which some elementary particles can change from one to the other, and the idea of 'weak interaction', a new type of sub-atomic force that causes radioactive decay. During the seminar Bohr quietly drew Fermi aside and broke protocol, telling him that the Italian was under consideration for that year's Nobel Prize. Normally such a breach would have been unthinkable but these weren't normal times and Bohr was well aware that Italy had stipulated that all its citizens must convert their foreign monetary holdings into *lire*. Bohr asked Fermi if he would prefer his name to be withdrawn until the Nobel Prize money could be used without restriction.

Faced with this dilemma, Fermi came clean and told Bohr plainly that he and his family wanted to leave Italy. And how good his instincts were. On the very day that Fermi's Nobel award was duly announced, it was also broadcast that racial laws were being introduced in Italy. Jewish children were excluded from public schools, Jewish teachers were dismissed, Jewish firms were dissolved, Jews were to have their passports withdrawn. Fermi's wife, Laura, was Jewish.

In November, the entire family managed to get to Stockholm for the Nobel ceremony, permitted to travel because they were bringing honour to Italy. But then, after the awards, instead of returning to Rome, the Fermis, with their children and nursemaid, went to Copenhagen to stay with the Bohrs until they could sail for America.

On the morning that the *Aquitania* docked in New York, the news that day was still dominated by the Teheran Conference, where Churchill had celebrated his sixty-ninth birthday and presented Stalin with the 'Sword of Stalingrad', to be passed on to the people of the defiant city who had withstood the tremendous siege there. In return Stalin had raised a toast to 'my fighting friend'.

As soon as the *Aquitania* tied up, several members of the security services came aboard, looking for Nicholas and James Baker. In New York, Bohr and his son were accorded special treatment and did not have to go through the normal immigration procedure. Their luggage was taken from them and hustled away. Although some of the most important papers on nuclear processes had been published on the eve of war – papers by Bohr himself, by Americans and by French scientists – since the fighting had started the professional journals had gone very quiet: no one was publishing anything. Therefore, should it leak out that Bohr, who had disappeared into the night from Stockholm, was in America, the Germans, Russians and Japanese would realise that something important was going on there.

Bohr was entertained by the fact that, as he and Aage moved around, and were passed from security man to security man, he and his son had to be 'signed for', showing they were 'in good condition' each time a transfer took place, as if they were parcels.[11] One hiccup in their smooth reception occurred in the form of a report that had appeared in the *New York Times* while Bohr had been in London, not long before he left for America. His movements were supposed to be secret but the *New York Times* accurately reported:

> **SCIENTIST REACHES LONDON**
>
> Dr. N. H. D. Bohr, Dane, Has a New Atomic Blast Invention
>
> LONDON, OCT 8 (AP) – Dr. Niels H. D. Bohr, refugee Danish scientist and a Nobel Prize winner for atomic research, reached London from Sweden today bearing what a Dane in Stockholm said were plans for a new invention involving atomic explosions.
>
> The plans were described as of the greatest importance to the Allied war effort.

Other reports appeared later in the *Evening Standard* and *Daily Sketch*, which added: 'Bohr has gone to the US on a special mission after consultation with Lord Cherwell. Prof. Bohr ... is an expert on explosives. We understand that this subject is connected with this trip, and that he is taking to the US some new ideas.' Finally, the *New York Daily Mirror*, of 20 December, reported that 'the Germans believe he has a vast knowledge of atomic warfare – the miracle that might save Germany'. This news, though inaccurate in some important respects, was hardly pleasing to anyone involved – Winston Churchill, the British prime minister, and General Groves in particular were sticklers for secrecy. (It was the last time Bohr's name was mentioned in the *Times* while the war lasted.)[12]

Moving on from New York, the Bohrs went first to Washington, where they stayed (still as the Bakers) at the Danish embassy. There, a discreet reception was held in their honour where, among the other guests was Felix Frankfurter, a justice of the Supreme Court. Frankfurter, a good friend of the president, Franklin Delano Roosevelt, knew Bohr of old, having met him in Oxford in 1933 and having seen him during earlier trips Bohr had made to America in 1939. Frankfurter was well placed, therefore, to work out – in general terms at least – why Bohr was in the United States. But nothing was said – not then.

Bohr met General Groves in Washington, a few days after Groves had briefed Peierls, Frisch, Fuchs and the other members of Tube Alloys. Groves was a military man: brusque, practical, direct, in a hurry and with a lot on his plate. Bohr was a famously careful man, who subscribed to a philosophy that one can never be accurate *and* simple at the same time. In order to be accurate, he said – and this applied in many areas, not just advanced physics – one had to bring in all sorts of qualifications and caveats, which invariably made it impossible to be clear and simple. And in his own speech he was also notoriously hesitant, roundabout, even long-winded. He would often pause for minutes on end as he sought the perfect phrase to encapsulate his thinking. Even so, Groves – not easily intimidated – was appreciative of Bohr's brilliance and the men soon achieved a modus vivendi.[13]

After their meeting, Bohr travelled west to Chicago, where he met up again with Enrico Fermi, who had been working at the university there. While Bohr was in Chicago, Groves turned up and the two of them travelled on together by train to Lamy, 18 miles from Santa Fe and the nearest station to Los Alamos. Groves was hoping to have Bohr to himself for the train ride and wanted to pick his brains before he joined the other nuclear physicists, whom Groves privately called his 'crackpots.' The general, as was his manner, emphasised for Bohr the rules he had established, in particular his notion of 'compartmentalisation'. He conceded that Bohr would of course be near the top of the pecking order but the general still emphasised 'what could be talked about and what was barred'.

As the train rolled across the Mississippi and the great plains of Texas, into the desert of New Mexico, the balance of the conversation began to subtly change. Bohr did more and more of the talking and, with his low-pitched tones, slow delivery and the click-clacking of the wheels over the rails, the Dane's voice took on a mesmerising quality.

Historians have in general been mesmerised by this journey too, skipping over it with nothing so much as a second thought. At one level this is understandable. In his papers at the National Archives in Washington, there is a note to the effect that the general adopted a policy of committing as little as possible to paper in the interests of preserving secrecy. So there is no documentation about this journey, either in his files or Bohr's. But, as we shall see in chapter 14, what happened on that train journey was central to the shape of the bomb story.

On the morning after their arrival at the Journey of Death, the 'Jornado del Muerto', as Los Alamos and the surrounding area was known locally (because so many had died of thirst there during the opening up of the West), Robert Oppenheimer, the scientific director, met Groves in the street and remarked that the general seemed a little stiff and was, apparently, limping. He asked what the trouble was. Groves, with deadpan humour (not his strong suit) replied: 'I've been listening to Bohr.'[14]

After their arrival on American shores, over the next few months Bohr and Fuchs, each in his own very different way, would try to shape events in regard to the development of the atomic bomb and the arms race that both could see on the horizon. The irony of what each was trying to do, simultaneously and in parallel, is brought out sharply in the pages that follow.

But their juxtaposed actions must also be understood against one all-important reality that has invariably been left out of all previous accounts and completely changes our understanding of what transpired.

For the fact is, by the time they arrived in America, both men knew that the original reason for building an atomic bomb – as a deterrent in case Hitler should get there first and hold the world to ransom – could be safely discounted. *Both men knew when he arrived in the United States that there was no threat from Germany.* They knew that German scientists were nowhere near building a bomb, and in fact weren't even trying to. So there was, in truth, no pressing need for Britain and America to build a nuclear weapon. The world did not need one.

Instead, hawkish figures among the higher echelons of the Allied administrations had developed new priorities which, they convinced themselves, would give the West an unrivalled dominance in the post-war world where Russia, then an ally, would become an adversary. This book will show that crucial intelligence, which confirmed that Germany had no viable nuclear weapons project, was deliberately withheld from the scientists at Los Alamos, so that they would keep working – as they thought – on a bomb to combat Hitler, when in fact the aims of the project were fundamentally changed without their knowledge.

PART TWO

The Overestimate of the Germans

2

The Taste of Fear: The Menace of Fission

The world first tasted the fear and threat of atomic weaponry in the early weeks and months of 1939. Not everyone did so, of course. The basic new knowledge was confined to begin with to a few score scientists, plus the politicians and military personnel they alerted. But, during the year – the year the Second World War broke out – the unease and the anxiety gradually spread.

Nuclear fission, the basis of the atom bomb, was discovered by Otto Hahn and Fritz Strassmann, at the Kaiser Wilhelm Institute for Chemistry in Dahlem, the suburb of Berlin often called the 'Oxford' of Germany because of the many academic outfits located there.

From the very first, nuclear fission was a worrisome discovery, because it was the culmination of several advances, all pointing with increasing menace towards the possibility of the explosive release of nuclear energy. The crucial entity of the nucleus, the neutron, the particle that made a runaway chain reaction theoretically possible, was discovered by James Chadwick, a Cambridge-based British physicist, in February 1932, barely a month before Adolf Hitler attempted his first (unsuccessful) bid to become German chancellor. Nuclear fission became a reality at Christmas-time 1938, a few weeks after Kristallnacht, and a month before Hitler's speech to the German parliament, the Reichstag, in which he felt emboldened enough to predict that a continent-wide war 'would lead to the annihilation of the Jewish race in Europe'.[1]

Between those 'bookends', scores of physicists, like hundreds of other scientists and scholars who were Jewish, or 'politically undesirable' in some way, were forced to leave continental Europe. However, thanks to their special talents, the physicists were able to play leading roles in the narrative of the bomb. The ironies in this story are endless but the crucial role played by the Jewish exiles is the most poignant – one might even say the most beautiful. In Britain, the German scientists were, to begin with, classified as 'enemy aliens' and so were not allowed to work on secrets directly related to the war effort, such as radar or jet engines. Instead, they were forced into more 'peripheral' long shots, such as atomic theory.

There was some disquiet in certain quarters even when Chadwick discovered the neutron. This was because the new entity, one of the three basic particles that make up our world, alongside the proton and the electron, lacked any electrical charge (hence its name), thus enabling it to be used to probe the nucleus far more intimately than before. And, thanks to Einstein, physicists knew that enormous amounts of energy were locked up in the nucleus.

One man who understood the implications of all this was Leo Szilard, a brilliant but maverick Hungarian physicist. In a famous epiphany, which occurred when he was waiting for a traffic light to change in London's Southampton Row, the year after Chadwick made his discovery, Szilard conceived the concept of the chain reaction, that some day the neutron could be used to disintegrate the nucleus and release the energy tied up, which in turn would throw out more neutrons to disintegrate more nuclei, releasing yet more energy, and so on in an ever-increasing – and explosive – cascade. So frightened of his own idea was Szilard that he assigned the patent to the British Admiralty, on condition that it was kept secret.[2]

In that same year, the Italian physicist Enrico Fermi, in Rome, burst on the scene with his theory of beta decay. This too related to the way the nucleus gave up its energy in the form of electrons. Although theoretical, Fermi's paper was based on extensive research, which led him to see that when a neutron enters a nucleus, the effects are, as someone else put it, 'about as catastrophic as if the moon struck the earth. The nucleus is violently shaken up by the blow.'[3]

More than that, though, Fermi showed that although lighter elements, when bombarded with neutrons, were transmuted to still lighter elements by the chipping off of either a proton or an 'alpha particle' (two protons and two neutrons bound together, the same as a helium nucleus), heavier elements acted in the opposite way. Their stronger electrical barriers *captured* the incoming neutron, making them heavier. However, being now unstable, they decayed to an element with one more unit of atomic number. This was totally new and raised a fascinating possibility.

Uranium is the heaviest element known in nature, the top of the periodic table, with an atomic number of 92. If *it* was bombarded with neutrons and captured one, it should produce a heavier isotope: U-238 (which has 92 protons and 146 neutrons, giving it a mass number of 238) should become U-239. Being unstable, this should then decay to a 'transuranic' element that was entirely new, never before seen on earth, with the atomic number 93.[4]

That was exciting enough but Fermi and his colleagues were faced with a major difficulty. At that time, atomic theory treated the nucleus as if it was one large particle. And that meant it must have a definite size, with a definite diameter. Experiments in Rome showed that a speeding neutron could cross from one side of the nucleus to the other in about 10^{-21} seconds – that is, a billion times less than a trillionth of a second. The neutron therefore would have to be captured within that minuscule window of time. But the Rome experiments also showed that the gamma-emission times the group had measured (essentially the time it took for the nucleus to react to being bombarded) were very much longer – slower – than what theory said they should be. In fact, the nuclei Fermi's group had experimented with took at least 10^{-16} seconds to emit gammas – 100,000 times too long.[5]

The problem was further complicated by the fact that neutron bombardment affected some elements more than others. To begin with, Fermi and his colleagues had loosely described that activation as 'weak', 'medium' or 'strong'. As time went by, however, that was no longer good enough – they needed a standard measure. For convenience's sake, they chose to investigate the reaction that neutron bombardment brought about in silver.[6]

The assistant given this task of standardisation immediately hit upon a fascinating surprise. He found that, for some reason, the silver cylinders were activated differently according to where they were located in the laboratory. Specifically, wooden tables produced much more activity than when the silver was laid on marble tables in the same room. How could that be?

Fermi was not known around the Rome laboratory as the 'pope' for nothing. His scientific intuition was near infallible and, acting on a hunch, he repeated the experiment with paraffin. He found an even greater effect. What Fermi had correctly intuited – and confirmed by experiment – was that the neutrons were colliding with the hydrogen nuclei in the paraffin and the wood. And that slowed them down.[7]

This was his crucial insight. Other physicists had taken it for granted that faster neutrons were better for nuclear bombardment, because faster protons and alpha particles had always been better. But that failed to take into account the neutron's neutrality. A particle that was electrically charged needed energy to force its way past the nucleus's electrical barrier and so its speed was an important asset. But a neutron did not. And so, slowing down a neutron gave it more time as it crossed the nucleus, and in turn that allowed more time for it to be captured.[8]

Fermi therefore had produced no fewer than three important advances: he had established the crucial significance of uranium, as the heaviest element in nature; the crucial difference between fast and slow neutrons; and third, their role in producing 'transuranic' elements. All three breakthroughs would come to matter.

Fermi and Szilard were both compelled to become émigrés. Szilard, more sceptical – and more scared – than most about the Nazis in 1933, had been forced into a hurried departure from Berlin by overnight train to Vienna on Thursday 30 March. Hardly more than twenty-four hours later, on the following Saturday, Julius Streicher, one of Hitler's notorious propaganda Gauleiters, directed a national boycott of Jewish businesses, with the result that Jews were attacked in the streets, and turfed off trains leaving the country. This was barely two months after Hitler had assumed power on 30 January. After that, for years, Szilard always kept two suitcases fully packed, wherever he was, 'just in case'.

Otto Frisch, as we shall see, was forced to escape *twice*, once from Hamburg to London, and then from Copenhagen to Birmingham. Several other Jewish physicists who were to take part in the atomic bomb project were also exiles: Rudolf Peierls, Hans Bethe, Franz Simon, Walter Heitler, Fritz and Heinz London and still others.

But of all the physicists who were forced to flee, none was faced with the dangers, the changes in fortune or the personal betrayal that confronted and humiliated Lise Meitner. Her story, which overlapped importantly with the worrying discovery of nuclear fission in an intimate way, was nothing less than harrowing, epitomising the taste of fear that was so widespread in the run-up to war.

Born Elise in Vienna in 1878, one of eight children, to parents who were ambivalent about their Jewish affiliations, she was never religious, never studied Jewish history and, like many assimilated Jews, thought of herself as purely Austrian. Her father was a politically conscious lawyer.[9]

She entered the University of Vienna in October 1901, a bluestocking, her nephew would judge later, 'a young woman who cared for nothing but study'.[10] So much so that her brothers and sisters used to tease her. 'Lise,' they would say, 'you are going to flunk [your exams], you have just walked through the room without studying.'[11]

She studied physics and mathematics, becoming only the second woman to earn a doctorate at Vienna, before she moved to Berlin, where Max Planck allowed her to become the first woman to attend his lectures. After a year she became his assistant but then found work with a colleague of his at the experimental physics institute in Berlin. That is where she met Otto Hahn in 1907.

Much the same age, Meitner and Hahn got on well from the start, making several discoveries. The friendship was solid and, throughout the 1930s, Hahn, who by then was director of the Kaiser Wilhelm Institute for Chemistry, helped protect Lise. Nonetheless, as early as 1933 her name appeared on a list of faculty to be dismissed from the University of Berlin. Planck wrote on her behalf but, despite his eminence, the ministry was not moved.

Life didn't improve. The Kaiser Wilhelm Institute for Chemistry had, like other similar institutes, a 'party steward' on the books, while others regularly wore their uniforms in the lab after joining the SA, the *Sturmabteilung*, or Brownshirts, whose main job was to rough up Nazi opponents. In 1936 Hahn and Meitner were nominated for the Nobel Prize by three distinguished colleagues, each a Nobel Prize-winner himself, Max Planck, Werner Heisenberg and Max von Laue, apparently in an attempt to protect their Jewish colleagues.* It didn't work and after the Anschluss, in March 1938, when Austria became part of Germany, she could no longer be protected and her fate turned even more dramatic. A fanatical Nazi chemist, Kurt Hess, who had worked next to Meitner for years, now snarled that 'the Jewess endangers this institute'.[12]

She watched as former colleagues were dismissed or left Germany on their own initiative, including Otto Frisch, her nephew.

A short time later, Hahn heaped pressure on her by insisting that she stop coming into the institute. She and Hahn had been colleagues and friends for thirty-one years by then and she was devastated by this betrayal. But she stood her ground and continued to turn up at the laboratory so she could write up the results of her recent experiments.

Feeling low, she approached Paul Rosbaud, whom she knew would have reliable inside information. Rosbaud was a remarkable individual who plays a major role in our story. They had been friends since the late 1920s when Rosbaud, a fellow Austrian, studied physics in Berlin. Sociable, with a ready wit, Rosbaud and his wife surrounded themselves with a lively mix of academics, actors, directors and musicians. Paul's brother Hans was conductor of the Frankfurt Radio Orchestra.[13]

In the early 1930s Rosbaud had become scientific advisor to Springer Verlag, who published scientific reference works and academic

* After Carl von Ossietzky was awarded the Nobel Peace Prize in 1935, no German was allowed to accept any of the Nobel Prizes.

journals. After Arnold Berliner, its Jewish founder, had been forced out in 1935, Rosbaud took over much of the editorial responsibility for *Naturwissenschaften*, Germany's premier science journal. This meant he travelled widely and established many contacts among scientists in universities, industry and the military. He loathed the Nazis but he cleverly cultivated friendships with those well placed in the regime, 'some with very low party numbers'.

Despite the inner access he had crafted, it was impossible for Rosbaud to assess how seriously Lise's position was threatened. Emigration was complicated. A new job outside Germany necessitated letters being written, contracts negotiated, visas arranged. The very idea paralysed Meitner with indecision for weeks.[14]

She was not entirely alone. Within forty-eight hours of the Anschluss, Paul Scherrer, a Swiss physicist who would become an Allied agent in the Second World War, sent a letter from Zurich inviting her to a Congress there. This was of course camouflage, an excuse for her to leave. Niels Bohr wrote a few weeks later from Copenhagen with a similar offer. But leaving the world she knew was a wrench.

Then, towards the end of April, she learned that her case was under review by the Ministry of Education. This was hardly good news and it finally dawned on Meitner that she must get out.

She opted for Bohr's offer, not least because her nephew Otto Frisch was in Copenhagen. But she was devastated all over again when, at the Danish consulate, she was refused a visa. Following the Anschluss, it now became clear, Denmark would no longer recognise her Austrian passport.

Ever more anxious, she prevailed upon Carl Bosch to write on her behalf to Wilhelm Frick, minister of the interior, arguing that she was an extremely distinguished scientist and wished to leave.[15] Bosch was himself an exceedingly distinguished figure, a chemist who had won the Nobel Prize in 1931 and helped found I. G. Farben, at one point the world's largest chemical company.

In May Dirk Coster wrote from the Netherlands. He was a physicist who had collaborated with Bohr, and he urged her to spend the summer with him and his family in Groningen. Paul Scherrer wrote once again, from Zurich, more forcefully this time. In June the Bohrs themselves passed through Berlin. Niels met Meitner and returned to Copenhagen very worried about her and, on her behalf, began trying to find somewhere for her to live and work in Scandinavia.

While all this was happening, or not happening, on 14 June she learned to her dismay that there were to be new restrictions on emigration from Germany. In particular, technical and academic people would be forbidden from leaving. It was a policy change seemingly aimed directly at her. Two days later a reply to Bosch's letter was received from the Ministry of the Interior: 'Political considerations are in effect that prevent the issuance of a passport for Frau Prof. Meitner to travel abroad. It is considered undesirable that well-known Jews leave Germany.'[16]

Not only was she forbidden to leave: she had lost her anonymity and her case had reached the notice of the Reichsführer of the SS, Heinrich Himmler himself.

Scherrer, Bohr and Peter Debye, a Dutch physical chemist, who had succeeded Einstein in 1934 as head of the Kaiser Wilhelm Institute in Berlin and won the Nobel Prize in Chemistry in 1936, rallied to her cause. In particular Bohr suggested that Manne Siegbahn, a leading Swedish physicist, who would soon have his own new institute in Stockholm, might step into the breach, while in Holland, Dirk Coster travelled personally to The Hague to petition the Ministry of Justice and the Ministry of Education on her behalf.[17] He told them he had found an unsalaried position for her at Leiden. He had in fact done this by hijacking a faculty meeting being held around that time and browbeating those present to approve the position, since it had no financial implications. He had also, by this time, collected enough funds to support Meitner for one year. It wasn't much, but it was enough.[18]

On 27 June, Coster sent Debye a short message in code. He was coming to Berlin, he said, to look for an assistant to fill a one-year appointment. Debye understood. That same day, however, one of Bohr's associates arrived in Berlin with news that Siegbahn's new institute in Stockholm would soon be completed and she *would* be welcome there. This removed some of Lise's anxiety and she decided on the Swedish offer. Quietly, she visited the Swedish consulate with her lawyer and arranged to transfer her assets.[19]

It was not a moment too soon because Bosch now contacted her to give her the inside information that the policy prohibiting scientists from leaving Germany was about to come into effect. This prompted her to get in touch with a former assistant, Carl Friedrich von Weizsäcker. His father, Baron Ernst von Weizsäcker, was a high official in the Foreign Ministry, and she asked if he would push through her application for a German passport. The reply was swift and devastating. No passport would be forthcoming. Worse, the rebuff meant that her case had now been drawn to the attention of two ministries.

Debye told Coster and Coster understood at once. But there was one final hoop to go through. He had to clear Lise's entry into Holland with the border guards. Maddeningly, Debye's message reached him on 9 July, which was a Saturday, when the border guard's office was closed. On Sunday all they could do was to wait.

On Monday morning the Dutch border guard gave their answer: Meitner *would* be admitted. Coster reached Berlin later that same day.

Only four other people knew of the plan to spirit her away: Debye, Hahn, von Laue and Paul Rosbaud. Hahn was involved because Rosbaud had 'shamed' him into changing his stance. On the Tuesday, 12 July, Lise worked normally at the institute until 8 p.m., correcting a paper to be published by a young associate. Then she left for home and packed two small suitcases, transferring to Hahn's house to spend the night (in case anyone should 'come looking' for her). She stayed there throughout the daytime hours of the 13th.[20]

After dark, Hahn made up for his having to dismiss her by giving her a large diamond ring that had belonged to his mother, 'so she had something to sell'. Then Rosbaud drove her to the station. 'At the last minute, overwhelmed by fear she begged him to turn back.' But Rosbaud wouldn't be deflected and Dirk Coster was waiting in the train. As agreed, they greeted each other as if by chance.[21]

The train left. They had agreed with Hahn on a code-telegram in which they would let him know how the journey ended. The chief danger consisted in the SS's passport control of trains crossing the frontier.

At the border, to Meitner's intense dismay and fear, they were confronted with a Nazi military patrol of five men. They took away her Austrian passport. Ten minutes elapsed. She said later that it felt like ten hours, because the passport was out of date and, technically speaking, invalid.

But the men returned and handed back the passport without a word.

Even then, it was a close-run thing. She didn't know it at the time but Kurt Hess, the fanatic who had thought that a Jewess endangered the institute, had spotted her sudden absence and sent a note to alert the authorities.

From Groningen in Holland, Coster sent the prearranged telegram to Hahn: 'the "baby" had arrived'.[22]

Two weeks later, on 26 July, Meitner's Swedish visa was granted and two days after that she flew from Holland 'with hidden money to Copenhagen ... fearful the whole time what would happen to me if the airplane should be forced to land in Germany'.[23] But the summer skies were clear and at Bohr's institute Frisch showed his aunt the new cyclotron then under construction. Afterwards, Niels and Margrethe Bohr welcomed her to their country home in Tisvilde, about 35 miles north of Copenhagen, on the coast. On 1 August she moved on to Sweden.

After the nerve-racking drama of her escape from Germany, Stockholm was something of an anticlimax. She would be free and safe there but never very happy. It took ages for her belongings to catch up with her and when they did it was clear they had been deliberately damaged en route.[24]

But, in the grand scheme of things, all that was overshadowed by the intellectual breakthrough that Lise Meitner and Otto Frisch were about to make. Moreover, subsequent developments took place on two levels: the obvious, surface level, and a hidden, semi-secret level in which individuals tried to shape events without others being too aware of what they were doing. The taste of fear was spreading.

On Monday 19 December, the day before the Christmas party of the Kaiser Wilhelm Institute for Chemistry in Berlin, Otto Hahn telephoned Paul Rosbaud at Springer-Verlag to tell him of a paper he had written with his colleague, Fritz Strassmann. The paper was, he said, especially important and needed to be published as quickly as possible. Rosbaud knew something of Hahn's work and reputation, of course, but he warned Hahn that, as it was nearly Christmas, he would need the manuscript in his hands no later than the following Friday, 23 December, for it to make the next issue of *Naturwissenschaften*. Hahn and Strassmann's paper was duly delivered to Rosbaud at the Springer offices on Linkstrasse in Berlin on Thursday 22 December.[25]

At much the same time as he had submitted the paper to Rosbaud, Hahn had written to Lise Meitner in Sweden, telling her of the bewildering results he and Strassmann had found that were the subject of their paper: that when he bombarded U-238 with neutrons he had found not radium, R-230, as he was expecting, which would imply that a few particles had been 'chipped' off the nucleus, but instead something 'very like barium'.

Meitner received Hahn's letter of 19 December in Stockholm two days later. She had meanwhile made arrangements to spend Christmas with a friend in Kungälv (King's River) outside Gothenburg, halfway up Sweden's west coast. She left Stockholm on Friday the 23rd, taking the letter with her.

Otto Frisch, who had been in the habit of spending Christmas with his aunt in Berlin, had been invited to Kungälv as well. He came up from Copenhagen by ferry across the Sound to Malmö, joining Meitner for breakfast at an inn in Kungälv the next morning. It was Christmas Eve.

By the time Hitler came to power in 1933, Frisch, who was the son of a Polish Jew born in Galicia, was working in Hamburg. He had never been politically conscious and in the early thirties did not pay much attention to the crisis atmosphere. 'When a fellow by the name of Adolf Hitler was making speeches and starting a Party I paid no attention. Even when he became Chancellor I merely shrugged my shoulders and thought, nothing gets eaten as hot as it is cooked, and he won't be any worse than his predecessors. There, of course, I was wrong.'[26]

Dismissed from Hamburg, a scholarship was found for Frisch with Patrick Blackett (Nobel Prize-winner in 1948) in London. The scholarship lasted for only a year and after that he took up the offer from Bohr to work in Copenhagen. He was still there at Christmas 1938.

When he arrived in Kungälv, he found his aunt still poring over Hahn's letter, as bewildered by the results as was Otto Hahn himself. Frisch was ready to discuss his own recent work, on the magnetic properties of neutrons, but Lise wouldn't hear of it. She insisted he read Hahn's letter before they did anything else.

Afterwards, the pair went for an expedition in the woods, Frisch on skis, Meitner struggling to keep up on foot. They turned the barium problem over in their minds as they moved between the trees. The fact is, until then physicists had considered that when the nucleus was bombarded, it was so stable that at most the odd particle could be chipped off (this is why Hahn was looking for radium, the nucleus of which has 230 particles compared with uranium's 238). Now, huddled on a fallen tree in the Gothenburg woods, Meitner and Frisch began to wonder whether, instead of being chipped away by neutrons, a nucleus could in certain circumstances be cleaved in two.

Though they dismissed this notion at first, gradually their confidence grew. What Hahn had produced was not a matter of 'chipping' protons or alphas off a nucleus. Rather, what he and Strassmann had found brought to mind Niels Bohr's 'liquid-drop model' of the nucleus. Like a drop of liquid, the nucleus has a certain temperature, surface tension and modes of vibration. The total energy, or mass, of the droplet is partly determined by several things – by the binding energy, modified by surface energy and an electrical energy between the protons. The surface tension holding the drop together is opposed by the electrical repulsion of the protons: the heavier the element, the more protons, therefore the more intense the repulsion. On this model the nucleus is a roiling bag of particles held in check, as it were, by a membrane that *just* manages to keep it all together.

When the deformed uranium nucleus reaches a critical point, the system breaks into two fragments of more or less equal mass, which fly apart with great energy. The energy forcing the fragments apart is explained by Einstein's $E=mc^2$, which is also known as the 'packing fraction'. The overriding fact is that the two smaller nuclei weigh *less* combined than the parent nucleus plus a neutron. This 'mass defect', as it is called, is small – only about 0.1 per cent. However, multiplying that by c^2 gives a very sizeable energy output.[27]

Meitner and Frisch had been in the cold woods for three hours. Despite the cold, and the roughness of the log they were perched on, she remembered how to compute the masses of nuclei from the so-called 'packing fraction'. The arithmetic was mind-boggling but straightforward. Lise multiplied the lost mass by c^2, the speed of light squared. It came out as more than 250 million electron volts. The calculation worked: the lost mass would supply the energy with which the drop would tear apart.

So fission had been discovered, described, explained and, up to a point, understood. But it was not clear yet whether energy would be released in sufficient quantities to sustain an explosion.

3

The Beginnings of a 'Strategic Game'

It was natural for Hahn and Strassmann to want their paper to be published as quickly as possible. They were well aware of its importance, even if they didn't know exactly *how* important, until Meitner and Frisch pointed it out.[1]

Rosbaud saw to it that the paper was published promptly, as it was his job to do. But there was more to it than that. Unbeknown to Hahn, Strassmann and Frisch (but perhaps not to Meitner), Rosbaud was a British spy – a very well-placed British spy, as it turned out. In fact, in Paul Rosbaud, Britain had a spy of unrivalled access, courage and tenacity, who was arguably the greatest hero in the entire atom bomb story. When it came to the bomb, Rosbaud 'was the main source of scientific intelligence for Britain'.

Rosbaud was born Paul Wenzel Matteus Rosbaud in Graz, Austria, on the evening of 18 November 1896. His mother, a music teacher, was the mistress of the choirmaster of Graz Cathedral and had three illegitimate sons by him, Paul being the third. She died of breast cancer in 1913, when Paul was seventeen. He never knew his father.[2]

In the First World War he enlisted as a private in a Styrian Regiment and saw constant fighting along the Isonzo River, which runs from the Dolomites into the north shore of the Adriatic. But for him almost the most important episode in the war was his outfit's surrender to the British. As he later wrote, 'My first two days as a prisoner under British guard were the origins of my long-time anglophilia. For the British soldiers, war was over and forgotten. They did not treat us as enemies but as unfortunate losers of the war. They did not fraternize, but they were polite and correct.'[3]

After the war, Rosbaud entered Darmstadt Technische Hochschule (technical university) to study chemistry, and he married Hildegard Frank, who was Jewish. (He also befriended a fellow student, Walter Brecht, whose brother Bertolt was a budding playwright.) Graduating from Darmstadt Rosbaud received a fellowship to study at the Kaiser Wilhelm Institute of Chemistry in Dahlem, Berlin, alongside Hahn and Meitner, where he did pioneering work on X-ray crystallography and earned an advanced degree. This did not get him an academic post, however, and when he was offered a job in Berlin with *Metallwirtschaft* ('Metal Industry'), a new weekly magazine covering developments in metallurgy, he leapt at it. It was in fact a plum job: he was a sort of editorial scout, looking out for ideas for articles, rather than a writing journalist, and it allowed him to travel all over Europe and meet scientists. He visited not only German destinations but also Oxford, Copenhagen and Oslo, forming long-term connections. Along the way he met and became friendly with Albert Einstein, Pyotr Kapitsa, Niels Bohr, Ernest Rutherford, Leo Szilard and Frederick Lindemann, later Lord Cherwell – all names that recur in our story. In the small world of elite science, he knew everyone who was anyone.

The one blot on this otherwise rosy picture was that the proprietor of his journal, Dr Georg Lüttke, was a fervent Nazi, which turned Rosbaud against the magazine he worked for. But when Dr Ferdinand Springer and his brother Julius, who headed the prestigious publishing house Springer-Verlag, sought Rosbaud out as scientific advisor to all the firm's publications, he found himself in an even bigger plum of a job, which gave him access to the best minds of the European scientific community.[4]

At more or less the same time, he made another contact, very different but no less important. Francis Edward Foley was born in Burnham-on-Sea in Somerset in 1884, educated in France and became fluent in French and German. (Later, he interrogated Rudolf Hess who observed that Foley spoke German without an accent.) After World War One, Foley had remained on the staff of the British Army of the Rhine, but was then moved to SIS (MI6) and posted to the legation in Berlin as a passport control officer, a cover for intelligence operations.

After Hitler's accession to power, Foley began to use his position to help Jews leave Germany on British passports. This was something Rosbaud was also engaged in and it may be how they met.*

But with the Anschluss in March 1938, life changed for Rosbaud and his family, no less than it had for Lise Meitner. Until then, his Jewish wife, Hilde, and his *Mischling* (mixed-race) daughter, Angela, had been protected by their being Austrian. But now they were German and the racial laws applied to them. Paul hurried to see Foley. British passports were issued immediately for Hilde and Angela. They quickly flew to London and found a small flat, with the aid of Robert Hutton, a professor at Cambridge and another of Paul's scientific contacts. Rosbaud's Anglophilia was set.

Until that point, scientific intelligence had not been a priority for the British. Economic, political and military information were what counted. That would change.

The change began with the publication of Otto Hahn and Fritz Strassmann's seminal paper on nuclear fission in *Naturwissenschaften*. Possibly there was more than met the eye in Hahn and Strassmann hurrying into print, natural as that was. Hahn was not a Nazi and at times his career suffered for it. Strassmann and his wife, we know, hid a Jew from the Gestapo in 1943 (and Paul Rosbaud, in a letter he wrote after the war, said that Strassmann was 'the most decent of all German scientists'). Author Ruth Sime has shown that Hahn's institute benefited in the war from its military links but there always remained in Germany, among some senior figures, a regret that Hahn and Strassmann's fission paper had been published openly (see below).[5]

Rosbaud's biographer, Arnold Kramish, argues that he was actually playing a 'strategic game'. To Rosbaud, Hahn was a pure scientist, someone like Faraday or Hertz who was not interested in

* It is estimated that Foley saved 10,000 German Jews by his actions as a passport officer. On 24 November 2004, the 120th anniversary of his birth, a plaque was unveiled in Israel at Yad Vashem, recognising him as 'Righteous Amongst the Nations'.

the applications his discoveries could be put to. But, 'probably earlier than any of the scientists, he [Rosbaud] realised the vast destructive potential of what Hahn, Strassmann, and Meitner had discovered, and he was acutely conscious that the fundamental research had been done in Germany. He wanted the rest of the world to know of the significance of the work at least as soon as the Nazi planners did. By rushing into print with Hahn's manuscript, he was able to alert the world community of physicists.' Paul Lawrence Rose, in his book on Heisenberg, concurs, saying that Rosbaud 'influenced' Hahn to publish early in his own journal. It was, says Gerard DeGroot in his history of the bomb, 'a calculated act of subversion'.[6]

Was this in fact the first episode in a series of actions by German physicists to compromise their Nazi masters? There are some grounds for thinking it was.

Hahn wrote to Meitner, in her exile, 'The day before yesterday, I have spoken with Rosbaud in detail, who will again travel to England and meet with Cockcroft.'[7]

It was a revealing admission, because John Cockcroft was a British scientist who took a close interest in any intelligence concerning atomic matters and who was himself in direct touch with Hahn. The Cockcroft-Rosbaud link would be crucial.

After their adventures in the Gothenburg woods on Christmas Eve 1938, Lise Meitner and Otto Frisch split up. Frisch returned to Copenhagen and told Bohr, who was just leaving for America, what his aunt and he had worked out. 'I had hardly begun to tell him,' Frisch wrote later, 'when he struck his forehead with his hand and exclaimed, "Oh, what idiots we all have been. But this is wonderful. This is just as it must be."'[8]

Bohr left the next day for New York, on the SS *Drottningholm*, where he had a blackboard installed in his cabin so that he could confirm Meitner's and Frisch's calculations. By the time he reached Manhattan he was convinced that, though there were some anomalies that still needed to be ironed out (important anomalies, in fact), aunt and nephew were right about fission. Frisch had coined the word 'fission' by borrowing the term from a friend who told him that was what biologists called the splitting of cells.

After Bohr introduced the idea to America (before Frisch's, Meitner's, Hahn's and Strassman's papers had actually been published), there was a rush to repeat the Berlin experiments. 'Rush' is the appropriate word here for, by all accounts, at the lecture where Bohr first told a group of assembled American physicists about fission, several of them (some in black tie) left their seats and hurried from the room before he had finished his remarks, to repeat Hahn and Strassmann's experiment. In a matter of days, the nucleus had been fissioned in several American cities.

The situation was hotting up, but there was one other breakthrough that needed to be made before a bomb could become a practical possibility. Bohr played the crucial role.

It took place early in the following month, February 1939, when there was still fresh snow blanketing the main quadrangle of the Institute for Advanced Study at Princeton. Bohr was in the Nassau Club, a fine old house on Mercer Street, discussing physics with Léon Rosenfeld, a polyglot Belgian physicist, and George Placzek, 'a Bohemian from Bohemia', who had lived in Russia for a time, but subsequently worked with all the great physicists of the twentieth century, and in Copenhagen.

During the exchange, Placzek asked how it was that slow rather than fast neutrons generally caused uranium to fission? And why should slow neutrons induce only a modest amount of fission in uranium?

As Bohr's biographer tells the story, at this precise moment Bohr's face suddenly went blank. He stopped speaking almost in mid-sentence, scraped back his chair and muttered to Rosenfeld, 'Come with me, please.' Placzek was left where he was, as if Bohr had forgotten him, a very uncharacteristic way for him to behave.

Bohr left the main building of the club and hurried across the snow to Fine Hall, where he had been assigned an office. He and Rosenfeld climbed the stairs in silence. 'Still without saying a word,' Rosenfeld said later, 'Bohr went to the blackboard and began to put down figures and symbols. He made some rough sketches, working rapidly.'

He stood for about ten minutes in total silence, save for the scraping of the chalk on the blackboard. Then he stopped and his face creased into smiles. 'He had the answer to Placzek's question, and to the major – the essential – problem posed by the fissioning of the nucleus.'[9]

What Bohr had realised, as he quickly told Rosenfeld, was that uranium actually exists as several isotopes. The ordinary isotope, U-238, comprises more than 99 per cent. U-235 is the rare form, making up only 0.7 per cent, and there are even more minute traces of two other isotopes. What Bohr had worked out, he said, was that U-238 merely captures slow neutrons. It does not undergo fission when a neutron penetrates its nucleus. Instead, it is U-235, the very rare isotope, that fissions with the intake of a slow neutron. This is why, overall, uranium undergoes only a modest amount of fission.

By now, Placzek and John Wheeler, an American physicist and a colleague of Bohr's, had crowded into the room in Fine Hall. Bohr and the group around him could not fail to grasp the implication of this. 'If U-235 could be separated from the abundant U-238 it would be highly fissile to slow neutrons. Fantastic amounts of power, or a tremendous explosion, could result if a reaction started.'* Bohr had found the explosive element at the heart of the bomb. Strangely, perhaps, the people in that Fine Hall room were elated. A new form of energy had been identified – at least in theory.

This was a great breakthrough, but there were still anomalies to sort out. Two in particular were important. Bohr, Wheeler and the others now paced the rooms and staircases of Fine Hall.

One anomaly, which led to another surprising advance, stemmed from their interest in 'transuranic' substances, first identified by Fermi. If U-238 does not fission when it captures a neutron, what then? If one neutron was added to U-238, it would bring its atomic weight to 239 (238 + 1) and 239, being an odd number, would be unstable. In seeking a return to stability, an electron would be emitted. If the loss of a negatively charged electron turned one of

* In order to construct a viable bomb, U-235 has to be enriched more than 100 times, from 0.7 per cent to about 85 per cent, to create 'Weapons Grade Uranium'.

the neutrons into a positively charged proton, as can happen (beta decay), the nucleus would have 93 protons and that extra proton would mean that the binding of an additional electron in the outer reaches of the atom would create a new element, one step beyond the table of elements, an element non-existent in nature.

This was another breathtaking idea but the excitement still didn't stop. The new element, Number 93, later to be called Neptunium, would itself be unstable. Through Fermi's beta decay, it too would emit an electron, and again the loss of the negative charge would convert a neutron into a proton. The addition of one more proton to the 93 already present would produce yet another 'transuranic' new element, Number 94 (93 + 1), which was eventually called plutonium. If the new, even-numbered element 94 should absorb a bombarding neutron, increasing its atomic weight from 239 to 240, it too might well undergo fission.

Which meant that, in theory, there might be *two* paths to a nuclear explosion.

The pitiful irony of all these exciting and potentially dangerous developments was that it was still early 1939. Hitler's aggression was growing, but the world was, technically, still at peace. The scientific breakthroughs we have been discussing were published openly. Many physicists thought that a bomb was unlikely, either because it would be too difficult to separate U-235 from U-238, their weights being too similar, or because too few neutrons were emitted during fission, meaning that Leo Szilard's idea of a chain reaction could never get going. Bohr in particular thought isotope separation would be such a superhuman task that it could never be achieved without fatally distending the economy of any country that went in that direction.

Others were not so sanguine. In particular, the émigré scientists in America, who had first-hand experience of what the Nazis could do and the efficiency with which they could do it, were especially alarmed. The individual most worried was Leo Szilard, the man who kept two bags always packed, 'just in case'. He had spent time in

Britain after leaving Vienna before moving on to the United States and while there he had written to Frederick Lindemann, head of the Clarendon Laboratory at Oxford, and suggested that the results of experiments in nuclear physics be circulated in only a limited way, to colleagues elsewhere in Britain, America and one or two other selected countries.[10] No one took him seriously, then.

Szilard had been in the United States for about a year when the discovery of fission was announced. If he had had his way, all fission research after the Berlin-Gothenburg breakthrough would have been kept under wraps. He became the most insistent advocate of secrecy.

4

The Struggles over Secrecy

Two days into 1939, while Bohr was still at sea on the *Drottningholm*, the SS *Franconia* docked in New York with Enrico Fermi and his family on board. (Fermi was delighted to be asked, as part of the immigration procedure, what half of twenty-nine was. He managed the correct answer, in English.)

Szilard was anxious to collaborate with Fermi who was going to Columbia University. Szilard did not have a formal position there but he was allowed to use the facilities on a 'freelance' basis. Before the collaboration could get going, however, Bohr arrived at New York Harbor, bringing with him the momentous news of fission, which he had been told about by Frisch but had not yet been published in the journals.

Szilard was immediately alarmed as, for him, the political implications were obvious: 'We were at the threshold of another world war.'[1] But Fermi was unimpressed – he was a cautious man, a confirmed experimentalist and, like Bohr, he repeatedly denied the grave implications of fission. He did accept that an extra neutron might be emitted during nuclear splitting, but considered the possibility so remote that he gave it little thought, then.

Szilard, though, 'the perpetually worried man', sent an urgent cable to the British Admiralty: 'REFERRING TO CP10 PATENTS 8142/36 KINDLY DISREGARD MY RECENT LETTER STOP WRITING LEO SZILARD.' He now believed that his original 1935 chain-reaction patent, which had occurred to him in Southampton Row, *would* create a violent explosion and he wanted the patent – like all fission research – kept secret.[2]

With the benefit of hindsight, Szilard's attitude was understandable. But, at the time, every major nuclear physicist group in the world was investigating the implications of fission, so that withholding publication on the subject in England, France, America or even Russia and Japan was far from simple.

And the possibilities and threat of fission were everywhere apparent. In Russia, for example, during the spring of that year, the physicist Igor Tamm asked his students, 'Do you know what this new discovery means? It means a bomb can be built that will destroy a city out to a radius of maybe ten kilometers.' The French, too, soon grasped where their experiments were pointing. Lew Kowarski later recalled jokes 'about whether we will get the Nobel Prize for Physics or for Peace, first because we made a war possible by discovering nuclear explosives, which obviously would make war impossible.'[3] Kowarski, born in St Petersburg but now a naturalised Frenchman, was, together with Hans von Halban, of Austrian-Jewish descent, part of a team at the Laboratoire de Chimie Nucléaire in Paris, working under the 1935 Nobel Prize-winner, Frédéric Joliot-Curie on neutron research.

Spencer Weart has reconstructed the relations between the Paris group and the Columbia group who, at this critical time, were the main teams outside Germany working on the explosive content of fission. He has shown how misunderstandings arose that led to more information being released than was perhaps in the circumstances advisable.

But Szilard wanted to bottle the atomic genie. Fermi, however, who was still enjoying learning English, told Szilard he was 'nuts'. Having mulled over what Bohr had told them Fermi did, however, now admit that there was 'a remote possibility' that a chain reaction could be made. 'What do you mean by a remote possibility?' he was asked.

'Well, ten per cent,' Fermi answered.

'Ten per cent is not a remote possibility, if it means we may die of it.'[4]

Such an exchange only made Szilard increasingly anxious, even after the British Admiralty assured him that the secrecy of the patent would be maintained. That went nowhere near far enough for the perpetually worried man. Szilard was still adamant that his colleagues should keep their own work secret. He was aware too that the French in particular, at the Laboratoire de Chimie Nucléaire, were working on aspects of nuclear physics that paralleled what was going on at Columbia, and so he wrote to Frédéric Joliot-Curie, Marie Curie's son-in-law, emphasising how the liberation of neutrons could create a chain reaction that might lead to dangerous bombs and suggesting that such research be kept secret. In the letter, Szilard also said that he would be sending a cable a while later when the Columbia team had discovered exactly how many neutrons were released in fission, which would determine whether a chain reaction was indeed possible. In an unfortunate coincidence, however, two days later Fermi also wrote to Joliot-Curie, telling him that he was trying to understand exactly what went on in uranium fission. He made no mention of secrecy.

Joliot-Curie had thus been informed by Fermi that the Columbia group were in effect rivals of the Paris group, and he knew from the difference between the Szilard and Fermi letters that secrecy was a private campaign by Szilard, not the official policy of Columbia, or the US come to that. So far as Joliot-Curie was concerned, therefore, there was every reason to believe that Fermi might publish first.[5]

Still frustrated by Fermi's caution, Szilard knew he could push things forward at Columbia only by carrying out some experiments of his own, as he had mentioned to Joliot-Curie in his letter, and towards the end of February he and Walter Zinn, a Canadian colleague, used a room on the seventh floor of the Pupin Building at Columbia to bombard uranium with slow neutrons, using a cathode ray oscillograph – like a small TV screen – to record any neutrons emitted as grey streaks. Their image on the screen would show what their energy levels were.[6]

The experiment was set in motion, and they waited. 'We saw the flashes,' Szilard said later. 'We watched them for a little while and then we switched everything off and went home. That night there was little doubt in my mind that the world was headed for grief.'[7]

Szilard and Zinn wrote up their experiment and applied for a patent. At the same time they found that Fermi had at last carried out his own uranium-fission experiment. His results had been less conclusive than theirs but he was now more inclined to believe that fission *could* produce a chain reaction.

Fermi and Szilard also had their thoughts sharpened, as did those of the other émigré scientists, by the news in early March, when Germany annexed Czechoslovakia. Like other physicists, they were only too well aware that the newly occupied country had at Jáchymov (Joachimsthal) Europe's richest uranium deposits.[8]

When news came that Joliot-Curie and his colleagues in Paris – who had easy access to plentiful radium thanks to the Radium Institute there, established after Marie Curie's ground-breaking discoveries – had published in *Nature* their finding that uranium did indeed free extra neutrons when it fissions, it put Szilard and Fermi under pressure to release their own findings. Szilard was adamant that *all* uranium work must be kept secret, but Fermi argued that such a ban went against hundreds of years of traditional scientific openness and that, moreover, with the publication of the latest paper by Joliot-Curie and his team, there was now in effect no secret to keep. However, in a remarkable but moving gesture, he then went on to say that, as he now lived in a democracy (as opposed to Mussolini's fascist dictatorship), he and his fellow scientists should take a vote. He lost to the Hungarians and for a while became an advocate of secrecy.

Szilard had taken publication of Joliot-Curie's paper in *Nature* to heart; he felt personally betrayed. But that didn't deter him. Later that March he, Eugene Wigner, another Hungarian émigré, and Victor Weisskopf began a concerted campaign of writing to their European colleagues, telling them of a new plan, that their papers on neutron emission had already been sent to the *Physical Review*, to established priority, but that the authors had agreed to delay publication, urging self-censorship and hoping that the Paris paper might be the last to appear.

Weisskopf, born in Vienna, had worked with several of the leading physicists in Europe. He cabled Patrick Blackett, whom Frisch had worked with at Birkbeck College in London, asking whether it would be possible for *Nature* and the Royal Society's *Proceedings* to cooperate in delaying publication of fission research in the same way, while Wigner wrote to Paul Dirac, who had visited Japan with Heisenberg and was a Nobel Prize-winning physicist (in 1933) at Cambridge, asking him to support Blackett. Blackett and John Cockcroft replied that they would support the secrecy plan and that *Nature* and the Royal Society were expected to co-operate.

Szilard, Edward Teller, a third Hungarian émigré, Weisskopf and Wigner also talked the matter over once more with Bohr, who was still in America. He remained sceptical that a bomb was possible, and thought moreover that enough had already been published to arouse interest in any military organisation. But he nonetheless agreed to go along with the attempt at secrecy and drafted a letter to be sent to his own institute back in Denmark.

The self-censorship campaign hit a number of snags. One telegram from Weisskopf to Joliot-Curie arrived in Paris on 1 April, with the result that at first it was thought to be an April Fool's joke. But Joliot-Curie did discuss self-censorship with his colleagues. Like Fermi, however, he strongly believed in the international fellowship of scientists. Another consideration was the natural one that if he and his colleagues failed to publish they might be eclipsed by people who did – they were extremely sceptical about everyone adhering to an unprecedented pact which, so far as they knew, was pushed forward 'only by two Central European refugees on the outskirts of the Columbia scientific community'.[9] (If Bohr had approached them, they might have felt differently.) Another reason was that if they were beaten to the punch in publishing, they might have trouble getting further funds to develop nuclear energy, which they felt was a more immediate prospect than a bomb. Finally, the French were aware that there had been quite a lot of press coverage of fission in the United States. Having been beaten to the punch once, they were not anxious to repeat the experience. Joliot-Curie replied to Weisskopf: 'QUESTION STUDIED MY OPINION IS TO PUBLISH NOW REGARDS JOLIOT.'[10]

Szilard was dismayed all over again but it only made him more determined than ever to reboot his crusade for secrecy. There followed intensive discussions at Columbia and in Washington about what – if anything – should be published. Those in favour of publication opposed secrecy on the grounds that it was not yet absolutely certain that a chain reaction would occur, and they also doubted fission could ever be used in a military weapon. Still others pointed to the scientific 'grapevine', arguing that a publication embargo was pointless, since word of the new findings would spread by informal personal communications that, they maintained, would have the same effect as publication. Then there was the argument that publication would bring about further developments in American laboratories that might not be carried out otherwise.[11]

The indecision continued for some time. But then on 22 April 1939, two days after the Eagle's Nest at Berchtesgaden was given to Hitler as a fiftieth birthday present, and two days before 5,000 works of 'degenerate art' were burned at the Berlin Fire Station, Frédéric Joliot-Curie, his wife and colleagues published in *Nature* their vital observation that, repeating Hahn and Strassman's experiment but standardising the measurements, nuclear fission emitted on average 2.42 neutrons for every neutron absorbed. There was now no escaping the conclusion that energy was indeed released in sufficient quantities to sustain a chain reaction.

The French had reinforced their scientific credentials.

But at what cost?

One immediate consequence was that Fermi reversed himself again and said he was now in favour of publication. His results, and those of others in America, were published openly in the *Physical Review*.

We shall never know if Szilard's plan for physicists to self-censor their work would have delayed the preparation of the atomic bomb. Szilard thought it was insane for the French to show the Germans* the way.[12] But we do now know that thanks to publication in *Nature*

* The Russians, too. Partly as a result of Joliot-Curie's article, Leonid Kvasnikov, director of the NKVD's scientific and technical department, sent a questionnaire to his agents in several countries seeking information on what those countries considered the 'military potential' of nuclear fission to be.

and in the *Physical Review*, the Germans learned that there was experimental proof that slow neutrons were more likely to fission U-235 and that neutrons with a certain energy were very likely to be captured by U-238, producing U-239. We now know that in July 1940 Rudolf Fleischmann at Heidelberg tried to separate isotopes using a method published by Harold Urey, the discoverer of deuterium, the basis of heavy water, and that in May 1941 Fleischman received a letter from Josef Mattauch about Alfred Nier's isotope separation apparatus at Minnesota, published in an American journal a year earlier. (Otto Hahn also saw this paper and was influenced by it.) In other words, the Germans were helped several times by open publication in the US. In Spencer Weart's words, 'The effort of Szilard and his friends, after coming within an inch of success, had failed disastrously.'[13]

We also know that the insistence on publication of Joliot-Curie's 22 April article provoked several governments to take immediate action. There was now no mistaking the menace of fission.

In London, for example, it was Nobel physicist George P. Thomson, at Imperial College, who saw in Joliot-Curie's article a probable new power source and he quickly alerted the British government. Only four days after Joliot-Curie's *Nature* article appeared, on 26 April, Britain's Ministry for the Co-ordination of Defence urged the Treasury and the Foreign Office 'to buy as much Belgian uranium as they could' – the Belgian Congo was known to be a major source. Edgar Sengier, a Belgian who was both a director of the Bank of Belgium and chairman of the Union Minière du Haut Katanga, the outfit that mined the uranium, refused an offer from the British to buy his metal (partly because they wouldn't explain why they needed it) but offered to sell it to Joliot-Curie in France (who did explain why). Francis Perrin, part of Joliot-Curie's team, had worked out that a uranium bomb would have a critical radius of 130 centimetres, approximating to 40 tons of uranium oxide and the French told Sengier about an idea for a secret test in the Sahara.[14] The swift collapse of Belgium and then France meant that the Sahara idea never came off and the uranium that Union Minière had stockpiled in Belgium was captured by the Germans, though Sengier saw to it that the uranium still in the Congo was quietly shipped to the USA.

In Germany it was much the same as in Britain. The same day as the Joliot-Curie article was published, 22 April, physicist Wilhelm Hanle described a 'uranium burner' (*Uranmaschine*) to a university physics colloquium in Göttingen where he was a professor. As Thomson had done in Britain, Hanle's superior immediately wrote to the Ministry of Education in Berlin, which quickly appointed Abraham Esau to convene a conference on uranium.

Then president of the German Bureau of Standards, Esau (not Jewish, despite his name) led a secret meeting called on 29 April (one week exactly after the Joliot-Curie paper was published) that paralleled the one held in London. The secret meeting urged the setting up of a 'Uranium Club' (the Uranverein) of selected nuclear physicists, and ordered the buying up of all the uranium stocks in Germany. Esau also wanted all exports forbidden and argued that secret contracts should be negotiated regarding the recently captured uranium mines at Jáchymov, Czechoslovakia.

At much the same time, in fact on Monday 24 April, in Hamburg, physical chemists Paul Harteck and Wilhelm Groth contacted the War Office in Berlin, drawing their attention to 'the newest development in nuclear physics', which 'will probably make it possible to produce an explosion many orders of magnitude more powerful than conventional ones,' and adding, 'The country that first makes use of it has an unsurpassable advantage over others.'[15] By the summer an office for nuclear research had been instituted in the Weapons Research Office of the Army Weapons Department. Even before the outbreak of war, German nuclear research had the backing of two government departments.[16]

Harteck, who had studied at the Cavendish in Cambridge in the 1930s, alongside Chadwick, Pyotr Kapitsa, a very talented Russian, and Rutherford, would remain a staunch nationalist throughout the war. But not everyone in Germany felt the same way, as we shall see.

In America, the reaction was very different to that in Britain and Germany. Despite Szilard's misgivings, a week after the Joliot-Curie paper appeared, the spring meeting of the American Physical Society took place in Washington and the proceedings – which included an open debate about chain reactions and isotope separation – were reported fulsomely in the press, including the information that the new explosives could destroy an area as large as New York City.

But not everything was made public and we shall now see how secrecy, in the end, was achieved. As mentioned, Fermi had discovered in Rome in 1934 that hydrogen was especially efficient at slowing down ('moderating') neutrons and, over the summer of 1939, the Columbia team, Szilard included, devoted much thought and study to which substances might be used as a moderator in a bomb, settling on heavy water and graphite, the former by definition having more hydrogen than normal water, and the latter being composed of carbon, which slows down neutrons less efficiently than hydrogen but is much more common in nature and therefore considerably cheaper.

Szilard thought that if a chain reaction could work in graphite and uranium then a bomb was likely and he assumed that, if he had thought of it, so too had the Germans. He believed a large-scale experiment was needed to settle the matter and it was his effort to raise funds for such an endeavour that led, indirectly, to the famous letter that Albert Einstein wrote to the president about the possibility of an atomic bomb 'using chain reactions in uranium'. Among other things, Einstein drew Roosevelt's attention to the fact that Ernst von Weizsäcker, state secretary in the German Foreign Ministry, was well placed to draw Hitler's attention to nuclear fission because his son, Carl, was a prominent atomic physicist.[17]

Those machinations did eventually result in the president becoming involved, and in the institution of a Uranium Committee, in October 1939, headed by Lyman Briggs, an uninspired and uninspiring soil scientist, with Szilard and several other physicists as members. Szilard hoped that this organisation would address the secrecy issue but the Briggs committee remained largely inactive, leaving everything to the physicists and the only response Szilard could get was a suggestion from Admiral Harold Bowen, director of the Office of Naval Research, present on the committee as an observer, that the scientists should impose on themselves whatever censorship they felt was necessary.[18]

This didn't really help and when Fermi and his assistant Herbert Anderson had worked on the details of how much the carbon in graphite slowed down neutrons and concluded that it would work very well, Szilard once again argued that this result should not be published. At which point, according to Szilard's own account, 'Fermi really lost his temper', thinking Szilard's caution was absurd. However, it then turned out that George Pegram, the head of the Columbia physics department, agreed with Szilard. Fermi caved in and reversed himself a second time.

And we now know that was just as well. As Spencer Weart put it: 'If the value for the cross section [of carbon] had been published [that is, the efficiency with which it slowed down neutrons], the course of World War II might conceivably have been changed.'[19]

This was because events in Germany were paralleling those in the US in some ways, but in other crucial areas they were very different. Walther Bothe, a future Nobel Prize winner, then director of the physics department at the Kaiser Wilhelm Institute for Medical Research at Heidelberg, would embark on much the same exercise as Fermi and Szilard had been engaged in – the study of graphite as a moderator. But in the course of it he made a serious – a momentous – mistake, one that shows that although it is highly likely that German physicists would have carried out identical experiments to those Joliot-Curie and his team in Paris performed, it is by no means certain.

In a separate paper to the one referred to above, Spencer Weart examined the comparative development of nuclear studies in different countries during this crucial time, when some results were published openly and others kept secret, and he has shown that while Joliot-Curie and his team, and the Columbia group, produced identical results, sometimes to within half a day of each other, the Germans followed a much more roundabout route (roundabout mathematically) and ignored a more efficient approach devised by Fritz Houtermans. He was a German physicist but was sidelined mainly because he was considered politically suspect. This shows that there is no guarantee that scientists will follow the same reasoning as their colleagues/rivals, even in the precise world of mathematics.[20]

There are two explanations for why Bothe made his momentous mistake, not necessarily mutually exclusive. According to Arnold Kramish, in June 1939 Bothe, then forty-seven, met Ingeborg Moerschner, thirteen years his junior, during passage to New York on the liner *Hamburg*. Ingeborg was going to San Francisco to work for Fritz Wiedemann, the German consul general there, who was a former adjutant to Adolf Hitler and a spy. Bothe was going to a meeting at the University of Chicago but he was also on a sort of spying mission himself, to see what progress the Americans were making in their fission research. (He concluded that very little of a military nature was going on though, as we have seen, Szilard and Fermi were working on graphite and heavy water at Columbia, something that would have interested him very much, had he discovered it.) In the course of his surveillance, however, Bothe and Moerschner visited the World's Fair in New York and saw the sights of San Francisco together and a close relationship began to develop. But she had to remain in California. 'In a highly infatuated state', Bothe returned to Heidelberg alone, and to his work measuring the nuclear properties of graphite.[21]

His diary recorded his emotional turmoil. On the first anniversary of their encounter, in June, 1940, he wrote, 'Ingeborg ... Tomorrow, it is a year that you came into my life ...' and he went on to refer to 'moonlight and dreams', concluding that he felt 'like a drunken teenager'. Two weeks later, in another letter, he confided that 'I have been speaking of physics the entire day, while thinking only of you.' And this is what has fascinated physicists and intelligence experts: the 'drunken teenager' was, at the time, measuring graphite the entire day – and it appears that the moonstruck Bothe made a serious error. He was using industrial graphite and concluded that it was 'unsuitable material' as a nuclear moderator, whereas in fact, as was discovered later elsewhere, pure graphite *does* work very well.

The alternative explanation is no less intriguing. There is no direct evidence that Bothe deliberately falsified his results but it has been pointed out that, soon after Hitler achieved power, Bothe had been hounded from his academic position because he was an anti-Nazi, hounded to such an extent in fact that he had to spend 'a long period' in a sanatorium. Even after he was discharged, he was still continually harassed, 'even to the accusation of scientific fraud'. Did he repay his tormentors in kind?[22]

Bothe's erroneous measurements, reached early on, were the reason the Germans turned to heavy water, a choice that, as we shall see, 'would doom their chances of making an A-bomb during the war'. By the time they discovered that graphite was a usable moderator, it was too late. Karl Wirtz, one of Heisenberg's closest associates, always blamed Bothe's error for the German failure to build a bomb.[23]

According to Paul Rose, and most experts agree with him, the most complete exposition of German nuclear understanding in this crucial initial period was for some reason not kept secret but published in June 1939 by Siegfried Flügge, who was now an assistant of Otto Hahn's in Berlin but had previously been a teaching assistant to Werner Heisenberg at Leipzig. So he was well integrated into the German physics establishment.

In his article in Rosbaud's *Naturwissenschaften*, published on 9 June 1939, the same month that Bothe met Ingeborg Moerschner, Flügge asked, and by no means rhetorically, 'Can the Energy Content of the Atomic Nucleus Be Exploited Technically?' As Rose puts it, in this article, 'Flügge tantalizingly estimated that one cubic meter of uranium oxide contained enough energy to lift a cubic *kilometer* [my italics] of water 27 km high.'[24]

Flügge went so far as to envision two types of explosion in reactor bombs, based on different materials and neutrons. The first type, he said, takes place within one 10,000th of a second and is brought about mainly by fast neutrons. 'In less than 10^{-4} seconds the whole of the uranium is converted. The liberation of the energy thus happens in such a short time that it produces an extraordinarily violent explosion.' The second type of explosion occurs, he said, in one-tenth of a second, and happens in uranium oxide mixed with a moderator and the agents are slow neutrons. 'The liberation of the energy proceeds certainly more slowly than with fast neutrons, but it is still nevertheless explosive.'[25]

The mathematical calculations in Flügge's article were, in fact, some way off. But, as Rose observes, there is no evidence that 'anyone in Germany' disagreed with the article's premise when it was published.

But another observation is even more intriguing. Why was the article published at all? On 29 April 1939, the Reich Ministry of Education, on behalf of the government, had called a closed-door conference at which it was agreed that secrecy be imposed on atomic research and that all uranium stocks in Germany and German-controlled territories be 'secured'. Otto Hahn was actually attacked there and then for publishing his discovery of fission at the beginning of the year, and a report of this reached Paul Rosbaud.[26]

Confirming how close he was to the main players in this story, Rosbaud later wrote, 'I do not deny that I was somewhat alarmed', when he learned (from Josef Mattauch, who had filled Meitner's old job in Berlin), that there was such a thing as the Uranverein and that Hahn had not been included. This was revealing in that Hahn was not a Nazi, so it suggested that the Uranverein was comprised of the more committed scientists, underlined by the fact that, as Rosbaud found out later, one of the administrators of the Uranverein even thought that a bomb might be ready *before* war broke out and was keen to make available the Wehrmacht's shooting grounds at Kummersdorf for experiments. But the scientists had dismissed this, saying such a development would take between five and thirty years.[27]

Fortunately, a week later, Professor Robert Hutton, the Cambridge man who had helped Rosbaud's wife and daughter settle in Britain after Foley arranged their British passports, was in Berlin. He had known Rosbaud since 1925, and was well connected with Britain's intelligence services, being also friendly with Eric Welsh, head of SIS's Norwegian desk, Charles Frank, deputy chief of Britain's secret outfit to assess new scientific developments for their military potential, and with Sir John Anderson, chancellor of the exchequer and the one cabinet minister Churchill brought into the atom bomb secret. Hutton was an ideal channel for Rosbaud's information.

In Berlin the two men had a brief but crucial meeting. In fact, it appears that Hutton was only in Berlin for a couple of hours, so it seems the meeting may have been specifically arranged for the exchange of information, in which Rosbaud made it clear that he knew who was and wasn't in the Uranverein, where its central location was – the Max Planck Institute – who had selected the members and who had criticised Hahn for publishing his original fission paper. After their conversation, Hutton quickly departed back to Britain. As Rosbaud put it, he 'had the kindness to transmit the information to Dr. J. D. Cockcroft, FRS'.[28] After the war, in a letter to Francis Simon, Rosbaud said that Hutton 'knows all about me'.[29]

Rosbaud was in London himself not long afterwards and met Cockcroft who took him for lunch in the neoclassical building of the Athenaeum Club ('scientists and bishops') just off Pall Mall. This time Rosbaud's masterly summary of the experimental results in nuclear fission impressed Cockcroft. Rosbaud was able to go beyond Hahn's and Strassmann's work, and he described other, as yet unpublished, experiments. These included that by Siegfried Flügge at the Kaiser Wilhelm Institute for Physics in Dahlem, 'aimed at determining whether atomic energy was practical', and that of Willibald Jentschke and Friedrich Prankl at the Institute for Radioactivity in Vienna, who were exploring how the energy of the split atom might be harnessed.[30] At the end of lunch, Cockcroft urged Rosbaud to keep sending reports. Scientific intelligence was now a top priority.

Not long afterwards, Flügge's article appeared and it seems that it too was a consequence of the disturbing April meeting that Rosbaud had heard about through Mattauch and passed on to Hutton. Flügge was so troubled by the setting up of the Uranverein that, as soon as he could, he published his article in Rosbaud's *Naturwissenschaften*, highlighting the dangerous potential of fission. In his June article, Flügge was flagging up the fact that the Germans were well aware of the military uses of nuclear fission – it was, in effect, a warning to the world. It was no more than natural for Flügge's article to appear in *Naturwissenschaften*, but even so, it was Rosbaud again who ensured that it was published in such a timely fashion. After the war R. V. Jones, the man who ran Britain's scientific intelligence collection and collation, asked Flügge if it was true that he had been trying to sound a warning with his article in *Die Naturwissenschaften*. Flügge said that it was.

And so, Rosbaud had already three times had a hand in ensuring that potential German secrets saw the light of day, when others would rather have kept them under wraps.

Of course, in light of how the war turned out, it might have become convenient for Flügge to say what he said; viewed in another way, his article could have been seen as threatening. But that too is unlikely. In a letter he wrote after the war, Rosbaud drew attention to the fact that one leading Nazi chemist had expressed the view, following Flügge's article, that 'the editorial staff of *Naturwissenschaften* were acting without any responsibility for the interests of the Fatherland in publishing the results of their [the physicists'] work'.[31] In the political atmosphere of the times, the fact that the article appeared at all suggests that Flügge told Jones the truth. And, as we shall see, it was part of a pattern among certain other German physicists.

Later that year, Rosbaud was in London again and Hutton travelled down from Cambridge to meet him. As the professor wrote in *Recollections of a Technologist*: 'He asked me to meet him in London as he had some important news. We found a safe spot in the Mall and he asked me to convey the valuable information to those most concerned with it. Apparently Hitler had considered the possibility of an atomic bomb as his secret weapon number 1, but this had to be put aside, because the only German physicists who could have given effective help refused to cooperate.'

On the face of it this wording has always confused historians. What exactly did it mean? In fact, what Rosbaud was trying to convey on this occasion was that although the Germans had their own uranium project, begun on 29 April as we have seen, it had run into trouble early on, mainly because the scientists concerned were squabbling among themselves. One bone of contention was between theoretical and experimental physicists – who should have the upper hand. A second problem was over which moderator should be used to slow down the neutrons. And a third problem was curiously Germanic: many scientists agreed to take part in the secret uranium project but they were less interested in its military uses than in the pure science aspects. They took part primarily because it would keep their younger scientists from being drafted into the army.[32] Paul Harteck put it this way in a letter: '[Wilhelm] Groth [a physical chemist] and a few other members of my institute in [the University of] Hamburg came formally into my office and asked me to propose a research project – since war seemed imminent – that would prevent them from being drafted.'[33] This is what Rosbaud was trying to get across. As he put it later, some of the scientists 'used the war to further their research'. At the war's end, when a raft of German physicists were held at Farm Hall, an SOE (Special Operations Executive) 'safe house' near Cambridge that was secretly bugged, one thing that became clear was that, as the Germans themselves admitted, they were incapable of collaborating like the Allies did. As one of them put it, 'real cooperation . . . would have been impossible in Germany. Each one said the other was unimportant.'[34]

Another early example of this ambivalence on the part of German physicists is arguably shown by the behaviour of Werner Heisenberg. Heisenberg was one of the more brilliant – and charismatic – German scientists, a full professor at the age of twenty-five, his 'uncertainty principle' being one of the foundations of quantum mechanics, putting him on a par with Bohr. He was included in the Uranverein and produced an early two-part report, headed *The Possibility of the Technical Acquisition of Energy from Uranium Fission*.[35] In the first part, produced in December 1939, he concluded that in a reactor fission could be controlled to produce energy rather than an explosion, that graphite and heavy water should be useful as moderators and that uranium could be used as the basis for a bomb of tremendous power if it could be significantly enriched. The second part of the report was ready in February 1940 when he expressed himself as more sceptical of nuclear fission, and in fact he omitted all mention of fission as a basis for a bomb, referring only to the engineering difficulties it would entail. While he acknowledged that

Germany possessed more than ample amounts of uranium, thanks to the Joachimsthal (Jáchymov) mines in Czechoslovakia, she lacked the means to process it on the industrial scale needed.[36]

Heisenberg's 'ambivalence' (or otherwise) would become an important issue in due course, but between 1939 and 1941 scientists at sixteen universities in Germany worked on technical aspects of atomic energy, some belligerent and some not. All the German research was secret. The physicists were not even allowed to keep copies of the reports they prepared.

Rosbaud's contribution to nuclear intelligence was already considerable and war had not yet broken out. Moreover, though he could have fled Germany with his wife and daughter, Rosbaud resolved to stay and do what he could to help the Allied cause in scientific matters. He could see that a war was coming. Being a thorough man, always thinking forward, he took the precaution, in the first week of September 1939, the very week war broke out, to again contact his old friend, Frank Foley, who was now a 'passport officer' in Oslo, where he had recently been posted, after fleeing Berlin. The ostensible reason for the contact was for Rosbaud to enquire whether he could obtain a visa to visit his wife in England. Foley went through the motions and passed on the request to London. After a few days, the reply came back: Rosbaud had left it too late. The Nazi–Soviet Pact, signed a few days before on 23 August, had closed all Europe down. In fact, Rosbaud's cover had been surreptitiously established: should anyone ask, he was persona non grata in Britain. Between them, Foley and Rosbaud had set up the most fruitful conduit for scientific intelligence to leave the Reich.[37]

On Friday 1 September, the same week that Rosbaud was in touch with Foley and the very day that Germany invaded Poland, Bohr and John Wheeler, an associate at Princeton, published a very influential paper ('a classic'). Entitled 'On the mechanism of nuclear fission', it was to be the last paper published in the spirit of openness that could not be maintained two days later, when a state of war existed between Germany, Britain and France.

In their paper, Bohr and Wheeler explained formally what had first been understood at that hectic meeting back in the snowy spring in Fine Hall at the Institute for Advanced Study at Princeton (see p. 42 above) – that it was primarily U-235 that fissioned explosively with slow neutrons.[38] This became the accepted orthodoxy and, since U-235 was so rare in nature, and so similar in weight to U-238, Bohr – and many others – continued to view the outbreak of war with less foreboding than was, in the event, justified. He and others like him remained confident that the separation of one isotope from another, which would be needed for a bomb, was so difficult that it was extremely unlikely to be achieved any time soon.

Not everyone shared Bohr's scepticism. In a letter written shortly afterwards, on 23 September 1939, addressed to Fritz Demuth of the Emergency Association of German Scientists Abroad, which was set up in Zurich to help physicists and others forced to leave the Reich, Francis Simon, an émigré scientist at the Clarendon Laboratory in Oxford, made it clear that every physicist in Germany, France, Britain and the United States would understand the potentialities of fission. He went on to say that he was sure the Germans were working on a bomb but insisted they would not know more than anyone else, adding that 'America does the most'. This shows the general climate in physics as the war got under way: the menace of fission was well understood among physicists everywhere.[39]

Almost everyone now knows that, across the winter of 1939–40 and on into the spring, there was a period of what the British called the 'phoney war' and the Germans called *Sitzkrieg*, when nothing much happened as regards real hostilities. During that same period, much the same happened in nuclear physics, at least in America. The Briggs Uranium Committee took time to produce its report (not finished until November 1939) recommending 'adequate support for a thorough investigation' of atomic bombs, but then the government sat on these recommendations until well into 1940. (This committee had its problems, not least the fact that not all its members were US citizens, something that could cause embarrassment in the event of a Congressional inquiry. Martin Sherwin makes the important observation that such bureaucratic fears would dog the bomb project throughout the war, to produce one overriding effect: 'The need to succeed would grow with the project's budget.')

Fission studies continued at several universities, but the prevailing view in America was still that a bomb was a very remote possibility. Research and thinking (and government action) were going ahead in the UK but in great secrecy, especially after the German invasion of Norway in early April 1940 (see the next chapter).

However, in January 1940 a major review article on fission had appeared in *Physical Review*, authored by Louis A. Turner, a Princeton physicist, who had surveyed nearly 100 papers that had been published since Hahn and Strassmann's original discovery a year earlier. These articles underlined the enormous impact of the Hahn and Strassmann experiments but Turner was bothered by something else. This was the fact that the Bohr and Wheeler report, confirming that slow neutrons did indeed fission U-235, had been allowed to be published. He wrote to Szilard – then becoming known as the great apostle of secrecy – saying he was puzzled as to the 'guiding principles' governing what was publishable and what wasn't.

More important still, his reading for the review article had produced some new thinking and on 27 May he again wrote privately to Szilard. In this letter, Turner enclosed a note he was intending to send to the *Physical Review* in which he developed the idea as to whether, when U-238 (and not U-235) was bombarded by neutrons, it might be transformed into a heavier element (later to be called plutonium), which might itself be fissionable. This possibility had been aired at the February meeting in Fine Hall but not taken further. Turner ended his letter to Szilard by asking if his paper should be withheld from publication because of its possible military value.

Of course this gave Szilard something else to worry about. He found the implications 'stunning . . . With this remark a whole landscape of the future of atomic energy rose before our eyes in the spring of 1940.' Szilard saw further than Turner, Richard Rhodes says, suspecting that bombs could more easily be made with plutonium than with uranium, a suspicion that was to prove correct.[40] It was also an idea that had occurred independently to Carl von Weizsäcker in Germany. His conclusions were of course kept secret.

Szilard wrote back to Turner to say that his own paper was secret, and implying that there was an official move under way to withhold papers. He prevailed on Turner to write to *Physical Review* and delay publication. 'It was as well he did so,' writes Spencer Weart. 'Turner's paper could have been an essential clue for the Germans and others.'[41] (He meant that it showed the Americans were thinking along the same lines as Weizsäcker and would surely have caused them to take the idea more seriously.)

Again, the timing was crucial. Barely two weeks later, on 15 June, two Berkeley physicists, Edwin McMillan and Philip Abelson, published in the *Physical Review* a paper in which they completed the fission jigsaw with a long letter where they described how they had used the Berkeley cyclotron to prove that they had in fact created a new 'transuranic' element, neptunium, element 93, which had a half-life of ≅ 2.3 days that would decay in turn into an isotope of element 94 which we now know as plutonium. They explained that this element – which does not exist in nature – was extremely stable with a half life 'of the order of a million years or more'. Experiment had confirmed theory and that fact had been published openly.

The implications were not spelled out – that it would be possible chemically to separate out plutonium, making preparation of fissionable material much cheaper and more practical – because McMillan and Abelson did not see the connection, as Turner and Szilard did, showing once more that the logical follow-through does not always come about, even at the highest levels.

But publication of this note was too much for James Chadwick, the discoverer of the neutron and professor of physics at Liverpool. Britain had been at war for nine months now and the phoney war was truly over. June 1940 was the very month he had heard that the Germans had bought large quantities of uranium from Canada just before war broke out. And he realised that all these experiments pointed to the fact that Bohr and Wheeler's theory was correct: that plutonium, element 94, was fissionable like U-235. Chadwick complained to the British authorities and they sent a letter of rebuke to the United States, urging secrecy. The US was of course not yet at war itself but the protest succeeded, in that the McMillan and Abelson paper was the last one published in America or Britain on uranium fission until 1946.[42]

From then on Szilard finally got his way, thanks mainly to Gregory Breit at the University of Wisconsin. Breit had known Szilard for some time and possessed a fellow feeling for secrecy. He was also in a position to do something about it. He had recently been elected to the National Academy of Sciences and been put in charge of the Division of Physical Science of the Academy's National Research Council. At a meeting of the committee he advocated censorship. The idea was far from popular but a committee on publications was set up to consider the problem and Breit was made chairman of a subcommittee specifically instructed to examine uranium. He immediately began writing to journal editors, proposing a voluntary plan in which papers relating to fission would be submitted to his committee before publication. Sensitive papers, he said, would be circulated only to a limited number of qualified individuals, though in due time the papers would all be published in book form, with their original date.

Not everyone was keen but the editors of the main journals did in the end agree. 'As recently as six months ago,' wrote Ernest Lawrence, the Nobel Prize-winning inventor of the cyclotron at Berkeley, to Breit, 'I should have been opposed to any such procedure, but I now feel we are in many respects on a war basis.'[43]

Within a few weeks, Breit had established a total censorship on American fission research. Papers were passed around by post and the innocuous ones were published; others were withheld. Long before the United States went to war, it was keeping vital information within its own borders, but there was no government participation.

In fact, the lack of government interest went wider than that. From 1939 to the end of 1941 the United States dragged its heels so far as atomic research was concerned.

As we have seen, the possibility of an atomic bomb had been first drawn to the president's attention at a meeting called at the White House on Wednesday 11 October 1939, a month into the war in Europe, when his old friend Alexander Sachs showed him what became a famous letter signed by Albert Einstein but actually drafted by Leo Szilard. This alerted the president to a new possibility that had arisen: a new and 'extremely powerful bomb, based on nuclear chain reactions in uranium'.

A government Advisory Committee on Uranium was convened immediately under the veteran Lyman Briggs, head of the National Bureau of Standards (America's National Physics Laboratory). Then aged sixty-five and a soil scientist by training, Briggs was pretty much the wrong choice and under him the American nuclear project languished (though the Navy Research Laboratory had been independently investigating isotope separation for some time). Things only started to move with the arrival in the capital of a smooth and urbane newcomer. Vannevar Bush – 'Van' to his friends – was the newly appointed president of the Carnegie Institution for scientific research. Before that he had been a professor of electrical engineering at the Massachusetts Institute of Technology, where his 'intellect and technical ingenuity' drew attention and a series of commercially astute patents provided a certain financial independence. (These included the self-justifying typewriter and a circuit for the self-dialling telephone.) At fifty, he was bristling with energy, and had the ability and requisite will to shake things up.[44] Like Sachs before him, Bush crafted a ten-minute slot in the Oval Office schedule, when he laid out his ideas for co-ordinating and consolidating scientific research on military matters. He had prepared in advance a single sheet of paper on which his plans were distilled. Roosevelt approved it there and then.

In no time Bush set up what became known as the National Defense Research Committee (NDRC). Its aim, from June 1940 – the time Chadwick was complaining about how much nuclear physics was being published openly in the United States – was to bring a new energy and direction to the application of secret science to warfare by American academics and military staff.[45] However, even at that stage an atomic weapon was not a priority. Bush was convinced of the bomb's 'great impracticability' and to begin with thought nuclear research would have medical and other applications.

By great good fortune, however, not long after Bush got organised, in August, Henry Tizard led a top-secret British mission to the United States, mainly having to do with an intellectual/materials exchange or deal, sharing with the Americans on the one hand air defence secrets and naval bases in the West Indies, in return for fifty First World War Lend-Lease destroyers for the British.[46] On the British side, the centrepiece was the cavity magnetron valve. For some, this – the basis of radar – was 'the "most important item in the whole Lend-Lease arrangement". So new and impressive was it that it played an important role in boosting trans-Atlantic co-operation by showing the Americans how advanced was much British wartime research.'

It was still the case, however, that the Americans were not really thinking about bombs. When Archibald Hill, the British scientific attaché in Washington, asked about uranium research, he was told that there were no practical results at present and that American scientists considered uranium research a waste of Britain's already strained resources.[47]

But the Tizard mission was such a success that, as a result, James B. Conant, president of Harvard and Bush's deputy on the NDRC, travelled to London to organise an office to help further the exchange of scientific information.[48] (According to a note in the Cockcroft archive at Churchill College, Cambridge, a meeting was held at the University of Liverpool on 30 June 1940, where Hans von Halban stated that Vannevar Bush had approached Joliot-Curie suggesting there be collaboration between the USA, the UK *and* France.)[49] Conant stayed at Claridge's for two months and what he learned while he was there would, in due time, change the course of the war.

5

The Midwives

While Leo Szilard and his colleagues at Columbia were just beginning to grasp the fearsome properties of plutonium, two sets of events in Europe, which occurred over the same brief forty-eight hours on Tuesday 9 and Wednesday 10 April 1940, would determine the ultimate outcome of nuclear rivalry.

Shortly before 5 a.m. on the Tuesday, the inhabitants of Oslo were shaken awake by the sound of heavy gunfire in the Oslofjord. News quickly spread that Norwegian forts on the coast were exchanging fire with an invading navy.

An emergency meeting of the cabinet was summoned. Ministers had barely begun their deliberations when Kurt Bräuer, the German ambassador in Oslo, interrupted the meeting with Hitler's demand for an immediate surrender. He was summarily rebuffed and records show that Bräuer was back in his own office by 5.52 a.m., cabling Berlin that he had been told that 'the struggle has just begun'. He scarcely took this response seriously and, together with colleagues from the legation, strolled down to the harbour to watch the arrival of the invading fleet.[1]

The ambassador and his team looked out to sea in vain for quite some time, shivering in the cold morning air. But Hitler's ships did not materialise, not then anyway. The Norwegian guns mounted at Oscarsborg, a fortress 15 miles down the fjord, had fatally holed one German cruiser and badly mangled another.

That bought time. Enough time at least for the Norwegian government and the royal family to make good their escape. Later that same day a train hurried them to the town of Hamar, 62 miles north.

The Nazis came back, of course, and in increased strength. But those opening exchanges set the scene. A small country, Norway could not resist Hitler's might for long, but they would do all they could to make life difficult for the occupying power.

The very next day, Wednesday 10 April, in the offices of the Royal Society in London, a small committee met to consider a brief memorandum by two émigré scientists working at the University of Birmingham.

If there was a single moment when an atomic bomb moved out of the realm of theory and became a practical option, it had occurred one night in the second week of March 1940, in discussions between Otto Frisch and Rudolf Peierls. They were the midwives who offered crucial help in the birth of the bomb.

As war approached, Frisch had grown more and more apprehensive. He was Jewish and had already been forced to flee Germany, losing his job in Hamburg. Every time he encountered an English physicist at Bohr's institute in Copenhagen he would badger them about the possibilities of work in Britain. Frisch was close to being a concert-level pianist, and his chief consolation was in being able to play. But then, in the summer of 1939, word must have got around because Marcus Oliphant, an Australian who was one of the inventors of the cavity magnetron (out of which radar was developed), and by now had become professor of physics at Birmingham, invited Frisch to Britain, ostensibly for discussions about his research. Frisch packed a couple of bags, as one would for a weekend away. Once in England, however, Oliphant made it clear to Frisch he could stay if he wished. While Frisch was in Birmingham war was declared, so that was that. All his possessions, including his beloved piano, were lost.[2]

Rudolf Peierls had been in Birmingham for some time. A Berliner, the son of a Jewish director of AEG (Allgemeine Elektricitäts-Gesellschaft), the General Electricity Company, founded in 1887 with a special emphasis on design excellence, he was a small, bespectacled, intense-looking man, and had the classic education of pre-war – even nineteenth-century – Germany, studying at several universities: Munich (under Arnold Sommerfeld, who taught various Nobel Prize-winners), Leipzig (Werner Heisenberg, Nobel Prize in 1932), Zurich (Wolfgang Pauli, Nobel Prize in 1945), later (on a Rockefeller scholarship) Rome (Enrico Fermi, Nobel Prize in 1938) and then Manchester (Hans Bethe, Nobel Prize in 1967). Peierls was in Manchester when the purge of German universities had begun – he could afford to stay away, so he did. He had a Russian wife, Genia, who did not speak German and, since Peierls did not speak much Russian, they communicated in the only language they both understood: English. Peierls would become a naturalised British citizen (after some difficulty) in February 1940, but for five months, from 3 September 1939, the day war broke out onwards, he and Frisch were technically enemy aliens.[3]

As a result of this, Peierls and Frisch were barred from working on war-related projects, such as radar or jet engines. Instead, they had to make do, concerning themselves with matters unconnected to the war: atomic theory.

Until Frisch joined Peierls in Birmingham, the chief argument against an atomic bomb, despite the understanding of fission that had been achieved, had been the amount of uranium needed to 'go critical', start a sustainable chain reaction and cause an explosion. Estimates had varied hugely, from 13 to 44 tons and even 100 tons.[4]

It was Frisch and Peierls, walking through the leafy streets of Edgbaston, the university quarter in Birmingham, and conferring in the Nuffield Building on campus, who first grasped that the previous calculations had been wildly inaccurate. Frisch worked out that, in fact, not much more than a kilogram of material was needed. Peierls's reckoning confirmed how explosive such a bomb would be. This meant calculating the available time before the expanding material separated enough to stop the chain reaction proceeding. The figure Peierls came up with was about four-millionths of a second, during which minuscule period of time there would be eighty neutron generations. Peierls worked out that eighty generations would give temperatures as hot as the interior of the sun and 'pressures greater than the centre of the Earth where iron flows as a liquid'.[5]

A kilogram of uranium, which is a heavy metal, is about the size of a large orange – surprisingly little. Later, Frisch and Peierls admitted they were 'quite staggered' by their results. They rechecked their calculations, and did them again, reaching the same conclusions. A bomb could be formed, they reasoned, by constructing two hemispheres of U-235, each half the critical mass, and firing them directly at each other so that a chain reaction would be set off. And so, as rare as U-235 is in nature (in the proportions 1:139 of U-238), they dared to hope that enough material might be separated out – for a bomb and a trial bomb – in a matter of months rather than years. A nuclear explosion, they said, would be equivalent to the blast of about 1,000 tons of dynamite and leave the area suffused with radioactivity for some considerable time, creating still more lingering deaths.

In the two-part paper that they prepared in secret (typing it themselves so that no secretaries were involved), they made the point that an atomic bomb would be so destructive that there could be no defence. The only precaution, as they quickly foresaw, was deterrence. If Hitler had the bomb, the only way to forestall his use of it was the threat of reprisal. And so they suggested the risk was too great to ignore, even though it would require an industrial-sized effort. 'Even if this plant costs as much as a battleship, it would be worth having.' At the same time, they pointed out that there would be considerable radioactive 'fallout' that could not help but have an effect on civilians, wherever it was exploded and might make its use 'inappropriate' for a democracy.[6]

Frisch and Peierls took their calculations, and their argument, to Oliphant. He, like them, recognised immediately that a threshold had been crossed. And it was this 'Frisch-Peierls memorandum' that was considered by a small government subcommittee, originally called the 'U-bomb subcommittee', which was brought into being especially for the purpose, and just happened to meet for the first time on that fateful Wednesday in April 1940. The committee came to the conclusion that the chances of making a bomb in time to have an impact on the war were good, and from then on the development of an atomic bomb became British policy.[7]

Momentous as it was, that wasn't the end of business that day, at the Royal Society. The same meeting also heard from Jacques Allier, a debonair French intelligence officer who, a month earlier, had masterminded a daring raid in Norway, in which he had stolen the world's only supply of heavy water from under the noses of the Germans. His presence at the meeting, and the story he had to tell, confirmed that the committee was meeting not a moment too soon. The Nazis *were* intent on developing their own nuclear weapon and Allier had the evidence (see the next chapter).

Amazingly, in retrospect at least, Frisch and Peierls received no acknowledgement that their report, which had been considered on that April morning, was being acted upon. As enemy aliens, they were not even allowed to know the name of the chairman of the Whitehall committee investigating their ideas and calculations.[8] Peierls discovered it on the physicists' grapevine, so he wrote politely but firmly arguing that it was madness that Frisch and he should be barred from working on their own idea. Common sense prevailed.

In late spring work got under way in earnest. Over the summer months, several sets of nuclear research were secretly launched at universities in Bristol, Birmingham, Liverpool, Oxford and Cambridge. Several of the individuals engaged in this research – quite apart from Frisch and Peierls – were technically enemy aliens.

Over the next several months the British scientists became increasingly confident that a weapon was feasible and by the late spring/early summer of 1941 a number of firm conclusions were emerging: that a critical mass of 10 kilograms of uranium would be large enough to produce a vast explosion; that a bomb that size could be loaded onto existing aircraft; and that the bomb could be ready in two years, and therefore could play a part in the war.

As the bomb emerged as a practical probability, it became all the more important to keep it secret. Two code words were introduced to stop outsiders wondering too much about what was going on. The committee itself was renamed the MAUD Committee. One version has it that the acronym stood for Military Applications of Uranium Detonation, but there appears also to have been an element of unwarlike whimsy in this. It had come about because, at one stage, a more or less garbled and incomprehensible telegram had been received from Lise Meitner in Stockholm. Meitner had sent the telegram to reassure British nuclear scientists that Niels Bohr was 'unhappy but safe' in Copenhagen now that Denmark was occupied (she was invaded on the same day as Norway). Meitner's cable ended with the words, 'TELL MAUD RAY KENT'. The British physicists fretted that this was a mysterious code but the truth was rather more prosaic, even banal. The Bohrs had had an English governess for their children, with whom they remained friendly. She was called Maud Ray and lived in Kent. The rest of her address had been left off the garbled telegram. Later, all references to nuclear weapons were coded as 'Tube Alloys'. It sounded scientific but was in fact nonsense and gave nothing away. A 'Tube Alloys Directorate' was set up in Old Queen Street, just off Birdcage Walk and 500 yards from Downing Street.[9] Literally, as well as figuratively, the bomb had moved to the heart of the British war machine.

6

The Strategic Sabotage of Heavy Water

Begun in April 1940, the fight to take Norway occupied the Nazis until June. In May, in Tromsø, 200 miles north of the Arctic circle, the Norwegian government held its last meeting for the duration on Norwegian soil. Then, 400 government ministers and diplomats, the king and the crown prince and other members of the royal family, escaped on HMS *Devonshire* for London.[1]

The Norwegian government and the Royal Family would see out the war in exile, in the (relative) safety of Britain. But Norway itself posed problems because of certain facilities at Rjukan and Vemork, about 80 miles west of Oslo. The Norsk Hydro factory was the only plant in the world capable of producing heavy water on anything like the scale needed for an atomic bomb. The importance of heavy water, to remind ourselves, lay in its being a moderator, especially suited to slowing down neutrons, so as to make fission more likely. That made Vemork of paramount interest to both the Germans and the Allies.

Moreover, the plant, though Norwegian owned, had two major foreign investors, one German, I. G. Farbenindustrie Aktien-gesellschaft, the other French, the Banque de Paris et des Pays-Bas. As war approached and the potential menace of fission became more apparent, the significance of Norsk Hydro grew.

Once Norway, with its crucial factory, was occupied, and while its next-door neighbour, Sweden, remained neutral in the war, Scandinavia couldn't help but become a vital route for war-related intelligence to reach Britain. And it was Jacques Allier's story, which he brought to the fateful meeting at the Royal Society in April 1940, that was the first real hard intelligence the British received about Germany's atomic programme.

Because of the French shareholding in Norsk Hydro, they had learned of a sudden heightened interest by the Germans, at the end of 1939, in buying up Norsk's stocks of heavy water, that they wanted the Norwegians to increase production by fully 500 per cent, and that the Norwegian management had been instructed to keep the matter secret. The French, when they learned about it, had taken this move by the Germans very seriously, to the extent that the French prime minister, the finance minister and Joliot-Curie met with Allier, who was both a director of the Banque de Paris et des Pays Bas and a member of France's Deuxième Bureau, the French intelligence service, and gave him resources and instructions to buy up all the heavy water that Norsk had.

Allier, as he told the Royal Society meeting, had travelled to Norway using a false passport, had established that the Norwegians were more sympathetic to French interests than German interests and, in an audacious coup, using specially prepared suitcases which would not destroy the chemical composition of heavy water, had stolen Norsk's entire stock from under the Germans' noses and secretly flown 185kg, or 410lbs, to Scotland and then on to London and Paris.

That was by no means the end of the adventures of these 185kg of heavy water. Following the Nazi occupation of Denmark, their sweep through Holland and Belgium and the dramatic collapse of France in May that year, the water was evacuated again, first to Riom, in the Massif Central, where it was hidden in the local prison. Then, as the full extent of the French collapse and the German occupation became clear, it was rushed ahead of the advancing forces to Bordeaux and smuggled on board a British coal steamer bound for Falmouth, together with two of the French scientists who had been planning to explore its possibilities, Hans von Halban and Lew Kowarski. Reaching Britain safely, the heavy water was again hidden in a west London prison, Wormwood Scrubs, before being transferred to the cellars of Windsor Castle. Halban and Kowarski were recruited into the British project and settled at the Cavendish Laboratory in Cambridge.[2]

It is not clear how much the Germans ever knew about Allier's coup during the war. But when Germany occupied Norway in June 1940, its physicists must have thought they had achieved a coup of their own, having gained control of the only heavy water plant in the world. However successful or otherwise their own nuclear research turned out to be, they had at least deprived the enemy of access to one of the most valuable ingredients on the way to any atomic explosion.

As it turned out, almost the opposite was true, and the plant at Vemork would become, before long, the Achilles heel for the German project. It would help provide, in effect, the nucleus of a network of agents by means of which, over the next two years, intelligence leaked out of the German project to the British. The authorities in London knew, from the experience of Jacques Allier, that at least some of the personnel on the Norsk Hydro staff were sympathetic to Allied aims, and of course he was able to provide specific names.

The central figures in what we might call the Norsk nucleus were Leif Tronstad, Jomar Brun and Eric Welsh, though they were surrounded by a raft of other individuals, equally brave, including Njål Hole, Sverre Bergh and Harald Wergeland.

Leif Tronstad (1903–1945) was born in Baerum, west of Oslo, and studied chemistry at the Norwegian Institute of Technology, graduating in 1927 with a paper so exceptional that, as was the Norwegian custom, it was drawn to the attention of the king. Tronstad later studied in both Berlin and Cambridge, becoming fluent in three languages. He was appointed professor of chemistry at NIT in 1936. One of the pioneers of research into heavy water, he was intimately involved in the creation of the Vemork plant.[3]

Having done his military service, he remained on the reserve and when war broke out and the invasion of Norway was completed, he formed part of the resistance, in particular providing technical intelligence to London. His main contribution, to begin with, was obtaining details of German warship movements in and out of Trondheimsfjord, a massive natural harbour. Before long, however, Tronstad started to supply information with an emphasis on heavy water. This brought him into regular contact with Jomar Brun at Vemork, a man he had known for years. Brun had been involved from the beginning in Norsk Hydro's interest in heavy water, which was a 1934 spin-off from the firm's main business of deriving hydro-electric power from the *Rjukanfoss*, or the 'foaming waterfall'. This fed electricity to the plant at Vemork, which manufactured explosives and ammonia for conversion into nitrates for fertiliser.

Gradually Tronstad widened his sphere of activities, eventually becoming involved in organising what became known as Milorg – the national military resistance organisation. All went well for a time, but then, in the autumn of 1941, the clandestine radio station that the underground used was discovered by the Germans, and Tronstad was forced to flee.[4]

In London he was immediately taken to the Secret Intelligence Services headquarters in Broadway near St James's Park, where he was introduced to Eric Welsh.

Welsh, British-born but married to a descendant of the Norwegian composer Edvard Grieg, was then in his forties and, like Tronstad, a trained chemist. He had once held a senior post with the Bergen firm Internasjonal Farvefabrikk A/S, which specialised in tiles and paints. This had brought him into contact with Jomar Brun at Vemork, who had asked his advice on certain corrosion problems associated with heavy water production. Welsh was therefore well informed about heavy water and Tronstad's many wireless reports, some about heavy water, had usually ended up on his desk.

Welsh confided that stopping the heavy water production at Vemork was now a top war objective. Brun was instructed to sabotage the heavy water production cells. He chose to do this at first by surreptitiously adding castor oil, which made the water foam and halted the production process.

This was not a satisfactory long-term solution, however. Moreover, there were an increasing number of intelligence reports from Brun showing that there were numerous visits to Vemork by scientists whom the members of the MAUD Committee suspected were directly concerned with the uranium programme in Germany. Welsh therefore came under pressure to do something about putting a permanent halt to Norwegian heavy water production.

This culminated in July 1941 when the war cabinet ordered the Vemork plant to be destroyed. Following this decision, Welsh arranged a dinner between Tronstad and John Cockcroft (on the evening of Thursday 6 August), when Cockcroft gave the other man an overview of the recent Allied decisions about nuclear matters, a technical briefing, and what the current opinion was on German progress.[5]

Pressed by his masters to do something about Vemork, Welsh looked beyond SIS, which had no capabilities for the kind of operation needed. The most obvious alternative was SOE, the Special Operations Executive, which Winston Churchill had created in July 1940 with the admonition, 'And now set Europe ablaze!' SOE specialised in relationships with resistance groups across Europe, most of which conducted sabotage of the German war machine.

Vemork was in western Norway, much of which lies on a great barren plateau, the Hardangervidda, in an area called Telemark. The landscape is windswept and wild; river valleys and lakes abound, as do waterfalls, which powered the plants at Vemork and Rjukan.[6] This harsh environment could not be avoided: any sabotage team would have to be dropped there.

Over the course of the war, there were no fewer than four attempts to sabotage heavy water production at Vemork. One commando raid on 19 November 1942, codenamed 'Freshman', was a disaster from the Allied point of view, with bombers and gliders being lost, and around thirty men killed. A later commando raid, codenamed 'Gunnerside', on 16 February 1943, was much more effective and succeeded in blowing up the plant with no loss of life for Allied forces. However, it put the plant out of action for rather less time than was hoped, and in November 1943 a 'precision' bombing raid by the US Air Force was tried. It too failed to knock out the plant but, despite this second failure, the repeated attacks persuaded the Germans that Vemork would never be left alone and so plans were made to transfer the remaining heavy water to Germany. Norwegian intelligence provided details of the date and route by which the water was due to be transported and on 20 February 1944 a third sabotage team blew up the ferry carrying the heavy water across Lake Tinnsø, sending its cargo to the bottom and killing two dozen Norwegians in the process.[7]

It should not be overlooked that the continued sabotage of Vemork had a strategic aim as well as a tactical one. The attacks had the deeper purpose of attempting to convince the Germans that the Allies attached far more importance to heavy water than was in fact the case, to keep them committed to heavy water when the Allies were going down a different route. In other words, it was a deception. By the second half of 1941, the MAUD Committee's research had shown that an atomic explosion was possible *without* a moderator. When he wrote the Final Report of the MAUD Committee, at the end of August 1941, Chadwick included a paragraph about heavy water. 'In the early stages we thought that this substance might be of great importance for our work. It appears in fact that its usefulness in the release of atomic energy is limited to processes that are not likely to be of immediate war value, but the Germans may by now have realized this, and it may be mentioned that the lines on which we are now working are such as would be likely to suggest themselves to any capable physicist.'

But the capable physicists in Germany never drew any such conclusion, partly because of Bothe's error over graphite, described on pp. 56–7, and partly because of the sustained campaign to sabotage Vemork.

In fact, following the creation of the Norsk nucleus, and the MAUD Committee – in effect on the same day – their fates were inextricably intertwined. Intelligence about the enemy's bomb was as central to the course of events that followed as were the advances in physics and engineering made by Allied scientists. That intelligence flowed to London rather than anywhere else. And it was the British failure to share that intelligence early, and the American reluctance to accept its conclusions, when it was at last shared, that was to cause the Americans, despite their considerable resources, to overestimate the German capacity in atomic weapons and saddle the world with a bomb that, strictly speaking, as we shall see, was not needed.

7

The First Glimpses of Germany's Nuclear Secrets

Sabotage of the heavy water plant at Vemork had begun in the late summer/autumn of 1941 and the first – catastrophic – raid was carried out by the end of 1942. As a result, the Germans must have realised then that the Allies were aware that they had some sort of bomb project in the works. This did not necessarily tell them that the British, or the Americans, had a similar project of their own and, as we shall see later, most German physicists remained convinced that they led the world in fission and that no one could catch up. Still less did they imagine that the bombing of Vemork was in part a secret decoy to keep them thinking that heavy water was more important than it was.

In fact, given the amount of effort and anxiety that went into keeping the existence and development of the Allied atomic bomb secret (chapter 3), it comes as something of a surprise to learn that there was quite a bit of press coverage of various aspects of the project throughout the war. In September 1944 General Groves's staff amassed a compendium of no fewer than 104 press references to the Manhattan Project 'or related subjects' over a 58-month period beginning in November 1939. Twenty-seven reports came before – and seventy-seven came after – a directive of 28 June 1943, when the US Office of Censorship issued a confidential letter to the nation's editors and broadcasters stipulating 'that nothing be published or broadcast about new or secret military weapons' or even 'experiments' involving atom smashing, atomic energy, atom splitting, heavy water, radioactive materials, cyclotrons, uranium, hafnium, protactinium, thorium, deuterium and so on. Certain decoy words were included, such as 'yttrium', to muddy the letter's real purpose.[1]

One very substantial report had been printed in the *New York Times* as early as 5 May 1940, by its science reporter, William Laurence, and headlined 'VAST POWER SOURCE IN ATOMIC ENERGY OPENED BY SCIENCE'. It was explicit about the enormous explosive power of uranium-235 and didn't shirk the implications it might have for the 'outcome of the European war'. Laurence also added that 'every' German scientist in the field had been ordered to 'drop all other researches and devote themselves to this work alone'. Though this latter claim was something of an exaggeration, Laurence thought it legitimate if it provoked US scientists into action.[2] A copy of this article was found in a German laboratory in Strasbourg at the very end of the war but not even this seems to have provoked too much disquiet among their physicists.

Press reports were published not only in the UK and the USA, and arguably the most fateful was one in the Swedish newspaper, *Stockholms Tidningen* at the beginning of September 1941, which referred to a report it had received from London to the effect that the Americans were working on a bomb 'of heretofore-undreamt-of power'. 'The material used in the bomb is uranium and if the energy contained in this element were released, explosions of heretofore-undreamt-of power could be achieved. Thus, a five-kilogram bomb could create a crater 1 km deep and 40 km in radius. All buildings within a range of 150 km would be destroyed.'[3] And it was this article that prompted Werner Heisenberg and Carl von Weizsäcker to visit Niels Bohr in occupied Copenhagen.

Heisenberg and Von Weizsäcker both played central roles in the German bomb project. Born in Würzburg in 1901, the son of a professor of Byzantine history, Heisenberg was a fervent nationalist and in the turbulent wake of the First World War took part in more than one street fight against the communists. An organisation called the Gruppe Heisenberg was formed at his school in Munich where, as the name shows, Werner was the leader, the group meeting in his home. Heisenberg was also a legendary chess player, who sometimes faced his opponents without his queen 'to give them a chance'.[4]

At Munich University, Arnold Sommerfeld was responsible for

taking Heisenberg to Göttingen to hear Niels Bohr lecture and where he made a crucial intervention, slightly correcting something Bohr said. Bohr, being Bohr, hadn't minded and after the lecture invited Heisenberg to accompany him on a stroll.

It turned into more than a stroll, for Bohr invited the young Bavarian to Copenhagen where, over the next few years, Heisenberg made his name for two fundamental ideas in physics: quantum mechanics and the 'uncertainty principle'. He was awarded the Nobel Prize for physics in 1932, aged thirty-one.

In 1941 Carl von Weizsäcker was twenty-nine, the son of a prominent diplomat, Ernst, and elder brother of Richard, who would become president of the Federal Republic in 1984. Carl made his name before the war for his work on nuclear fusion in the sun. He would win the Goethe Prize in 1958. Heisenberg and Weizsäcker played chess together, in their heads, without using a board.

As a result of the article in *Stockholms Tidningen*, these two visited Copenhagen in September 1941, the same month that the MAUD report was finalised. Heisenberg saw Bohr alone and it has always been a mystery as to whether Heisenberg was informing Bohr (without actually saying so, as that would have been treasonable) that Germany did not have a bomb, or whether he was trying to find out from Bohr what the Allied plans were. The timing of the visit, so soon after the report in the Stockholm newspaper, is certainly suggestive of the fact that Heisenberg was sounding out Bohr. America was not yet in the war, so contact between Bohr and American physicists was not out of the question. Chadwick thought that the purpose of the meeting was for Heisenberg to 'throw Bohr off the scent', but there are also good grounds for believing Heisenberg's post-war claim that he was trying to tell Bohr that Germany was not working towards a bomb. Either way, Bohr felt no need to alert the Allies.

Sabotage, denying the enemy knowledge of the Allied bomb, or misleading him about it, was one thing. Obtaining intelligence of German plans was something else entirely.

The intelligence reaching Britain about nuclear matters was, to an extent, different from that reaching the United States. Moreover, when the crucial information was received, relations between the two Allies were at an all-time low, meaning that certain vital details were not shared, with consequences that have been largely underplayed until now.

The British decided early on that they would not send intelligence agents *into* Germany to find out what they could about the enemy's nuclear preparations. They concluded it was too risky because any agents that they did send would have to be sufficiently familiar with Allied plans to know what to look for. Should they be captured, and interrogated or tortured, they might well reveal what they knew.

But in any case, so far as nuclear intelligence was concerned, Britain was aided by two factors. One was the Norsk nucleus, aided and abetted by individuals in Sweden. The other factor lay in the form of Paul Rosbaud, a spy of unrivalled access, courage and tenacity. His importance to the history of the atomic bomb is incalculable.

Despite these assets, the worry of a German bomb had not proliferated within the British scientific community until their own research began to show that a device was indeed a practical possibility. This happened following the Frisch–Peierls memorandum and the setting up of the MAUD Committee early in 1940. Before that time, Tizard, although he was always one of the biggest sceptics about whether a bomb could be built, was still worried enough in May 1939 to commission a study of the organisation and quality of British scientific intelligence. This was after the discovery of fission and three months before war broke out but around the time Rosbaud was alerting John Cockcroft to the existence of the Uranverein in Germany. Uranium supplies and atom bombs were on Tizard's mind but to begin with no one was to know that. He recruited the young scientist Reginald V. Jones, telling him only that he wanted to know why he was receiving so little information about German developments in air warfare and how the situation could be rectified.

Then in his late twenties, Jones, a Londoner, was a physicist who had worked in Oxford's Clarendon Laboratory. It was through one of his Oxford friends that he first learned – informally, accidentally – about atomic bombs. Just over a month after his appointment, but before he had actually taken up his new role, in July, he was waiting at a bus stop in Oxford High Street with his friend James Tuck, who worked as an assistant to Frederick Lindemann. From Tuck, Jones learned that, following the discovery of fission some months before, atomic bombs were now a distinct possibility. 'Reginald,' Tuck said, 'one day there is going to be a BIG BANG!' Tuck then referred to the article that Siegfried Flügge had published the month before in *Die Naturwissenschaften* as evidence that the Germans were already working on their own weapon (see above, p. 57). He too thought that Flügge might have been sounding the alarm.[5]

Jones turned out to be an inspired choice in his new role. One of his early achievements was to place in context a worrying speech by Hitler delivered in the newly conquered city of Danzig on 19 September, only sixteen days into the war. In his speech, the Führer threatened to employ a '*neue Waffe gegen die es keine Verteidigung gibt*' – a 'new weapon against which no defence would avail'. This caught the attention of the prime minister, Neville Chamberlain, who wanted to know how seriously to take this threat. Jones had the good sense to have made a new translation of Hitler's speech, culled from a recording held at the BBC, which showed that the speech had been reported out of context. What Hitler actually said was that he would soon have a weapon (*Waffe* in German) to rival the one that Britain had so much confidence in – her navy. He was referring to Germany's 'airweapon' – in other words, the Luftwaffe. Jones was therefore able to offer the prime minister a reassurance that Hitler had no new secret weapon such as an atomic bomb. At least not yet.

This cool reappraisal won Jones the trust of SIS and that trust would pay off as the war dragged on. But he was not the only one in London thinking about nuclear secrets on both sides in the war. On 11 June 1940 James Chadwick had written to Cockcroft, saying that he had heard from Otto Maass, a Canadian physical chemist, that the Germans had bought 'very large quantities' of uranium ('presumably pitchblende') from Canada just before the war. A top-secret document prepared by the Ministry of Supply in the same month – for G. P. Thomson's benefit – was more specific, saying that it had been 'recently discovered' that the Germans had bought 25 tons of uranium ('probably uranium oxide') from Eldorado Mines near Great Bear Lake, in Saskatchewan, Canada, again just before the war. The same document said that the Germans were understood to have purchased 25 grams of radium 'since the war', to which should be added about 3 grams per annum from Joachimsthal (Jáchymov), which had been increased to 5 grams a year. The Germans had tried to buy as much as 10 grams from Union Minière prior to the invasion of Belgium but the Belgians had refused. The document went on to 'conjecture' that Holland and Denmark, now occupied, could supply 'as much as 40 to 50 grams'.[6] All the stocks of radium and uranium in Belgium were evacuated via Bruges, and 20 grams of radium were got out via Bordeaux, like the heavy water.

Although R. V. Jones was in charge of the collection and collation of scientific intelligence in general, Cockcroft seems to have had an especially prominent role in the gathering and assessment of atomic intelligence, or else he became a sort of clearing house as news came in.

At one point he led the way in trying to track down any employees of Norsk Hydro who might be in Britain (there weren't any), and he also sought to follow up a rumour that had come via Peierls that a number of towers were being built in Berlin for thermal diffusion experiments. 'These rumours started in Holland, linked to the leaving of Debye.'[7] Debye, the Dutch Nobel Prize-winning physicist and head of the Kaiser Wilhelm Institute in Berlin, who played a prominent role in Lise Meitner's escape, had emigrated to America at the beginning of 1940, and there were those who thought he might have been sent as a Nazi spy to find out about America's atomic research. One rumour that circulated in London was that, on the contrary, Debye had left Berlin because he had refused to take part in experiments on isotope separation but Oliphant thought this might be a front.[8] On 20 July 1940, he sent a secret note to John Cockcroft: 'I am very worried about the position of Debye in the American work and I think that before the uranium question is discussed at all in the USA it must be definitely assured that people like Debye, who are still technically employed in Germany, are excluded rigorously from any discussion of the problem.'[9] Cockcroft agreed and said he would alert the British scientific liaison office in Washington so that the whole question of interchange would be 'carefully gone into'.

Again, it was Cockcroft whom N. B. Mann, of Imperial College, contacted when he too came across the rumour that isotope separation towers were being built in Berlin; it was Cockcroft who was alerted by Jacques Allier to the fact that Joliot-Curie, who knew Frisch was working in Birmingham, feared that he might be an (inadvertent) security risk because of his links to his aunt, Lise Meitner; it was Cockcroft who, with Chadwick, prevailed on the Ministry of Aircraft Production to warn the scientists in America not to publish their latest results in *Physical Review*; it was Cockcroft whom Oliphant asked to follow up Lise Meitner's garbled telegram about Bohr and Maud Ray (he approached the British legation in Stockholm to contact Meitner); and of course it was Cockcroft whom Rosbaud was regularly in touch with and saw most often. Although he seems never to have had any official title, Cockcroft did play a central role in the collection and collation of wartime nuclear intelligence.

But the first real initiative, so far as active nuclear intelligence gathering was concerned, appears to have occurred virtually simultaneously in the minds of Rudolf Peierls and James Chadwick. In a letter from Francis Simon, at the Clarendon Laboratory in Oxford, to Chadwick in Liverpool, written on 19 September 1941, the month the MAUD report was completed, the Oxford man said that Peierls and Fuchs had gone to London the previous week 'to go through the German periodicals that have arrived lately', to see whether German physicists were still working in their normal places of work, on what subjects they were publishing, 'or whether there was an ominous silence that would suggest they were collaborating on a Nazi atomic bomb programme'. In reply, Chadwick said he had had much the same idea and had mentioned it to Lord (Maurice) Hankey, as chairman of the cabinet's Scientific Advisory Committee. This idea may have originally been suggested by a document from Jacques Allier, the man who had stolen the heavy water in Norway from under the Germans' noses. He sent a note (also to Cockcroft), dated 26 May 1940, as the invasion of France was being completed, headed 'Lieu de Travail jusqu'en Septembre 1939' in which he gave the places of work of Hahn, Strassmann, Flügge, Weizsäcker, Bothe, and others.[10]

In London, Peierls and Fuchs consulted copies of the *Physikalische Zeitschrift*, a German journal which, they knew, produced each semester a list of which physicists were teaching what and where. (During the war, such journals arrived in Britain mainly via Switzerland or Sweden.) As Peierls wrote in his report, 'This showed that nearly all the physicists were in their usual positions, teaching their usual subjects, a very different picture from that which would have been shown by a similar list for the UK or USA.'* The list of thirteen names they drew up – which included such individuals as Heisenberg, Weizsäcker, Wirtz and Harteck – was spot on, insofar as all these individuals were indeed involved in Germany's uranium project.[11] At the same time they observed that both nuclear fission and isotope separation were being mentioned freely as late as the summer of 1941, though Peierls worried that this was a double bluff.

Lord Hankey passed on these lists to Brigadier S. G. Menzies, head of the SIS and known as 'C'. After some delay he wrote back confirming that most of the scientists were still in their usual places of work, and providing the first intelligence that Heisenberg had visited Bohr in 1941. This told Cockcroft and his colleagues that the Germans were interested in Allied plans, but they knew that Bohr was out of the loop, and they knew from the fact that he had made no attempt to contact them that Heisenberg and Weizsäcker hadn't told him anything of consequence either. Menzies also reported that Bothe had taken over Joliot-Curie's cyclotron in Paris, that Hahn had also been to Paris 'but refuses to do war work', that Heisenberg still taught in Leipzig but 'goes three days a week to Berlin to the Kaiser

* There is no evidence that the Germans had the same idea in reverse, as it were, and looked for the whereabouts of Allied scientists. Paul Rose argues that this was because the German scientists simply could not believe that their Allied colleagues/competitors could catch up with them. Chadwick worried at times that the British physicists should publish *some* work to avoid the obvious fact that everything had gone very quiet. But this never seems to have happened.

Wilhelm Institute (KWI), where he had been made director of atomic research . . . that [Klaus] Clusius [another member of the Uranverein], who is still in Munich, has been entrusted with the separation of uranium-235 from the mother substance, of which large quantities are received from Joachimsthal in Czechoslovakia.'

These inquiries found that everyone was in their regular place of work. There had been a small amount of movement but that was normal, Peierls and Fuchs calculated. They did consider the possibility of 'camouflage', that physicists were intentionally continuing to publish in their normal fields while still being listed in their normal places of work, while actually engaged in secret work. They concluded it would be 'difficult to do'. The fact that Joliot-Curie's cyclotron had been taken over in Paris told them that the Germans had no machines of that kind of their own, which they would need for a bomb (America had twenty).

They noted mention of a lecture by Gerhard Hoffmann, a professor at Leipzig, at a meeting of the Deutsche Physikalische Gesellschaft, in which both fission and separation were mentioned 'in one sentence'. 'This, of course, does not prove that the importance in this connection of fast neutrons is being realised, but if it were one would expect it to be kept secret.' In fact, fission and separation were mentioned 'freely' in the German literature, they found, as late as the summer of 1941, 'but this may be intentional'. Furthermore, in his article Hoffmann asked for more support, 'suggesting his work was not being taken seriously but again is that the intention?'

It was quite 'feasible' at that point, they concluded, that 'Germany is at the same stage as in the UK'. In March 1941, when a German parachute mine exploded in Liverpool, Chadwick asked that it be checked for radioactivity, just to be on the safe side.[12]

Regarding specifics, Peierls and Fuchs observed that Heisenberg had not published anything since 1939, 'although he used to publish regularly'. They added that Heisenberg was 'almost certainly' engaged in war work, partly based on his changed publication schedule and partly on account of his 'frequent absences from Leipzig'. Weizsäcker had not published since 1939, except on philosophy, and continued to lecture in Berlin. Nevertheless, they thought he was 'likely' to be involved. Wirtz had published on the 'Clusius-Dickel' method of separating liquids and so was also likely to be involved in isotope separation. Flügge was attached to Hahn's department and had published on fission but not since 1939. Hahn was at his normal place of employment and was publishing normally, so was probably not involved. Bothe was publishing normally, but on cosmic rays and electron scattering. Clusius and (Gerhard) Dickel published from time to time on thermal diffusion and Clusius on other matters. Gustav Hertz (the son of Heinrich Hertz) was still director of research at Siemens and publishing on sound pressures, while Harteck had given a lecture to the Bunsen Society in October 1940 on isotope separation by thermal diffusion.

Peierls and Fuchs felt that this overall picture would be hard to fake, that while some nuclear work was undoubtedly going on in Germany, there was no evidence to suggest a focused 'crash' programme that had reached the point where a central team of leading physicists had been commandeered and sequestered into a large-scale project.

Towards the end of 1941 and in early 1942 censorship was quietly introduced in Britain in regard to fission. The editors of six journals – *Nature*, the *Proceedings of the Royal Society*, the *Philosophical Magazine*, the *Proceedings of the Physical Society*, the *Cambridge Philosophical Society* journal, and the *Transactions of the Faraday Society* – were asked to clear any papers on nuclear physics or chemistry with Lord (Maurice) Hankey's Cabinet Scientific Advisory Committee, in effect the top body advising the government on science.[13]

A further report on the activities of German physicists was made by Fuchs and Peierls in spring 1942, who this time visited the Science Museum library and the Carnegie Science Library at the National Physical Laboratory in Teddington. They concluded there was 'not much new', or as James Chadwick put it, 'I don't think the new evidence outweighs the other evidence we have of interest in our problem', and that the annual report of the KWI for Dahlem for 1941–2 showed that Max von Laue was in charge.[14] Through

Meitner and Paul Rosbaud they knew that von Laue was an anti-Nazi, and would have nothing to do with a bomb, but even so the information that his institute was given over to *spezialaufgaben* ('special tasks') implied that some war research was going on. A list of papers published – or in the course of publication – was given, including some in which, it was said, 'nuclear transformation could be produced under neutron bombardment'.[15]

Fuchs further discovered an article on isotope separation by Fritz Houtermans, in the *Annalen der Physik*, published in November 1941. Since isotope separation was central to the production of a bomb, and one of its most difficult procedures, Fuchs, Peierls and indeed Jones felt that Houtermans would not have been allowed to publish such a paper if work on a German bomb was going ahead.[16] The Houtermans article calculated the minimum energy necessary for separating isotopes and went on to state explicitly that the calculations had been made 'with the view to a "possible future isotope separation for technical purposes"'. The paper considered two methods – distillation and rectification (a cascade of eighteen separate processes). As Fuchs observed, this did not seem to have any practical value, other than helping to provide a theoretical standard.

The significance of this publication, as the British researchers realised, was that it was small scale – merely an experiment – rather than showing any industrial commitment. Also relevant was the fact that Houtermans was not regarded as a reliable Nazi, having lived for a time in Russia. If he was being allowed to engage in such research, it did not seem likely that physicists in the Uranverein were working along similar lines.

Perhaps the most important item of information in this survey was a paper on fission neutrons, the work for which had been finalised before May 1940. However, the paper was dated March 1942 and shown as being received by the editor of the journal concerned, *Physikalische Berichte*, on 30 April. No explanation was given but, as Fuchs and Peierls observed: 'One that suggests itself is that they originally did not intend to publish but are now anxious to establish priority as against secret work now proceeding.' A similar argument applied to a paper by Erich Bagge in *Physikalische Zeitschrift*, where the work was finished in November 1940 but not received by the journal until April 1942. Then there were two articles published in *Naturwissenschaften*, also in 1942, one by Hahn and Strassmann, and the other by Kurt Starke, on element 93 and other transuranics. Paul Rose says, 'One may discern here the hand of Paul Rosbaud, who was probably trying to alert the Allies to German interest in plutonium.'[17]

In fact, Rosbaud was doing rather more than that. One man who realised what was going on was R. V. Jones, who summed up by saying that these comprised a number of papers that were kept secret 'while the Germans decided to go for a bomb or not'. That the papers were now being published showed that the German programme was a non-starter. Also, this was a process difficult to fake or camouflage. As Jones put it, 'What would be the point of giving away important information?' Publishing the dates on which the papers had been received by his journal, and the dates when the work had been carried out, was normal practice but at the same time Rosbaud was signalling important intelligence to the British.

In addition to these reports from academic journals, there were the intelligence reports, sent in the early years of the war from Norway and Sweden, by Jomar Brun and others. Apart from the raw details of what was happening, and when, which would provide the basis for when and where and how sabotage of the heavy water facilities at Rjukan and Vemork could be carried out, there was the all-important detail provided by Brun that, between October 1941 and summer 1942, Norsk Hydro had delivered 850 kg of heavy water to Germany. This, according to calculations carried out by Allied scientists, was only 17 per cent of the amount judged sufficient to ignite a self-sustaining chain reaction.[18] The Allies couldn't know it but Heisenberg made exactly the same calculation.

Through 1941 and into the early months of 1942, therefore, a tentative view began to form among intelligence personnel and scientists in Britain that Germany might not be working too hard on an atomic bomb. In America, the view was very different.

8

The 'Crown Jewel' of Secrets

For General Groves, the 'crown jewel' of secrets was the very existence of the Manhattan Project itself and his concept of compartmentalisation was specifically designed to maintain a total blackout. But, whether or not Groves fully realised it, or properly assimilated it, that secret had leaked out more than once quite early on, as we have seen. German atomic scientists convinced themselves that no one else could match them when it came to fission research, but even so two senior physicists, Heisenberg and Weizsäcker, had visited Bohr in the wake of the account in a Swedish newspaper that the Allies were working on a bomb of 'hitherto-undreamt-of-power'. Whether or not they were intent on conveying to him that the German bomb project was not serious, the timing of their visit surely suggests that they were interested in finding out what progress the Allies had made. The crown jewel of secrets was not secret at all.

In fact, the real crown jewel of secrets – by far the most important piece of atomic/nuclear intelligence of any kind, and a crucial turning point (or it should have been) – reached the British from Rosbaud in June 1942.

The quality and timeliness of Rosbaud's intelligence throughout the war were extraordinary. Brynjulf Ottar, one of the original creators of XU, the Norwegian resistance underground, told Rosbaud's biographer Arnold Kramish that in the autumn of 1939 – most probably early November – Rosbaud had travelled to Oslo and warned Odd Hassel that Germany was going to invade Norway. Hassel, a physical chemist who had studied in Berlin and Dresden, and a future Nobel Prize-winner, had not believed Rosbaud – not then – but we now know that the German decision to invade Norway was taken on 10 October 1939, which gives an idea of how precise and up to the minute Rosbaud's information was.

On his same visit to Oslo, Rosbaud asked Hassel to make a delivery for him to the British legation in the Norwegian capital. He couldn't make the delivery himself because the legation, at Drammensveien 79, was within sight of the German legation, at number 74 in the same street. Hassel, however, was a member of the Norwegian Academy of Sciences, also located on Drammensveien, so it wouldn't be out of place for him to cross from the NAS to the British legation.

Rosbaud's package that Hassel delivered that day contained three things. The first thing it contained was a book, with a rather dry title, *Technology of Magnesium and its Alloys*. Its author was Dr Ing. Adolf Beck, director of magnesium operations for the giant I. G. Farben complex at Bitterfeld (the major shareholder in Norsk Hydro), though Rosbaud had had a hand in the book's production. Part of the book's text covered the way magnesium could be used to facilitate heavy water operations and in explosives, and Rosbaud saw, as perhaps Dr Beck himself did not, that shortly after its publication the Luftwaffe might order it to be withdrawn from sale (as indeed they did). By then a copy was safely inside the British legation.

The second item in the package was a sealed glass tube which, on closer examination, proved to be a proximity fuse, designed to provoke explosions when missiles are close to – but not actually in contact with – a target such as an aircraft.

The third item was a number of sheets of paper, with writing and drawings on them, one of which was entitled 'Electric Fuses for Bombs and Shells'. The papers also contained information about a rocket-propelled glider that could be launched from aircraft against enemy ships, information about an experimental station at Peenemünde, the mouth of the Peene, near Wolgast, which was the first mention of Peenemünde in an intelligence report. (Peenemünde was where the V-1 and V-2 rockets, Hitler's 'vengeance weapons', in effect the world's first ballistic missiles, designed to attack British cities, would be developed.) The Oslo Report, as it became known, also contained information about two types of radar system, a pilotless aircraft, an acoustic-homing torpedo and details of new infantry tactics.[1]

Although most of this was intelligence of pure gold, it was at first dismissed on the grounds that it was too good to be true, that no one source could have so much information at once. To begin with, therefore, and despite the fact that the material was clearly written by someone with a scientific background, it was discounted. The presence of the proximity fuse was explained on the grounds that *some* real information had to be transmitted among all the hoax material to give it credibility.[2] Only later did attitudes change. In his war memoir, R. V. Jones continually refers to material in the Oslo Report that came to pass.

In one of the papers he wrote after the war, Rosbaud hinted at how he got information out of Germany. In one case he used a Dutch physicist who was able to travel to Norway. At other times he passed information to Sverre Bergh, a Norwegian student at Dresden Technical High School (and an Olympic skier), and in yet another case he informed Odd Hassel directly about a new and highly reactive compound being used in incendiary bombs. Because of his editorial work at *Naturwissenschaften*, he knew that there was a Dutch chemist, who had escaped to London in 1940, who was an expert on the chemical and might be able to help. So as well as warning about the new device, he was also able to suggest a possible way to counter the new threat. In yet another case, he persuaded a Dr Hjort to leak information to the Swedish press.[3]

In the second of the two books he wrote about scientific intelligence in the war, Jones said that in fact the author of the Oslo Report was not Rosbaud but Hans Ferdinand Mayer, a radio engineer who worked for Siemens but who had good links with the General Electric Company in England, and who became very disturbed by what was happening in Germany. Following publication of Jones's second book, Kramish wrote to him with the news that, because of her death, he was now free to reveal the name of Rosbaud's long-time mistress, and she had told him that Mayer was a regular in their household and that Paul had let it be known that Mayer and he had worked jointly on the Oslo Report.[4]

Despite this disagreement over sources in this one case, Jones did confirm in his memoirs that Rosbaud 'succeeded in getting information passed to London'. He admitted that he saw 'the most crucial of Rosbaud's reports that were successfully transmitted', adding that he had provided key information about the activities in particular of Heisenberg. Most important, Jones said that because Rosbaud's reports 'helped us correctly to calculate that work in Germany toward the release of nuclear energy at no time reached beyond the research stage, his information thus calmed the fears that might otherwise have beset us'.[5]

Rosbaud's most audacious coup came in June 1942. Some time between the 4th and the 10th of the month Rosbaud was present when several members of the German Physical Society met at the Restaurant Orient, on Fasanenstrasse, near the Ku'damm* in the fashionable Charlottenburg area of Berlin.[6] The gathering took place a short while after a momentous meeting between physicists and the country's military leaders – generals and admirals and, not least, Albert Speer, the minister of armaments and war production. The meeting was held at the conference centre of the Kaiser Wilhelm Institutes.[7] This was in fact the third – and the climax – of three similar meetings held that year between physicists and military leaders to discuss nuclear research and the possibilities for an atomic weapon.

At the meeting in the Ku'damm cafe, Rosbaud learned that a momentous decision had been taken just days before for the Germans *not* to proceed with atomic bomb development on an industrial scale, which would be necessary if an actual bomb were to be built. The advice that Heisenberg and the other scientists had given Speer and the military leaders was that nuclear energy held great promise for the future but that there was no chance that bombs could be constructed within such time as to have an effect on the current war. Germany did not at that stage have the right equipment (such as cyclotrons) and so the development of nuclear weapons contravened

* Ku'damm is the colloquial short form for Kurfürstendamm, in effect the Champs-Élysées of Berlin.

Hitler's policy of concentrating only on weapons that could be of immediate use.

Almost as important, Rosbaud also learned that it was the considered opinion of the German scientists that the Americans and British did not have the required expertise to manufacture a bomb either.

At this, 'a rather intoxicated' Rosbaud seems to have broken cover for a moment, in that he burst out, 'If any one of you knew how to make the bomb, he would not hesitate a minute and tell your Führer how to destroy the rest of the world in order to get the highest order of the Iron Cross.' The scientists were stunned into an 'icy silence', evidently frightened that Rosbaud might denounce them to the Gestapo. They dispersed quickly and, thankfully, no one denounced *him* to the authorities.

As a result of this meeting in the cafe on the Ku'damm, according to Arnold Kramish, Rosbaud flew by military aircraft to Oslo a short time later on 10 June. He had hoped to stop off in Copenhagen on the way to visit Niels Bohr but permission was denied and he was forced to spend eight days in a flat belonging to Victor Goldschmidt. Goldschmidt was Norwegian-Jewish, a heavily built brilliant geochemist who had done work on the relative abundance of elements in the earth, and some atomic research. An academic dispute had led him to resign his Norwegian post, after which he went to Göttingen in Germany, where he became friendly with several leading scientists and took out German citizenship. Throughout the 1930s he met Rosbaud countless times and it was the latter who, when the pressures on Jews in Germany became intolerable, arranged for Goldschmidt to reclaim his Norwegian citizenship in 1935. Rosbaud thought Goldschmidt's work deserved a Nobel Prize and, when he died, in 1947, Goldschmidt left his friend some money in his will.[8]

Rosbaud made the trip to Oslo in a German uniform, conceivably a Luftwaffe uniform (the service sponsoring his visit). His main mission was to pass on to Eric Welsh, via XU – the Norwegian underground – what he had heard in the Ku'damm cafe, but he was also in touch with Bohr. On 3 July he wrote, 'It would have been very important for me to see you again in order to discuss several questions that probably interest you as well as myself.'[9]

This wording is interesting. The questions he wanted to discuss were in the plural and that raises the further question as to just what Rosbaud knew at this time and what he passed on via XU.

Rosbaud was an experienced spy, and he had important contacts. He saw Otto Hahn once a week. He had 'a close personal relationship' with Walter Gerlach, another prominent physicist and also not a Nazi sympathiser, meeting him 'nearly every week' and providing him with news from the BBC. For Rosbaud, Gerlach was incorruptible, a man who loved Germany but not the Nazis. More to the point, Gerlach was apparently 'very talkative' and in that way helped the Allies – via Rosbaud – 'keep abreast' of how backward the German bomb work was. He went so far as to tell his friend that, 'I don't intend to make any war physics' even though he was administrative head of German nuclear research under the Reich Research Council, overseeing Heisenberg's work, and was Speer's deputy for nuclear matters.[10] As Rosbaud put it in a letter to Francis Simon after the war, Gerlach 'was a close ally of mine in many things'. Towards the end of the war Gerlach phoned Rosbaud and told him quite openly that he was leaving Berlin with 'the heavy stuff', meaning heavy water. ('He was always careless on the telephone.')[11] Rosbaud was also on good terms with Flügge and Mattauch.[12]

Rosbaud, who was bombed out twice in the course of the war, had been in touch with the British several times before the Ku'damm meeting, providing prompt information and swift action in particular on nuclear material.[13] We know from reports he wrote after the war was over how measured he was in his judgements. For example, although Pascual Jordan, a mathematical physicist, became a fervent Nazi (and was excoriated for it by émigré physicists), Rosbaud found him always 'decent' in his personal dealings and not anti-Semitic. And although Rosbaud sympathised with Heisenberg after he won the Nobel Prize yet was vilified by students for being a 'white Jew' (meaning he refused to denounce 'Jewish physics' – Einstein's relativity), he noted that the newly honoured physicist always made the most of his family's connections with Himmler. And Rosbaud warmed to Wolfgang Pauli when he mocked Heisenberg's ego. At one stage Pauli quipped that one could imagine Heisenberg saying something like: 'I can paint like Titian. Look: □ only technical details are missing!'[14]

Rosbaud was also creatively brave. At one point in the war he arranged for two French prisoners of war to be released from captivity and for them to do paid work in an electrical company. Later he arranged for these two men to wear civilian clothing, then had them moved out of the camp, one of them living with him. One reason he did this was because Piatier, one of the prisoners, was in touch with sources who leaked information on new weapons, which he passed to Rosbaud. In return Rosbaud arranged for a clandestine communications link between the French prisoners of war and their relatives back home. As Rosbaud phrased it later to an American colleague, throughout the war he made 'attempts to inform Blackett and Cockcroft via Sweden on valuable material'.[15]

And he was scientifically literate. It is therefore extremely unlikely that he would have travelled to Oslo without first trying to find out more details and background in connection with the meeting between the physicists and the political/military leaders. (Flügge, Hahn and Heisenberg were all at the meeting.) What else might that have been?

An informative clue comes from Mark Walker, professor of history at Union College, New York. He examined 138 of the relevant German nuclear research documents, which were top secret at the time (their authors were not allowed to keep copies). Eighty-two of those Walker examined were completed and written up before the crucial 1942 meeting. He shows how, in early December 1941, Erich Schumann, a military acoustic physicist and a powerful figure in Wehrmacht Ordnance, set into motion a reappraisal of Germany's nuclear power project and informed the leading scientists in the fission group that continued support of the project could be justified only if the military application of the phenomenon was expected in the foreseeable future. The changing fortunes of war exacerbated both the military demand for manpower and the limited availability of raw materials. The German offensive in the east had ground to a halt, the subsequent Russian counter-attack had brought the 'lightning war' to a definitive end, and the Third Reich was on the defensive, doubly so as the United States was now in the war.[16]

In view of these changing fortunes the Army Ordnance's own scientists began arguing that the time was now ripe for the large-scale industrial exploitation of nuclear power to be attempted. The Army Ordnance scientists believed that, contrary to what the nuclear physicists said, the problem was being investigated in America and other enemy countries, and so atomic research must be supported 'by all possible means'.[17] 'In particular, they wished to take the considerable and consequential step from laboratory to industrial scale research and development.'[18]

That was in late 1941. By early 1942, however, the army began to have second thoughts and was in fact now 'leaning toward' relinquishing control of nuclear power 'because they were doubtful of its likely utility in the war'. As a result of this change in thinking, a meeting was called in February 1942, attended by, among others, Schumann, Albert Vögler, president of the Kaiser Wilhelm Society and General Wilhelm Leeb, the actual head of Army Ordnance. The results presented to the February meeting showed that some progress had been made: uranium isotope separation had been uniformly disappointing but heavy water production had made modest advances. As the supply of moderator increased, the prospect of a uranium machine improved. The most exciting results had been obtained in Leipzig by Robert Döpel, a colleague of Heisenberg. His heavy water machine with uranium powder had come very close to neutron production: 'The first uranium machine seemed in sight.'[19] Siegfried Flügge, Otto Hahn and Fritz Strassmann also discussed the isolation and chemical properties of element 94 (i.e. plutonium), stressing that it should be relatively easy to fission.

At the same time, says Walker, the meeting underlined the scope of the nuclear project 'at its height', which showed that there were never more than seventy scientists connected directly or indirectly with nuclear power.[20]

After being told of these results, and assessing the relatively meagre dimensions of the programme, General Leeb argued that, as far as the German army was concerned, it had become 'convincingly clear' that the practical application of nuclear fission in Germany could not be achieved in time to help in the present war. And so confident of its scientists was the army that it was felt the enemy could not have a bomb either.[21] In view of these 'formal conclusions', Army Ordnance officials believed that sponsorship of the project should be transferred elsewhere and that the Kaiser Wilhelm Society was the natural destination.

This decision, Walker says, 'was final'. No one – not the military, German industry, the National Socialist government, or even the academic and military scientists themselves – ever revisited this conclusion.

At the same time as this purely professional conference, a more popular lecture series was organised by the army and the Reich Research Council, designed for a restricted audience of military personnel, party officials and selected industrialists. The importance of this popular conference for us is that it was written up in the newspapers, albeit carefully. Published under the title 'Physics and National Defence', the words 'atomic', 'nuclear', 'energy' or 'power' were not used but the thrust of the report clearly showed how modern physics would be important to the future – not current – defence of the country and to its economy.

One effect of this was that the minister of education, Bernhard Rust, decided to take the nuclear power project away from the Kaiser Wilhelm Society and give it to his own Reich Research Council. In so doing this moved nuclear research even further away from the army, and into the realms of pure academe.

This then is the involved background to an even more important – and climactic – meeting held in the Harnack Haus conference centre at the KWI on 4 June, attended by none other than Albert Speer, which was the immediate cause of Rosbaud's expedition to Oslo.

Utterly loyal to Hitler, as least until the very end of the war, Speer was a tireless worker and drove the German wartime economy to astounding feats of production despite ever-tightening shortages of raw materials. He won the trust of leading businessmen and established a close friendship with Erhard Milch, the Luftwaffe commander.

The latter friendship was important because by the time Speer was promoted, after the death in an aircraft explosion of Fritz Todt in February 1942, the air war had turned through several phases. After the bombing of Berlin in August 1940, the Battle of Britain and the London Blitz had ensued in September 1940–May 1941, with Hitler eventually being forced to back down in the face of heavy aircraft losses. Then, after the British forces had recovered, toward the end of March 1942, the RAF tested a new theory of bombing, using incendiaries in addition to high-explosive bombs. They began with Lübeck, a medieval city purposely chosen because of its narrow streets and ancient timbered houses. The new system worked beyond expectations – Lübeck burned and, for the first time, the German dead and injured in one bombing raid numbered in four figures.[22] Rostock was the second victim of a fire raid. Hitler seethed and demanded revenge and it fell to Milch to fulfil the Führer's wishes.

In the middle of all this devastation Speer met with General Friedrich Fromm, the military officer in charge of armaments. Over lunch in late April 1942, in a private room at Horcher's, a Berlin restaurant popular with Nazi officials, Fromm told the minister that although Germany's situation was then by no means desperate, he thought the country's best hope of winning the war was to develop a devastating new weapon and he passed on an account of atomic bombs which, he said, could wipe out entire cities. Fromm urged Speer to meet the scientists working on the weapon.

Speer was naturally intrigued by the possibility of a new and terrible explosive. He therefore arranged another conference at Harnack Haus, on 4 June, when a large contingent of military leaders would meet the nuclear physicists.

It was, in theory at least, a most propitious meeting for the physicists. Speer was the one man with the authority to put the full weight of the German economy behind a bomb programme.[23] Moreover, by then circumstances were changing. It was becoming clearer that the war would not be over quite so soon as Germany had originally thought but she still controlled vast swathes of Europe and could command those conquered territories to help in all sorts of ways militarily, materially and economically.

Speer's party was extremely distinguished. It included Ferdinand Porsche, the Austrian designer of the Volkswagen, General Fromm, the military armaments chief, Generals Emil Leeb and Erich Schumann of the Heereswaffenamt (the Army Weapons Office), Field Marshal Erhard Milch of the Luftwaffe, and Admirals Rhein and Witzell of the Navy. Among the scientists were Werner Heisenberg, Otto Hahn, Fritz Strassmann, Hans Jensen, Karl Wirtz, Carl Friedrich von Weizsäcker, Erich Bagge, Walter Bothe, Klaus Clusius, Manfred von Ardenne, Arnold Sommerfeld, Kurt Diebner and Paul Harteck, about fifty individuals in all who crowded into the Helmholtz Lecture Room at Harnack Haus.[24]

Heisenberg spoke first. Speer's office diary records that the discussion that day covered 'atom smashing and the development of the uranium machine and the cyclotron'. Speer's memoir written after the war also suggests that Heisenberg focused on nuclear research that day 'as a purely scientific enterprise'. This wasn't the purpose of the meeting at all and Speer wasn't so easily sidetracked. After Heisenberg ended his talk, Speer asked him bluntly 'how nuclear physics could be applied to the manufacture of atomic bombs'. Use of thc word 'bomb' apparently caused an uneasy stir through the lecture room, as many people had never heard the word used in connection with nuclear fission. According to Speer, Heisenberg declared, 'to be sure, that the scientific solution had already been found and that theoretically nothing stood in the way of building such a bomb. But the technical prerequisites for production would take years to develop, two years at the earliest.'[25]

Heisenberg further explained that German progress was hindered by the lack of a cyclotron, that he thought the Americans had several and that the one in Paris, to which they theoretically had access, was also hampered by promises made earlier in the war not to use it for belligerent purposes. When Speer countered that he could sanction a cyclotron, Heisenberg argued that they would need to test ideas on a small machine first.

Questions continued. How big must a bomb be, asked Milch, to wipe out a city like London? Heisenberg cupped his hands in mid-air and said, 'About as big as a pineapple.' This also created a stir among those present.

Milch then asked how long it would take for the Americans to build a reactor and a bomb. Heisenberg replied that even if the Americans 'pulled out all the stops' they could not build a working reactor before the end of that year, 1942, and that a working bomb would take at least another two years. Germany need not fear an American bomb before 1945.[26] To his credit, he was right on all counts. (It also appears he knew something about Fermi's activities in Chicago – see chapter 10.)

Speer next asked the physicist how his ministry could best aid Heisenberg's work. Heisenberg said they needed money, new buildings and ready access to scarce raw materials. When pressed to name a sum needed for immediate expenditure, Weizsäcker suggested a figure that was 'substantial' only by pre-war university standards – 40,000 marks. (Between $12,000 and $16,000 then, $211,000–280,000 now.)

The military men present couldn't believe what they were hearing. Schumann alone had already spent two million Reichsmarks on atomic research. As Milch observed, 'It was such a ridiculously low figure that Speer looked at me, and we both shook our heads at the artlessness and naïveté of these people.' The words Speer actually used when told a German bomb was years off, were: 'There's not much music in the thing.'[27]

After the conference Speer bitterly criticised Vögler for dragging him to an occasion that considered a project on such a 'paltry' scale. Stung, Vögler demanded that Heisenberg revise his financial demands, so they could accomplish something 'really useful'. A week later Heisenberg drew up a rough revised budget, increasing his requirements to 350,000 marks. But this meagre increase didn't impress Speer any more than the first figure did and after the Harnack Haus conference, 'Speer gave the possibility [of an atomic bomb] no further thought.'[28]

9

Intelligence Blackout: The Fatal Mistake

Rosbaud's realisation that the army had given up control of the bomb, and that an all-important decision had been taken *not* to develop nuclear explosives on an industrial scale – plus the intelligence that the Germans didn't think the Allies could produce a bomb either – were momentous developments.

Nor was Rosbaud's report an isolated event that summer. In London – and quite independently, as mentioned earlier – R. V. Jones noticed that, beginning in 1942, the Germans 'allowed a number of relevant papers to be published on nuclear work done in the previous two years, which seem to have been kept secret while they decided to go for the bomb or not'.[1] This too signalled that the Germans didn't think the Allies could make use of such results either.

Moreover, at much the same time, during the summer months soon after Heisenberg's meeting with Speer and the top brass, a revealing raft of reports reached British intelligence authorities concerning the German bomb programme. All carried a similar thrust, all the more impressive because they came from German scientists involved in nuclear research. 'There is nothing quite like this series of reports in the whole history of intelligence organisations,' says Thomas Powers in his book about Heisenberg. 'Together these messages reveal an unprecedented level of disaffection among scientists at the heart of the German bomb programme.'[2]

Between the summer of 1940 and the last months of the war, several German scientists made visits to Norway. In doing so they developed the tactic of stopping in Stockholm on the way there or the way back, where all manner of desirables – from pickled herring to

French and American cosmetics – could still be had for ready money, long after they had vanished from the shops in Berlin or Munich. But neutral Sweden also became a convenient listening post for Allied intelligence organisations as well as the unofficial seat of resistance groups operating with British help in Norway and Denmark. This helps explain how, in the summer of 1942, several German physicists managed to pass on intelligence from the Harnack Haus conference.[3]

A central figure was Jomar Brun, the engineer who had helped the initial design of the Rjukan plant and had been good friends with Karl Wirtz since before the war. Wirtz was one of those involved in heavy water research. In January 1942 Brun had been called to Berlin, where he met Wirtz and Kurt Diebner, scientific leader of the army's weapons research office, to discuss yet more increases in heavy water production. He was very close to the German nuclear scientists.

In July 1942 Brun received a visit from Hans Suess, Austrian-born, a future Nobel Prize winner, and at the time another of the German experts on heavy water. Suess was very friendly with Brun (and with Rosbaud) but in his dealings with the Norwegian he had to be cautious. Brun's premises might be bugged, whether Brun knew it or not. Suess was apparently aware that Brun had a line to the British but that was not without its problems. If the Germans were to win the war – as was still possible in 1942 – and then capture British records, those records might contain proof of his incipient treason.

Suess was, nevertheless, the first in a series of visitors who each brought much the same message, though cloaked in language designed to protect them should they make a false contact.

Suess started to talk after Brun referred to rumours he had heard, that heavy water played a role in biological warfare and new poison gases. Suess dismissed this and then added that the real purpose of heavy water was to function as a moderator in a chain-reacting pile. But, he added, the German research 'wouldn't produce anything useful for many years'. He himself guessed it would take at least five.

Brun naturally asked why Germany was devoting time and money to such a long-range project. Suess tactfully gave two answers. 'Those who believed in a quick victory surely were hoping for its peaceful application after the war; those who expected a long war were thinking that they must have some knowledge of all the possibilities that might result from such research.'[4] This was careful wording but Rosbaud confirmed much later that 'when he [Suess] was in Norway during the war he did his bit to help the English'.[5]

Suess wasn't exactly an emissary of Heisenberg but he was, in effect, outlining the position Heisenberg had given to Speer at Harnack Haus: nuclear research heralded a new source of power at some stage in the future; no one knew exactly when, but it would definitely be after the war. The German military *were* interested in nuclear research in theory, Suess told Brun, but his own work with heavy water was pure science, at least for the time being.

Brun duly reported this exchange to British intelligence. But Suess was not the only source of intelligence that summer to reinforce the Rosbaud message. In June – the same month as the Rosbaud report – the Swedish theoretical physicist Ivar Waller, a professor at the University of Uppsala, sent a letter to a colleague in London saying that nuclear research was being conducted at a number of laboratories in Germany, directed by Werner Heisenberg, and that their research was being carried out on nuclear fuels that 'might be used to create a chain reaction, especially uranium-235'. And, as he phrased it, 'results cannot be excluded'. The British official intelligence history says Waller 'may have got his information from Niels Bohr in Denmark', but Lise Meitner was a more likely source. Waller saw her occasionally on his trips to Stockholm and she too was in contact with friends in Germany and in Britain, such as Max Born in Edinburgh. The Waller letter mentions 'fission', 'chain reactions' and 'uranium-235' but there is no reference to a bomb.[6]

Karl Wirtz also visited Norway in that momentous summer of 1942. He arrived a week or so after Suess, in July, and moved between Oslo and Rjukan. During the war and, except for Weizsäcker, no one was closer to Heisenberg than Wirtz. Both Jomar Brun and Harald Wergeland were also close to Wirtz. In a memoir that he wrote after the war, Wirtz had this to say: 'I was able to make it understood relatively quickly that I was not a passionate Nazi and that as far as possible it would be best if also in the future there would be a certain scientific collegiality.' He confirmed, without putting it into words, that everyone knew that the Germans wanted to buy heavy water for use as a moderator in a nuclear reactor. But he insisted he got across the message that the goal of research was a power-producing machine, not a bomb. 'I made it clear,' he insisted.[7]

Harald Wergeland, a Norwegian physicist and chemical engineer, played a crucial role here. Wergeland had studied with Wirtz under Heisenberg in Leipzig in the 1930s. Wirtz had had many meetings with Wergeland and had told Rosbaud as much. As we have seen, Wergeland was a member of XU, and Wirtz was aware of this. Wergeland apparently went so far as to suggest to Wirtz that he escape to Britain, and said he could arrange it. Wirtz declined but, while he chose to remain in the Reich, he sought to make it clear that, as he himself put it, 'no danger was coming from Germany'.

The fullest piece of intelligence, after Rosbaud's, reached the British also by way of Scandinavia, and also in the summer of 1942. It came from Hans Jensen, known for his work on heavy water alongside Paul Harteck at the University of Hamburg. In his thirties, Jensen was described by colleagues as a 'socialist-going-on-communist' and was very different politically from Heisenberg. Despite their differences, however, they had often discussed the bomb project with the result that Heisenberg urged Jensen to visit Bohr in Copenhagen, in the hope that it would make up for his own unsuccessful visit in September 1941. This account was confirmed after the war by Jensen who said that Suess also took part in the discussions.[8]

In Copenhagen, Jensen talked to Bohr and his assistant, Christian Møller. Møller confirmed to Stefan Rozental, a young Pole then at Bohr's institute, what Jensen had to say.

After Jensen had left Copenhagen for Norway, Møller confirmed that Jensen had been very blunt about his work in Hamburg: he was experimenting on heavy water; he was travelling to Norway in an attempt to boost supply; and finally that the goal of this effort was 'a chain reaction producing nuclear power'. But he also made it clear that 'he was quite sure his work would be of no use in making a bomb'.[9]

Jensen assumed that his message had gone down well and said

as much to Heisenberg on his return to Germany. Bohr had made emollient noises about how the German physicists were coping with their moral dilemmas, but in truth, although Bohr thought Jensen sincere, he was far from certain how much he knew about Germany's bomb preparations.[10]

What Bohr didn't know was that Jensen confided not only in him that summer. After Copenhagen he moved on to Norway where he made a point of addressing a colloquium of Norwegians in Oslo. It was a brave thing to do: except for himself, everyone in the room was connected to the Norwegian underground. Jensen described the progress of nuclear research in Germany, again underlining that it posed no threat of developing a bomb. Even a power-producing machine, he said, would not give practical results until well after the war.

Notes of the meeting were made by Harald Wergeland, and a précis was passed on to Brynjulf Ottar, who later remembered:

> Heisenberg's opinion was, according to Jensen, that Germany would be unable to make the bomb. Wergeland made a detailed report from this meeting and what Jensen told him privately. The next day or so, one of my co-workers in the XU organisation . . . told me I had to go and see Wergeland, because he had some important papers that had to be sent to England as soon as possible. I visited Wergeland in his house . . . and I think Jensen was there, too.[11]

There is no question but that this report reached Britain with all due speed. F. H. Hinsley, in the official history of British wartime intelligence, says: 'In August [1942] a German professor who had left Germany for Norway sent a message that Heisenberg was working on a U-235 bomb and a "power machine". Heisenberg was said to be doubtful about the former but certain of the latter and satisfied with progress.'[12]

So news of Jensen's visit to Bohr, which must have occurred after 10 June, had reached Britain by August, a matter of a few weeks. Rosbaud's account of Heisenberg's meeting with Speer and the generals would certainly have crossed the North Sea no less rapidly.

After the war Wirtz tried to explain the difficulty that he, Heisenberg and the others had been in: they didn't want Hitler to get the bomb, but they didn't want to be disloyal to Germany either. One way or another, Wirtz insisted that his Norwegian friends come to understand that German work was aimed at a reactor, not at a bomb.[13]

Of course, these various reports from German physicists – from Suess, Jensen and Wirtz – could all have been deceptions, a deliberate conspiracy to lull the Allies into a false sense of security about the apparent German lack of progress in building an atomic bomb. It was in fact what several Americans did think (see below). And it might well have been how British Intelligence assessed the information coming out of Norway and Sweden in the summer of 1942. But, at the same time, the British also had the detailed report from Paul Rosbaud, on whom they knew they could rely. In London, Tronstad also had access to Brun's intelligence. And Brun had signalled by the summer of 1942 that the heavy water shipments went to one of just two addresses: 'Kaiser Wilhelm Institute for Physics, Boltzmannstrasse 20, Berlin-Dahlem', or 'Schering (Quinine Factory), Leipzig'. MAUD Committee scientists knew from Paul Rosbaud that the centre of the Uranverein was in Berlin and that Werner Heisenberg was in Leipzig. Brun's intelligence, as we have seen, also included the amount of heavy water that had been transported to Germany from Vemork – 850 kilograms – which, experts calculated, was only 17 per cent of that needed to sustain a chain reaction. The fact that the heavy water was going to two locations, not one consolidated facility, confirmed the exploratory, experimental nature of the German programme.

And this put a totally different perspective on the situation. The reports of Suess, Wirtz, Jensen and Brun dovetailed *exactly* with what Rosbaud had said, so that each report reinforced the others, making the overall assessment even stronger. And it fitted with what R. V. Jones was observing independently and simultaneously in London, that publication of previously secret experiments was now permitted.

There *was* one report that summer that, at least in theory, was somewhat alarming, though in truth it highlights the difference in quality between UK and US atomic intelligence. This report came from Leo Szilard in Chicago. As we have seen, among the émigré nuclear physicists none were more nervous about the threat of a Nazi bomb than the Hungarians Leo Szilard, Edward Teller and Eugene Wigner. It was, therefore, unfortunate that, in the late spring of 1942 (and so *before* the meeting between the physicists and Speer), Szilard took delivery of a telegram that Fritz Houtermans had managed to despatch from Switzerland.[14] The actual text of the telegram has been lost but Wigner later remembered the wording 'as very general'. It was almost bland, something like, 'They are getting organised.' Houtermans said no more than that Heisenberg had been made director of atomic research at the Kaiser Wilhelm Institute in Berlin. This was true, as we have seen, but Houtermans then added a sentence about his own work, to the effect that a nuclear chain reaction (to create plutonium) was a better route than the separation of isotopes. The words 'better route' could be construed as a sign of progress and Kramish says, 'For the perpetually frightened man, this meant the worst.' Szilard lost no time in circulating this news, in the process drumming up huge excitement, so much so that it eventually helped create a rift between Britain and the United States.

At much the same time, Szilard had also picked up 'through the scientific underground' another troubling hint about recent German activity, which again caused the Americans to overestimate Nazi capabilities in the atomic field. Thomas Powers says that it is impossible now to know 'how many hands' this rumour had passed through before Szilard got hold of it. But it appears that his assistant, John Marshall, relayed to him a conversation that allegedly took place between two German physicists, Friedrich Dessauer, recently arrived from Switzerland, where his father Gerhardt had emigrated. According to this exchange, the father had learned in Switzerland that 'the Germans got a chain reaction going'.[15] This was mid-1942, six months before Enrico Fermi achieved the first self-sustaining chain reaction in Chicago. As a result, 'The permanently worried man got it into his head that the Germans were a whole year ahead.'[16]

From these scraps, Kramish says, 'Szilard built his own Potemkin village', in the course of which he managed to convince the director of the Chicago laboratory, Arthur Compton, that he should relay this 'intelligence' to James Conant, Vannevar Bush's deputy. Compton, originally from Ohio, had won the Nobel Prize in 1927 and, before the war, was on good terms with Heisenberg, who lectured in Chicago. As Compton put it in July 1942, in his cable to Bush, 'We have become convinced that there is a real danger of bombardment by the Germans within the next few months using bombs designed to spread radio-active material in lethal quantities ... Apparently reliable information has reached us to the effect that the Germans have succeeded in making the chain reaction work. Our rough guess is that they may have had the reaction operating for two or three months.'[17]

So concerned were the Americans that, not long afterwards, on 23 July, the US embassy in London sent a paraphrase of Conant's cable to the British scientific authorities. By this time, says Kramish, Szilard 'and the entire American atomic hierarchy' had transformed Houtermans's vague wording into 'hot' intelligence that the Germans were already operating a large-scale nuclear power plant and that an attack on the United States by radioactive fission products was 'imminent'.

To doubt that the Germans were ahead seemed complacent to the Americans. Conant thought that they might be ahead by as much as a year, and Wigner thought they might have a bomb by Christmas 1944. Samuel Goudsmit, the physicist appointed to head up the Allied team (codenamed 'Alsos') sent in to capture Germany's nuclear scientists towards the end of the war, went so far as to say that the thought of German superiority drove some people 'almost to panic' and they 'sent their families into the countryside'.[18]

The British obviously needed to take such a threat seriously (or appear to) and called a full-scale meeting of the Tube Alloys Technical Committee on 18 August. The committee was not convinced by Szilard and Conant's alarums. They had their own sources, they said (without identifying them), and replied to Conant that same day, saying that 'information that has just come to us indicates that research is still in progress'. This was somewhat muted usage, but the rapidity of the response confirms the confidence they had in Rosbaud and the rest of the intelligence they were receiving at exactly this time. It would also appear to confirm that Rosbaud's intelligence had arrived. The British went on to firmly insist that *their* informants had said nothing about a *working* reactor. They confirmed that they had a good source in place but that was as far as they would go.

The British were content with this but the Americans always retained a fear that, even if a bomb was not being built by the Germans, they might still produce deadly radioactive materials and, says Kramish, they inflated that fear 'out of all proportion'. General Groves said later that he was surprised the Germans didn't take this route, because it was the easiest way to a weapon of some sort, but we shall see later how the British thought this through more practically.[19]

There were to be two consequences to these differences between the Allies. One was that, to placate the Americans, the British put one of their best scientific analysts on the problem. This was Alan Nunn May. Like Fuchs, Nunn May was a Soviet spy. Thus, the Russians knew in 1942 about the Americans' fear of fission products, and had a good idea of how Allied thinking was progressing. Possibly more important, they were also aware of the second consequence of Szilard's alarums, the widening rift between the entire British and American nuclear programmes. At the president's retreat, Hyde Park, in June, Roosevelt and Churchill met for the third of their twelve wartime summits and agreed on close co-operation. But before the year was out, as we shall see in the next chapter, Roosevelt's advisors were trying hard to distance the British and diminish their role.

In a way, this made sense from an American point of view. However, what the Americans did not grasp was that by excluding the British on technical matters, they also isolated themselves from the first-rate intelligence Britain was receiving on the German atomic programme. 'Until the end of 1943, the British did not pass on atomic intelligence of any importance to the United States.'[20] According to some accounts Groves was not given the full intelligence picture by the British until December 1943.[21]

The British never briefed the Americans on their contacts with

German scientists, nor did they bring Groves or his staff into the secret of the Enigma code-breaking machine. Although there were large gaps in the Enigma decrypts, the British knew how reliable the Enigma traffic was in general and that there was no reference at all in that traffic to atomic bombs, or to the scientists working on the German bomb project. This was in marked contrast to the several references to the rocket programme and other new weapons.

This blackout, the failure by the British to share fully the intelligence they had, and its quality, was arguably one of the most egregious errors of the entire atom bomb story. It was a contributing factor, and an important one, which led the Americans to overestimate the Germans' nuclear capability, and in turn helped propel them unnecessarily into the Manhattan Project.

One other source of intelligence for the British, almost certainly, was Lise Meitner in Sweden. To her friends, the idea of Lise Meitner being an agent of any kind in the war is 'preposterous'. They insist that she always refused point-blank to have anything to do with bomb work. But as her biographer Ruth Sime says, her 'innate openness may have made it impossible for her *not* to pass on what she knew to people she trusted'. She was in regular contact with Max von Laue and from him she learned bits and pieces. 'It is unlikely that Laue's information was ever very technical – he was not a member of the uranium club [though he was in charge of the KWI in Berlin-Dahlem] – but it may have helped the Allies form a general picture of the scope and progress of the German fission project, the fact that it was important enough to be moved to southern Germany when Berlin became too dangerous.' But von Laue made various trips to Sweden during the war and, in addition to meeting Meitner, wrote letters from there to friends in Allied countries describing the situation inside Germany.[22]

Sime thinks it unlikely that Paul Rosbaud sent much vital data through Meitner. He had other more robust and direct routes, one of which, we know, was Njål Hole, a young Norwegian physicist who was also in Manne Siegbahn's institute in Stockholm. Although the details of Hole's contributions are not known, what we do know is that after the war he received a British OBE. So Siegbahn's institute did seem to be sympathetic, at the least, to the Allied cause (and see chapter 13). We also know that Meitner received a number of technical books from Rosbaud who often used book codes (a natural choice for a publisher's agent). After the war, Meitner expressed her gratitude to Rosbaud for the books 'he had so kindly selected'.[23]

In the Meitner archive at Churchill College, Cambridge, her diaries show that she was in touch with Rosbaud in April 1940, June 1942 and December 1943, and had four meetings with Hahn in September, October and November 1943. The Rosbaud encounter in the week of 6 June coincided exactly with his travels following the Ku'damm cafe rendezvous. We know that during at least one of Hahn's meetings in Sweden he gave a lecture on fission, and that he also gave much the same lecture in Rome, even earlier, in 1941.[24] Although Hahn was known to be anti-Nazi, in fact especially *because* of that, and because he was in the forefront of uranium experiments, and had attended the meeting with Speer and the generals at Harnack Haus in June 1942, British intelligence felt that if any large-scale work was going on in Germany, the Germans would not have allowed Hahn to travel, 'as they would have had to assume that however secret the work was, Hahn would have known about it'.[25]

The Meitner archive also shows that she was in touch by letter frequently with both Paul Rosbaud and his wife in Britain. She was, in effect, a go-between who could help them keep in touch with each other on personal and domestic matters when direct communication was obviously not possible. On a number of occasions she thanks Paul for sending her books, usually books on physics, such as cosmic rays. This may have been no more than the truth, but it does not contradict the suspicion that Rosbaud was sending out messages in a book code.

One very unusual intelligence leak – if it can be called that – occurred, seemingly out of the blue, early in 1943 with the astonishing news from Rome that Pope Pius XII had referred to a bomb in an address to the Pontifical Academy of Sciences, delivered on 21 February. The actual words the pope used were: 'The thought of the construction of a uranium machine cannot be regarded as merely utopian. It is important, above all, however, to prevent this reaction from taking place in a chain reaction ... Otherwise a dangerous explosion might occur.' British officials attached to the Holy See advised the intelligence services that the phrase 'uranium machine', which was German usage ('*Uranmachine*'), strongly suggested that the original source of the pope's remarks was almost certainly Max Planck, a member of the Pontifical Academy visiting Rome at the time of His Holiness's speech. How can we explain such an extraordinary breach of secrecy? Was Planck building on what Suess, Jensen and the others had said, but also signalling that the Germans knew something of the Allied nuclear plans? Or was he using the pope's moral authority to prevail on the consciences of the Allied physicists not to develop a bomb?[26] Since so many of the Allied scientists were Jewish this seems doubtful.

The argument that Hahn would not have been allowed out to lecture in Sweden also applies to Heisenberg's visit to Zurich in November 1942, where he gave a lecture on his theoretical work and on cosmic rays, and afterwards sat up until 12.45 a.m. at the home of Paul Scherrer, discussing modern and 'degenerate' art. Scherrer was a British informant and told MI6 about the visit, perhaps including the important detail that Heisenberg, along with Weizsäcker, once spent four weeks in Switzerland. Heisenberg would surely not have been allowed to travel abroad had he been running a viable bomb project.[27]

It would be quite wrong to give the impression, however, that nuclear research had stopped totally in Germany. Paul Harteck continued his work on isotope separation, a heavy water pilot plant was mooted near Munich, a subcritical pile was built in Berlin, there was talk of a nuclear reactor capable of giving a submarine (U-boat) an 'action radius' of 25,000 miles for the consumption of a kilogram of uranium. Even Speer agreed to give nuclear physics the coveted *kriegswichtig* (important for the war effort) security rating, if not *kriegsentscheidend* (decisive). The production of uranium was eventually put on an industrial basis, and one small group even worked on the idea of thermonuclear fusion – in other words, the theory behind a hydrogen bomb. Among the documents that the German academic Rainer Karlsch found in Russia in researching his book, *Hitlers Bombe*, published in 2005, was one that showed that German scientists had carried out a 'hitherto-unknown nuclear reactor experiment' and tested 'some sort of nuclear device' in Thuringia in March 1945, as a result of which several hundred prisoners of war and concentration-camp inmates had been killed.[28]

But never, at any point, did the Germans obtain anywhere near enough heavy water to make a go of things. Several purely academic exercises were run as so-called military programmes so as to keep the physicists involved away from being drafted, continuing a practice that had begun early on.[29]

Gradually the War Office and Göring lost faith with even the minimal academic projects that were under way. Like others before him, the Reichsmarshal lamented the fact that Hahn had published his fission paper openly in the first place. Although several reports of American progress reached Germany, such as the fact that a number of physicists and chemists were being collected together, German progress continued to be slow. There was a lack of urgency for at least two reasons, other than the fact that the Speer conference had killed any idea of an industrial project stone dead. First, German physicists couldn't believe that anyone could overtake them. And second, they were hamstrung by Hitler's policy that priority had to go to ideas that might produce weapons in the current war. Enquiries made after the war ended confirmed Jones's assessment during it – German nuclear physics research peaked in 1942; and the Army Weapons Department had commissioned a 144-page document, 'German secret reports on the exploitation of nuclear energy from the years 1939–1942', which also marked 1942 as the time when the army passed over the work to a more research-minded institution because a weapon was not seen as practicable within the duration of the war.[30]

It had in fact become 'the settled opinion' of British intelligence and British scientists attached to the Directorate of Tube Alloys by the summer of 1943 (and well before Fuchs was sent to America) that, while the Germans were *researching* atomic energy, their primary objective was not the production of a weapon but the development of power, and that 'the danger that they could acquire an atomic weapon before they were defeated could be discounted'.[31] As a Directorate of Tube Alloys report underlined on 5 January 1944, 'All the evidence available to us leads us to the conclusion that the Germans are not in fact carrying out large-scale work on any aspect of TA [Tube Alloys]. We believe that after an initial serious examination of the project, the German work is now confined to academic and small-scale research, much of which is being published in current issues of their scientific journals.' As we shall see more than once, this was spot on. As Margaret Gowing put it in her official history of the British role as 'midwife' in the building of the atom bomb, 'The correctness of their 1943 assessment and their success in thereafter building upon it "an almost uncannily accurate picture of the German effort" were remarkable.'[32] In fact, of course, the British intelligence services had first learned there was no German bomb a whole year earlier, in June 1942. This discrepancy is explained later on.

Given the chronology of intelligence leaks considered in this chapter, it is an extraordinary irony that, at exactly the same time, in fact on 17 June 1942, the very same week that Rosbaud was in Oslo, Vannevar Bush and James Conant reported to the president of the United States that, with respect to nuclear power, it seemed likely that, granted adequate funds and priorities (much the same wording as Heisenberg had used with Speer), full-scale plant operation could be started soon enough to be of use militarily in the current war. And by the same token, as Mark Walker adds, 'It also appeared certain [to Conant and Bush] that the desired end result, nuclear weapons, could be attained by the enemy, provided he had sufficient time.'[33] Only a short while later, in July, Conant reported, as mentioned above, that 'We have become convinced that there is a real danger of bombardment by the Germans within the next few months using bombs designed to spread radio-active material in lethal quantities ... Apparently reliable information has reached us to the effect that the Germans have succeeded in making the chain reaction work. Our rough guess is that they may have had the reaction operating for two or three months.'[34]

But of course, this wasn't true. It wasn't anywhere *near* true. This was *the* fatal overestimate and the British had the intelligence to prove it. But it wasn't shared, not then.

Although the Bush/Conant report for the president was composed on 17 June 1942, General Groves was not appointed to run the Manhattan Project until 17 September. Thus there was clearly a window of time, three whole months, when the Rosbaud report could have – should have – been considered at the highest levels on both sides of the Atlantic.

One might argue that the Rosbaud report was not enough on which to base such an important piece of policy. Despite his impressive record, he might have been wrong in this one case. But the Rosbaud report was not a one-off. It was, as we have seen, reinforced over the same few weeks by several other reports from German scientists working on nuclear research and by R. V. Jones's observation that the Germans were now allowing research reports on nuclear matters to be published, which must have meant not only that they were not proceeding with a bomb of their own, but that they didn't think Allied scientists would benefit from the publications either.

All these developments reinforced one another in a coherent and therefore convincing whole. Jones himself had a most impressive record of interpreting scientific intelligence, so much so that he had the ear and confidence of Churchill. We don't know what Churchill was told, or when. The matter is not discussed in either Churchill's memoirs or Jones's.

Simply put, in the summer of 1942 there was no need for the Allies to embark on building a nuclear weapon – not if the main reason for building a bomb was to counter a Nazi threat, because there was no Nazi threat.

At this point, the Manhattan Project had not got under way to any appreciable extent. Very few funds had been expended and there was no real industrial momentum towards a bomb. It would have been relatively easy to call the whole thing off. Since the Nazi bomb didn't exist, there was no urgent military need for the Manhattan Project.

Conant's worries – that the Germans had succeeded in making a chain reaction work – were directly countered by the British. The British were much closer to Germany than America was, far more exposed in the event that Conant was right.

Had the issue been considered properly, without question the course of history would surely have changed in a momentous way. In the summer of 1942 the world was failed by a handful of British and American personnel who kept to themselves what they should have shared.

We now know that there was one other vital development during that all-important second week of June 1942. On the fourteenth of the month a coded radio message was sent from Moscow Centre to NKVD *rezidents* in Berlin, London and New York. It concerned a project they had codenamed, fittingly, Enormoz.[35]

TOP SECRET

REPORTEDLY THE WHITE HOUSE HAS DECIDED TO ALLOCATE A LARGE SUM TO A SECRET ATOMIC BOMB DEVELOPMENT PROJECT. RELEVANT RESEARCH AND DEVELOPMENT IS ALREADY IN PROGRESS IN GREAT BRITAIN AND GERMANY. IN VIEW OF THE ABOVE, PLEASE TAKE WHATEVER MEASURES YOU THINK FIT TO OBTAIN INFORMATION ON:

- THE THEORETICAL AND PRACTICAL ASPECTS OF THE ATOMIC BOMB PROJECTS, ON THE DESIGN OF THE ATOMIC BOMB, NUCLEAR FUEL COMPONENTS, AND THE TRIGGER MECHANISM;
- VARIOUS METHODS OF URANIUM ISOTOPE SEPARATION, WITH EMPHASIS ON THE PREFERABLE ONES;
- TRANSURANIUM ELEMENTS, NEUTRON PHYSICS, AND NUCLEAR PHYSICS;

- THE LIKELY CHANGES IN THE FUTURE POLICIES OF THE USA, BRITAIN, AND GERMANY IN CONNECTION WITH THE DEVELOPMENT OF THE ATOMIC BOMB;
- WHICH GOVERNMENT DEPARTMENTS HAVE BEEN MADE RESPONSIBLE FOR CO-ORDINATING THE ATOMIC BOMB DEVELOPMENT EFFORTS, WHERE THIS WORK IS BEING DONE, AND UNDER WHOSE LEADERSHIP.

While much has been made of the British atomic spies – not just Fuchs and Nunn May but John Cairncross, who was Lord Hankey's secretary at the Scientific Advisory Committee – the American spies, Morris Cohen and his wife Lona, both communist sympathisers (he fought in the Spanish Civil War), had formed a fruitful contact with a well-placed physicist, only ever identified as 'Perseus'. Whoever Perseus was, he – or she – clearly had as ready access to American policy-making as Cairncross had to the MAUD Committee deliberations. The information about Roosevelt's green light for the Manhattan Project reached Moscow in a matter of days.

And so, across the second week of June 1942, the atomic age was ushered in at the very time that its original raison d'être collapsed. It was an extraordinary coincidence or irony. The Battle of Stalingrad broke out, while the German scientists were reassuring the British intelligence services that there was no German bomb. On 23 August some 600 German bombers killed 40,000 Russian civilians and three weeks later the ground assault began. Russia lacked cyclotrons and, following the damage done during the German invasion, none would be up and running until the summer of 1944. It seems highly unlikely that, faced with such terrible exertions, the Russians would have embarked on a weapon of their own that could have had no part in the war. The Russians developed a bomb because the Allies were developing theirs. By not acting as Allies, the Allies kick-started the nuclear arms race we inherit today.

10

Fall Out

In this narrative of ironies and coincidences, the very greatest without question was that which occurred in the second and third week of June 1942 when, as we have just seen, the Germans killed off their bomb project while, at exactly that time, that very week, President Roosevelt gave the American project the green light. The irony was made all the more fateful by the fact that the British knew about the German decision but didn't tell the Americans. Of all the episodes in the atomic bomb story, apart perhaps from the fact of fission itself, this failure to share information must rank as of seminal and tragic importance. Why did the British not tell the Americans what they knew?

There were, conceivably, four reasons. One was the natural reluctance of intelligence organisations everywhere to share what they know with rival outfits, in case they compromise their sources. But this doesn't really bear a moment's reflection. The news from Rosbaud and the others was so important that ways should have been found to circumnavigate the reservations of the intelligence services, and that cannot have been beyond the wit of man. A second reason may well have been to prevent any information being circulated that would deter physicists from joining the bomb project now it was known that there was no Nazi threat. Some of the scientists shared the conviction of James Chadwick that a bomb was inevitable but many, as we shall see, did not. This is one aspect of the bomb story that has hardly been explored. Is this really why the British did not broadcast the Rosbaud intelligence more widely, or the reports of the German scientists in the summer of 1942? This is hardly a small point, raising the unpleasant prospect of a cover-up.

The other two reasons were more complex. One reason is the subject of this chapter. The other is considered in chapter 12.

In a nutshell, one important reason the British did not share what they knew with the Americans was because between the spring of 1942 and the autumn of 1943, during which time the crucial intelligence became available, the Allies had in effect fallen out so far as an atom bomb was concerned, and collaboration between British and American nuclear scientists had been put on hold.*

The roots of the problem lay in the fact that Britain was at war for a full two years before America joined after Pearl Harbor. As a result, British scientists thought harder about nuclear matters than did their American counterparts to begin with, and made most of the early intellectual running. The Americans were slower off the mark, because they were non-belligerents, but, after being attacked by the Japanese at the end of 1941, and as they saw Russia emerging as a superpower following its victory at the Battle of Stalingrad at the beginning of 1943, the bomb, originally developed as a counter-threat to a Nazi bomb, now took on a much wider significance. At the same time, having been under arms for more than two years, Britain was increasingly strapped for treasure and resources.

We left James Conant, Bush's deputy on the NDRC, in chapter 4, at Claridge's where he stayed for two months. He was in general much impressed by what he saw in Britain, and returned home very pro-British (at least up to a point). No less important, however, one of his meetings in London was with Frederick Lindemann, now Prime Minister Churchill's personal advisor on scientific matters, who took Conant to dinner at the Athenaeum Club on Pall Mall, where he talked openly about nuclear chain reactions. Conant was horrified,

* See CHAD IV 12/5, in the James Chadwick Archive at Churchill College, Cambridge, declassified in 2006, for a Cabinet Office letter, dated 23 December 1952, setting out the exact dates for the 'uncooperative' American attitude.

believing that the subject was too highly classified to discuss, even in private. Yet when Conant countered Lindemann's argument by using Bush's line – that a bomb was impracticable, at least during the current war – Lindemann shocked him still further by telling him about Frisch and Peierls's work.[1]

What Lindemann didn't say was that about now the first practical problems were occurring on the British side. By that stage, Rudolf Peierls, enemy alien or not, had emerged as the MAUD Committee's leading mathematical physicist. It fell to him to coordinate and check the calculations of his fellow theoreticians, working closely with experimentalists. The sheer size and complexity of the problem had forced him to cast around for help and, in the spring of 1941, he managed to find one excellent individual among the foreign-born scientists who had not moved on to America: this was the young applied mathematician Klaus Fuchs, a refugee from Nazi Germany, who had impressed his professor, Max Born, another German émigré, at the University of Edinburgh.[2] There was a brief flurry of excitement when MI5 discovered intelligence that Fuchs had been an active communist as a student in Germany. But there was no evidence to support the idea that he was still one and he was cleared inside days so that by the end of May he had a desk in the Nuffield Building in the University of Birmingham, and was living with Rudolf and Genia Peierls in Edgbaston.

Despite the mass of data from scores of fission experiments, Peierls and Fuchs's calculations could not guarantee that a uranium explosion would come off as planned. Without actually building a real pile, no one could be sure that enough neutrons would be emitted quickly enough.

And that is where the practical difficulties now began to surface in a serious way. One of Lindemann's Oxford colleagues was Francis Simon, still another German émigré (Simon had left Berlin in 1933 at Lindemann's invitation). Simon now argued that the best chance of going forward would be to isolate U-235 using a system of isotope separation known as gaseous diffusion. This involved forcing uranium hexafluoride gas through fine membranes, which separated U-235 from its slightly heavier sister, U-238. But to manufacture enough fissile uranium to produce an explosion would necessitate a large and complex industrial plant. Simon worked out that in fact it would require about 40 acres and cost anything up to five million pounds. That was roughly a tenth of Britain's weekly expenditure on the entire war effort, though in the long run the cost of a uranium bomb, then estimated at £8,520,000, was considerably less than the cost of an equivalent amount of TNT (£14,150,000).[3] Even so, could Britain really afford to build a bomb?

Faced with Simon's frightening calculation, it became increasingly clear to some that an accommodation with the Americans was a chief priority. But only to some. Differences began to emerge between the scientists that were to have serious long-term consequences. Oliphant, Tizard, Blackett and Cockcroft were strongly in favour of collaboration, while Chadwick and Lindemann were against.[4]

These disagreements were doubly unfortunate in view of a note received at this juncture, in early August 1941, from Charles Darwin, grandson of the great naturalist and himself a theoretical physicist then serving as director of the British Central Scientific Office in Washington DC. In his note, which he sent to Hankey, Darwin insisted it was high time for Britain and America to decide whether they were serious about nuclear bombs, sharply asking whether 'our own prime minister and the American President ... are willing to sanction the total destruction of Berlin and the country around' (confirming Germany as the target). Americans needed to get moving and the British needed to decide where a bomb should be built. More than that, he said he had been conferring with Bush and Conant who, following the latter's visit to London, had floated the idea that the two countries 'should go beyond coordinating their research on the Bomb and regard it as a joint project of the two governments'.[5]

This was an extraordinary offer, extremely generous and calculated to make the most of what each ally could offer. But Hankey wouldn't be hurried. He decided instead on a careful review process, which would tap the views of all the British scientists involved. But they, as we have seen, were divided on just this question. Lindemann also played a role in delaying things. Three weeks later, at the end of August, he got in touch directly with the prime minister, pointing out that research had shown that the weapon was now 'extremely likely'. But he suggested the government should fund research for another six months, 'when a final decision would be possible'. He himself was strongly in favour of building the bomb 'in England or at the worst in Canada'.[6] 'However much I may trust my neighbor and depend on him, I am very much averse to putting myself completely at his mercy.' Britain, he thought, 'should not press the Americans to undertake this work', but simply 'continue exchanging information'. He was doubtful a weapon could be produced in two years, as some believed, but he was now certain it should go ahead. 'It would be unforgiveable if we let the Germans develop a process ahead of us by means of which they could defeat us in war or reverse the verdict after they had been defeated.'[7]

The MAUD Committee's report – summarising and evaluating the research carried out over the previous sixteen months at half-a-dozen secret locations – was finished about a month after that, on 24 September 1941. Importantly, the report (its final version being drafted the previous month by James Chadwick in Liverpool) concluded that a 'sufficiently purified' critical mass of U-235 could fission even with fast neutrons, and that no moderator would be needed. The report dismissed plutonium production, thermal diffusion, centrifuges and the electromagnetic method of isotope separation and opted for gaseous diffusion on a massive scale. Its main administrative (as opposed to scientific) conclusion was that the project should be moved from the Air Ministry to the Department of Scientific and Industrial Research, which was the natural home for the government's science projects. No less important, the report recommended that the gaseous diffusion plant should be built in Canada, with the Americans treated as consultants.* There was still this reluctance to go the whole hog and collaborate: 'The British and American project would therefore be separate but linked.'[8]

A twist was added to these developments, when Sir John Anderson was appointed as lord president of the Council and later Chancellor

* In the event, a group was assembled in Canada, whose most notable scientist was Hans von Halban, regarded as *the* expert on heavy water. The group, which at one stage was led by John Cockcroft, worked on an alternative bomb design, using the heavy water method of manufacturing the highly fissionable isotope, plutonium-239, at Trail, British Columbia, right on the US border. Canada, besides being part of the Commonwealth, and a safe distance from the Luftwaffe, was also a source of uranium and its engineers were experienced in refining metals.

of the Exchequer, and then the cabinet minister responsible for science during wartime. Then approaching sixty, Sir John was already at the head of the rest of the government's science research and he was an established career civil servant: hard working, sensible, unexciting and unexcitable, an administrator without equal and an unparalleled negotiator, known throughout Whitehall as 'Jehovah'. By a happy coincidence he had been a student at Leipzig University, where he wrote his thesis on the chemistry of uranium.[9]

To begin with, like Bush and like Tizard, Anderson had been 'quite sure' that the uranium bomb was not feasible but lately his doubts had receded and he had changed his mind. Still, he sided with Blackett and others in his view that the MAUD report had badly underestimated the amount of time needed to develop a bomb.

One man who had no doubts was Mark Oliphant, Frisch and Peierls's professor at Birmingham. He knew that Lyman Briggs had been receiving the monthly minutes of the MAUD Committee, as well as the final report, with its important conclusions. But in the summer–autumn 1941, America was still a non-belligerent in the war and the Briggs committee was inactive. Oliphant flew to America (in an unheated bomber) with a cover mission about radar but in reality to see why the United States was ignoring the MAUD Committee's findings. When Oliphant called on Briggs he found that 'this inarticulate and unimpressive man' had put the reports in his safe and had not shown them to anyone else.[10]

Incensed, Oliphant then made a thorough nuisance of himself. He met with Briggs's full American Uranium Committee, which had a new member, Samuel K. Allison, an experimental physicist like Oliphant himself. At the meeting, as Allison recalled later, Oliphant 'said "bomb" in no uncertain terms. He told us we must concentrate every effort on the bomb and said we had no right to work on power plants or anything but the bomb. The bomb would cost $25 million, he said, and Britain didn't have the manpower or the money, so it was up to us.'[11]

Allison's colleagues on the Uranium Committee were shocked to find they had been left in the dark by Briggs. On that trip Oliphant also contacted Ernest Lawrence in Berkeley and Fermi in New York and his nuisance-mongering paid off: he persuaded them to influence Vannevar Bush to bypass Briggs.[12] Like his colleagues in Birmingham, Frisch and Peierls, Oliphant in this sense was also a midwife to the bomb.

The difference between the British and American views at this stage is brought out clearly in a letter from James Chadwick to a colleague in the Ministry of Supply, on 18 October 1941, the month after he had completed his report:

> What is being done in America? Are they only interested in the boiler [the reactor]? Their MAUD Committee has not seriously considered a bomb, in spite of our reports to them . . . Americans are not aware how seriously we are taking this and if we divulge we must have assurances about secrecy. There has been so much loose talk both here and in America that the enemy must be well aware that we are engaged on the uranium problem. We must take every possible step to prevent them from learning that we hope to proceed to a manufacturing stage. This is a matter that may be of vital importance and we cannot afford to ignore it.[13]

While the British were bickering among themselves about collaboration, draft copies of the MAUD report did finally reach Vannevar Bush and Conant, with its specific plans and detailed calculations for manufacturing a bomb on an industrial scale.[14]

Following Conant's trip to London, and his worrisome conversation with Lindemann in the Athenaeum Club, Bush – like Anderson – had been forced to start reconsidering that his original doubts about the practicability of a bomb were mistaken. The MAUD report further underlined this. Bush digested its contents and then had a long conversation with the president and vice president, Henry A. Wallace. As a result, Roosevelt finally instructed Bush to set the American wheels in motion and have his own physicists confirm the British results, in the process assessing that if the report's conclusions were substantiated, what a bomb was likely to cost. At that stage Roosevelt was unwilling to go further. He stipulated that there was to be no further practical development without presidential authorisation.[15]

But that was not the only action the president took.

On Sunday 12 October 1941, what turned out to be an all-important message from Roosevelt landed on Winston Churchill's desk. Normally, Churchill attended to the president's notes assiduously. As he quipped after the war, 'No lover ever studied every whim of his mistress as I did those of President Roosevelt.'

'My dear Winston', the president wrote. 'It appears desirable that we should soon correspond or converse concerning the subject that is under study by your MAUD committee and by Dr Bush's organization in this country, in order that any extended efforts may be coordinated or even jointly conducted.'[16]

As Graham Farmelo has pointed out, the benefits of coordination or joint action to Britain were potentially enormous. The president's note reflected the conversation Bush and Conant had had with Darwin but amplified it. The British would be able to make the most of the lead in theory that they had built up over the Americans, benefit from the Americans' huge scientific and financial resources and, not least, be able to manufacture the weapon at a safe distance from the attentions of the Luftwaffe.

But, just as Hankey had failed to respond promptly to Darwin's initiative, Churchill also failed to strike while the iron was hot. The prime minister had shown himself wary of giving technical secrets cheaply to the US and still felt that Roosevelt should be somewhat more generous in his support of Britain's war exertions. And this appears to be why he did not respond immediately to the president's overtures where Tube Alloys was concerned.

Many of the British nuclear scientists would have urged faster action on the prime minister's part, had they known about it. Instead the British situation got worse. While Oliphant was nuisance-mongering in Washington, the British government decided to hand over the running of the MAUD Committee and its research efforts to a brace of officials from Imperial Chemical Industries (ICI). The company was vast in its reach (it made everything from paints to food to fragrances) and its research director, Wallace Akers, a chemist, was also an experienced administrator of science projects. He would be in charge, assisted by Michael Perrin, a company administrator.

To be fair, the government had a reason for this, especially if Britain were to follow its own path to a bomb. ICI had already helped with some of the preparations and, mindful of the fact that nuclear fission could provide power for industry, as well as for a bomb, it made sense for the government to have one eye on the future. But, in the context of the war, it was a major miscalculation. The majority of scientists involved could not believe this news.

Still, all need not have been lost at that point. On 6 November, the NDRC committee – under Arthur Compton at the University of Chicago – confirmed the basic conclusions of the MAUD Committee. A transportable fission bomb was feasible at the cost of $50–100 million, spent mainly on isotope separation.[17] The numbers were new but, that apart, the British and American scientists were on the same page.

Churchill finally replied to Roosevelt's initiative on 21 November, fully seven weeks after Roosevelt had written to him, and three weeks after the Compton Committee confirmed the MAUD findings. But even then the prime minister's response was merely a peremptory cable, registering the fact that, following the president's suggestion, he had delegated Anderson and Lindemann to explore the matter with Frederick Hovde, the American scientists' representative in London.[18]

This was an almost fatal misjudgement and, in retrospect, an embarrassment. Anderson and Lindemann retained an exaggerated opinion of British progress and felt they were holding all the aces. In doing so they badly overestimated the ability of Britain's scientists to sort out the scientific, technological and industrial problems posed by the bomb. Worse, after being told that Bush and Conant were 'anxious' for a fuller collaboration and that the president had ordered it followed up 'with all possible speed', Anderson put up yet more obstacles, arguing at one point for example that the US must improve its security so it was on a par with Britain's own. This was high-handed, an unconscious irony as it turned out and a mean-spirited response to a fulsome offer of nuclear partnership.

To some, it might have seemed justified as a negotiating ploy to secure for a cash-strapped Britain more material rewards in the midst of dire circumstances. The British aircraft carrier *Ark Royal* was sunk that month, and the RAF took heavy losses in the night bombings of Berlin, the Ruhr and Cologne. But these tragic events were overshadowed and overtaken and the situation transformed irrevocably overnight when, on 7 December, Japan bombed Pearl Harbor. The US was now at war itself (Hitler declared war four days later) and that changed everything. This brought about a marked change in the American attitude, and the view swiftly gained ground that the bomb's rapid development was 'the single most important necessity of the war', because the Germans, it was felt, had had a two-year head-start.[19] Martin Sherwin says that American scientists in particular 'entered this race convinced that the outcome of the war depended upon their ability to recover the time lost'.[20]

From then on the country was to pursue the bomb with an intensity that matched her size. In March 1942 the president insisted to Bush that he wanted the project 'pushed not only in regard to development, but also with due regard to time. This is very much of the essence.' More than that, America now had no interest in giving Britain any more of a role than was strictly necessary to help the Americans acquire the weapon as quickly as possible. Churchill had lost the opportunity to be a full partner with the Americans over the bomb. It was arguably his greatest misjudgement.

Following Pearl Harbor, and a visit Churchill made to Roosevelt about a week later, when the president seemed to think – quite wrongly – that Churchill did not regard the bomb as especially significant (both men did have a lot on their plate), relations between the Allies on nuclear matters sank like the *Ark Royal*. On the British side, despite a modest lead that they still held on nuclear theory, the main administrators of the project – Akers, Lindemann and Anderson – had no real understanding of American industrial and organisational capacity.

The situation was not improved when, during the winter of 1941–2, Akers changed his mind about the plan to build a plant in Britain. At the beginning of 1942 he had been invited to the United States to compare notes with his counterparts in the American project. He was accompanied by Francis Simon and Rudolf Peierls, who knew as much about isotope separation as anyone on the Allied side, and by other scientists. It was immediately obvious to Akers that the sheer scale of the American effort dwarfed that of the British. The Americans were trying several methods of isotope separation, for example, not one, and also concentrating a lot of effort on newly discovered plutonium.

Akers returned to London in March quite convinced that collaboration was the way forward. But, for a time anyway, he had a problem convincing Anderson. This, too, was unfortunate for, as the weeks and then months went by, it became ever clearer that the Americans were pulling ahead.

It also became increasingly clear to Akers that Bush and Conant had changed in their attitudes and were now 'stringing him along'.[21] On the surface they were reassuring that they were in favour of collaboration. But behind his back these two were scheming to have the president reverse his policy of collaboration. During the late spring and early summer Bush sent a string of 'long, courteous and factually accurate' letters to Sir John Anderson, which nonetheless always played for time. At one point, Conant went so far as to praise Bush for his 'masterly evasive' replies.[22] (In a letter from Margaret Gowing to James Chadwick, dated 11 July 1963, as she was preparing her book on the British role in the atomic bomb, she wrote that Conant's attitude to the British 'was in general rather unworthy'.)[23]

The underlying ice in Bush's stonewalling communications may have done as much to convince Anderson to come round to Akers's change of heart as Akers himself, though at one point the ICI man did tell Lindemann and Sir John bluntly that to refuse a merger with the Americans would risk giving the Germans the chance to get to a bomb first.[24] (This was early 1942, before the Rosbaud report in June.) Faced with this ultimatum, on 30 July that year Anderson finally wrote a 'very urgent' memo to the prime minister, in which he was no less explicit: 'We must . . . face the fact that the pioneer work done in this country is a dwindling asset and that, unless we capitalize it quickly, we shall be outstripped. We now have a real contribution to make to a "merger." Soon we shall have little or none.'[25]

This was an interesting exchange. Had Anderson been told about the Rosbaud report? Was he of the view that an atomic weapon was needed, come what may?

Churchill finally saw sense at the end of July 1942 by which time it was too late. The British had missed the boat, for the time being at least.

In addition to now having their own urgent reasons for moving ahead as rapidly as possible, the Americans had acquired a suspicion, or pretended to a suspicion, of British motives, believing – or pretending to believe – that the presence of Akers, difficult as he was to dislike, and of Perrin, showed that their ally was most concerned to position itself for the commercial exploitation of nuclear energy after the war. In contrast, from September 1942 the Americans turned their programme over to the military.

The sheer momentum of American progress was brought home in the very fact that, barely three months later, on 2 December, now working in Chicago, Enrico Fermi produced the first chain reaction in a former squash court under the west stands of the University's Stagg Field stadium. After Fermi's experiment had ended, Szilard, who had hoped against hope that a chain reaction would prove unviable, shook Fermi's hand but said softly, 'This will go down as a black day in the history of mankind.'[26] He was more right than he knew. The British already had evidence that there was no German bomb, and that this all-important experiment, which conclusively proved that a chain reaction was viable, was not needed.

Akers, who had visited Fermi at his laboratory in Stagg Field only the day before the first chain reaction came off (but was not invited to stay on for the 'main show'), was in the United States again in the New Year, in Washington. As we have seen, there had been much toing and froing between Bush and Anderson over the nature of the collaboration between the Allies and, during the harsh winter months, the attitudes of the Americans had chilled. More than that, the president had been brought round to Bush and Conant's view.

All of which came to a head on the morning of 13 January 1943, when Akers was asked to visit Bush and Conant at their offices on 32nd Street in north Georgetown. It was a cold and blustery day and the mood was very different from the warm welcome Akers had received almost exactly a year before. Bush and Conant wasted no time in delivering the blow. Bush was present as a witness, while Conant read Akers a single-page note and then handed it to him.

The note covered seven aspects of the project, including the various methods of separating U-235 from U-238. In almost all cases, the wording was identical, brief and to the point: 'No further information to be given to the British or Canadians.'[27] The British researchers were being shut out, except in one or two areas where their expertise might be useful. Bush and Conant impressed on Akers that the decision had the full approval of Roosevelt.

The British felt aggrieved. They had done the early thinking, the early *motivating*. But for the British, the Americans almost certainly would not have had any kind of project by the time Pearl Harbor came along. Now the United States was unilaterally abrogating what had until then been a fruitful collaboration.[28]

In approving the new policy of 'restricted interchange', Roosevelt had, as Martin Sherwin puts it, 'rearranged' America's priorities. The desire for sole post-war control of the atomic bomb had suddenly become as important – and maybe more important – than the need to build the bomb quickly. Bush's aim, of course, was for America to keep for itself alone a weapon that, as he put it, 'may be capable of maintaining peace in the world'.[29] He shared with the secretary of war, Henry Stimson, the view that if America possessed the best weapons available, she would never need to use them.[30] As part of this he now agreed with Conant that the rapid development of the bomb was no longer the primary aspect of policy: atomic energy was a military secret 'in a different class from anything the world has ever seen', and sharing it with the British could not be defended to Congress.[31]

This was extraordinary because 'the new policy was adopted despite the possibility that it might "slow down" work on the bomb'. Harold Urey, the Nobel Prize-winning discoverer of deuterium, was just one who thought 'this re-arrangement' of priorities would delay the bomb by six months and that this might give the Germans the edge. The Americans were still heavily overestimating the German capability and the British did nothing to correct them. But Conant and others took the view that though a delay would occur it would not be fatal or a serious hindrance. The post-war monopoly was now a – if not the – major consideration.[32]

Of course, this position needs more attention than it has so far received. Because a moment's thought will see the argument as hollow. Unless the Allies got to the bomb first, before Hitler, they would likely never be in a position to use an atomic monopoly to enforce their 'peace' on the world. And this raises the issue as to whether Bush and his colleagues were *really* in the dark as far as the British atomic intelligence was concerned. Would they *really* risk a delay in the production of a bomb without knowing that, in fact, a German bomb was a non-starter?

All the senior figures in Washington agreed with the Bush-Conant line, in particular Henry Stimson. He had long been a friend of Britain 'but firmly believed that it was America's destiny to lead the world'.[33] And this general agreement was maintained, according to Graham Farmelo, '*even though they knew that putting an end to collaboration would delay the project and might conceivably mean that Hitler got hold of the weapon first*'.[34] Looked at in this way, it was a remarkable decision.

What no one mentioned openly, but it must have been on all their minds, was that, as Akers and Bush and Conant met on 32nd Street, Operation Ring, the last episode in the Battle of Stalingrad, had been under way for seventy-two hours. The German forces were encircled, their tanks and guns iced up, and the battle had only a few days to run. After Stalingrad, Russia would still be an ally but she would also be – as anyone could see – America's main post-war rival.

After the deadly Bush-Conant note the British fought back. British scientists were forbidden from attending meetings in the United States, or providing information on certain specific projects they specialised in. Clearly this was not ideal but the fall out continued for months. One especially heady skirmish took place in London in July

1943, when Bush was in London to attend the War Cabinet's anti-U-boat subcommittee at 10 Downing Street. Or he thought that he was. Once inside the building, he was directed into the cabinet room. And there sat Winston Churchill, incandescent with rage. The prime minister was just up from his afternoon nap, and he could barely contain himself. He laid into Bush, growling about the failure of the interchange agreement while struggling to light his cigar, tossing one match after the other over his shoulder in the general direction of the fireplace. Bush could not get a word in.

Churchill had only grasped the full state of the British nuclear weapons project at the beginning of April that year. And only then did he realise that Britain was already in serious jeopardy of being shut out of America's plans. Not only that. As Lindemann minuted, 'The principles and possibilities are known to scientists throughout the world ... Can England afford to neglect so potent an arm while Russia develops it?'[35] This did little to comfort Churchill.

At a further meeting a few days later, with Stimson also present, and with Lindemann and Anderson on the British side, Bush aired the American suspicions of Akers and Perrin's involvement, suggesting that Britain was trying to retain for itself the option of pursuing nuclear power commercially after the war. Churchill countered formally that he had no interest in commercial matters.[36] He thereupon made a proposal that would become the basis of the Quebec Agreement that would be signed when the two leaders met later that year in August. 'First the bomb would be built as a joint venture with the free exchange of information; second, neither country would use the weapon against the other; third, neither would pass information to other countries without the other's consent – a move that would remove any chance of Roosevelt sharing the science with Stalin, with whom the President was more sympathetic than Churchill thought wise.'[37]

Bush privately thought Churchill's proposal so poorly worded that its intentions would be easy to deflect. While he was still in London, however, Bush received a telegram from Roosevelt 'To review' the full exchange of information with the British. He was unperturbed but, ironically, the cable had been wrongly transcribed: the actual wording was 'To renew' the agreement. The Fall Out was about to end, whatever Bush might want.[38]

This was the back story to the meeting between Churchill and Roosevelt held at Quebec in August 1943. Just before the meeting took place, Sir John Anderson outlined the British position to the Canadian prime minister, Mackenzie King. He said that the British knew that 'both Germany and Russia were working on the same thing' and that the bomb 'would be a terrific factor in the postwar world as giving an absolute control to whatever country possessed the secret'.[39] His memoirs fail to make clear what exactly he had heard about German and Russian plans.

The most important decision at the Quebec Conference had nothing to do with the bomb. It was that Churchill agreed finally to open a second front in the war – an Allied landing in France that would take place in the following year under American command. At the same time, Roosevelt finally gave way on the bomb and confirmed that collaboration would be resumed. On 19 August, the two leaders signed an agreement, drafted by Anderson but substantially the same as Churchill had discussed with Bush in Downing Street.[40]

There is one important point to be made about the Quebec Agreement. It included a clause that the speedy completion of the project was 'vital to our common safety'. This was a reference to the danger that Hitler might get to a bomb first. *Yet Churchill in particular now knew, from the intelligence reports he was receiving, that this was unlikely. The point of the American bomb was beginning to change from beating the Germans to the weapon, to winning global dominance after the war.*

For many of the nuclear physicists involved, especially the émigré German and European Jewish scientists, the threat of a Nazi bomb was their chief reason for being involved in the Manhattan Project. For many of the politicians, military personnel and science administrators, however, certainly for Roosevelt, Bush, Conant and Groves, and for Churchill and Cherwell, the prospect of a bomb of hitherto unimaginable power was too mouth-watering to resist. John Moore-Brabazon, minister of aircraft production, on reading the MAUD report in the autumn of 1941, had described 'the unprecedented political clout' that would accrue to any nation if they were the sole possessor of the atomic bomb. In the words of Lord Cherwell, it would mean that 'whoever possesses such a plant should be able to dictate terms to the rest of the world'.

For them, and others like them, the prospect of the atomic bomb offered, as they saw it, an unprecedented opportunity to coerce other nations – in particular the Soviet Union – after the war was over, to establish a post-war settlement on their terms. This may seem obvious to us, now, but as we shall see it was anything but obvious to many of the scientists themselves at the time.

Moreover, evidence was beginning to accrue that the Russians were not entirely unaware of what the US and UK were jointly embarked upon. And this time it was the Americans who did not share this intelligence with the British.

11

'Hard Evidence' of Soviet Spying

The intelligence predicament so far as the Americans were concerned was somewhat different from that confronting the British. They were, geographically speaking, relatively distant from Germany and, from 7 December 1941, pitched more directly against Japan, across the Pacific, rather than the Atlantic. In addition, America was where the bomb was being constructed. So they were concerned with preventing secrets from leaking out, as much as or more than obtaining secrets about the enemy's plans.

The marked difference that existed then is shown by the fact that, in June 1942, the very month the British received the Rosbaud intelligence about Speer's decision not to proceed with a German bomb, Arthur Compton, the director of the atomic energy project at the University of Chicago (and yet another Nobel Prize-winner), was so upset at the American rate of progress that he urged a programme for researching and developing 'counter measures' against a German weapon. The following month Oppenheimer was in despair that the war could be lost before an answer was found and in September Szilard headed a critical memo, 'What is wrong with us?' in which he too lamented the slow rate of progress in finding out about Germany's plans.[1]

The first time the atomic programme and espionage collided (albeit unknowingly then) had come in March 1941, when the FBI opened a file on Robert Oppenheimer.[2] Julius Robert Oppenheimer, aged thirty-seven in 1941, had been born in New York City, the son of a German textile importer who emigrated to America in 1888. He had studied in Cambridge, Göttingen and Leiden and worked or studied with Heisenberg, Fermi and Franck, plus Paul Dirac in Cambridge (Nobel Prize in Physics in 1933, for his discovery of 'anti-matter' and quantum electrodynamics) and Wolfgang Pauli in Zurich (Nobel Prize in Physics in 1945 for his discovery of the 'exclusion principle', that no more than two electrons can inhabit any orbit around the nucleus). Oppenheimer's fields of interest included nuclear physics, astrophysics and relativity. At the time he was appointed scientific director of the Manhattan Project – the ranking physicist at Los Alamos – he was professor of theoretical physics at Berkeley and very close to Ernest Lawrence.

Oppenheimer came under surveillance accidentally, because he attended a meeting of a discussion group at the San Francisco home of Haakon Chevalier, an assistant professor of French literature at Berkeley. Born in New Jersey, with a Norwegian mother and a French father, there was something of the Viking adventurer about Chevalier: in his twenties he had enjoyed a stint as a deckhand on a four-masted schooner during a round-the-world voyage before settling in San Francisco – and attending Communist Party meetings. The FBI had been primarily interested in other people at the meeting at Chevalier's but wrote down the registration number of Oppenheimer's car, which led to him.

San Francisco was regarded by the Soviets as a particularly fertile area for recruiting subversives. This was because union activity was especially strong there, going back to the West Coast waterfront strike of 1934. Aware of this, the FBI had been surreptitiously listening to the telephone calls of several known and suspected communists as part of its campaign against the Soviet Union's so-called Comintern Apparatus.[3] Although Oppenheimer was probably never a member of the Communist Party himself, he did attend rallies in support of the waterfront strike, and was vocal in support for the Republican side in the Spanish Civil War and other radical causes. Both his long-time women friends, with one of whom he had two children, were members of the CP.

The head of the FBI in San Francisco was a lawyer, Robert King.

He had been interested to learn, through bugging, that the meeting at Chevalier's in March 1941 was to be of 'the big boys'. These turned out to be Thomas Addis, a physician at Stanford, and a recruiter for the Party; Isaac Folkoff, a sixty-year-old Latvian émigré, who owned a clothing repair shop, who was known as a 'dogmatic Marxist' and collected Party dues; and Oppenheimer.[4] These three continued to meet, and to be bugged, until, in October that year, Folkoff asked Oppenheimer to connect with a certain Steve Nelson, known to King as the Bay Area's top communist – another 'big boy'. At this point, the FBI were bugging Oppenheimer because of his apparent importance in local CP matters; they knew that he was a senior scientist but they knew nothing about the Manhattan Project.

Nelson was colourful too. Born Stefan Mesarosh in Croatia in 1903, he had entered America illegally but managed to become a naturalised citizen only a few years later. He had joined the Communist Party in Pittsburgh and fought in the Abraham Lincoln Brigade in the Spanish Civil War.[5] Oppenheimer and Nelson had first met at a Spanish Civil War relief party, held at the home of Louise Bransten, a wealthy San Francisco socialite who was known to the FBI as an intimate friend of Gregory Kheifetz, an agent for the NKVD (forerunner of the KGB) working undercover at the Soviet Consulate in San Francisco. Kheifetz was known to be cultivating scientists.

Shortly after America's entry into the Second World War, Nelson had become chairman of the San Francisco branch of the Communist Party, so that Oppenheimer's many meetings with him made both of them of natural interest to the Bureau.

Requests by the local San Francisco FBI to bug Oppenheimer's phone were turned down. At the same time, the army refused him security clearance. This was appealed and the Provost Marshal's Office overruled the decision. This gives a measure of the confusion surrounding Oppenheimer: no one could be exactly sure who he was, an uncertainty that, in some quarters, was never to go away entirely.[6]

Then, on 10 October 1942, after some months of eavesdropping, the FBI bug in Steve Nelson's office picked up an exchange between Nelson, Lloyd Lehman – a Bay Area labour organiser affiliated with the Young Communist League – and another person, unidentified. According to the Bureau record of the conversation, 'Lloyd told Steve about an important weapon that was being developed, and indicated that he was on the research of it.' They then went on to discuss someone else they knew who was in the project at Berkeley, who was 'considered a "Red"', who had been involved in the Teachers' Committee and the Spanish Committee, 'but whom the government allowed to remain because he was such a good scientist'.[7] This description pointed to Oppenheimer. Equally alarming were remarks caught by the bug on Nelson and Lehman about one Rossi Lomanitz. Lomanitz, it was said, was working on the weapons project but was considering quitting. 'Nelson said the project was extremely important and Lomanitz would have to be an undercover Party member, since it was important to have knowledge of such discoveries and research developments. If he quit his work they did not know what political work he could do [afterwards] as he might be drafted.'[8] But still the FBI did not know fully what it had. It was even then in the dark about the Manhattan Project.

The FBI was not alone in its suspicions, however. No less aware of communist sympathisers in San Francisco was the army – which by now did know about the Manhattan Project – and they also took action. They set up a dummy business office – the 'Universal Subscription Company' – in a building just off Market Street as a base for undercover agents, and installed their own wiretaps. By arrangement with the FBI, the army concentrated on university employees under contract to the Manhattan Project, while the Bureau concentrated on known or suspected communists with connections to the Radiation Laboratory. For this it rented a two-storey house near the Berkeley campus and conducted wiretaps on selected Rad Lab personnel.

The plans for Oppenheimer's surveillance were even more elaborate than they were for others. The FBI had been forbidden by Roosevelt's attorney general from tapping his phone so they were confined to examining the log of phone calls out of and into his home, as provided by the telephone company. This showed one thing of interest – that Oppenheimer had suddenly begun to travel and on a fairly frenetic schedule. Why?[9] At that point, Robert King, the one-man 'commie squad' in San Francisco, had approached the army fairly formally to find out what they knew. When the formal channels failed to reveal anything of real substance, King turned to Lieutenant Colonel Boris Pash, the head of counter-intelligence at the Presidio, the military base at the northern tip of the San Francisco peninsula.

Described by Gregg Herken as 'an intense and pugnacious bantamweight', Pash had by then been head of G-2 – military intelligence – for the Western Defence Command for two years, and was the 'Fourth Army's foremost defence expert'.[10] He had been born right there in San Francisco where his father was a bishop in the Russian Orthodox Church in America, and Boris had himself fought the Bolsheviks in Russia's Civil War (1918–20). When the White armies were defeated, he had returned to California and had a middling career as an athletics coach.

Following Pearl Harbor, however, his life changed again and, after being called up, he was assigned the running of a network of army undercover agents infiltrated into Mexico's Baja Peninsula to intercept saboteurs landed by Japanese submarines. He never did encounter either submarines or saboteurs but in the course of his exertions Pash realised that the army in general – and spy-catching in particular – was his calling.[11]

King had had Oppenheimer followed but Pash went further. He assigned to the physicist a pair of 'bodyguards', on the ostensible grounds that Axis saboteurs might try to assassinate him, but in reality to keep an eye on him.

The chief evidence for Pash's suspicions, apart from Oppenheimer's pre-war communist sympathies, was a second tap on the phone of Steve Nelson. The conversation took place in the early hours of Tuesday 30 March 1943. Someone identified only as 'Joe' on the wiretap had arrived at Nelson's house the previous evening, insisting to Nelson's wife, Margaret, that he had some important information to pass on. In fact, it was so important that he was prepared to wait right there until Nelson got home, however long it took. As Nelson discovered when he did get home, the information Joe had was indeed of great importance and was bound to be of enormous significance to the Soviet Union: people engaged in the new weapon project (which Joe at the time thought would include him) were about to be relocated to a remote spot where experiments on explosives could be conducted in secret.[12]

The wiretappers could sense that Joe was nervous, fully appreciative of the enormity of what he was doing. Confessing that he was 'a little bit scared', he spoke in only a whisper when going into technical details. The two men then discussed 'the professor' – Oppenheimer – who, Nelson complained, was 'very much worried now and we make him feel uncomfortable'.

'You won't hardly believe the change that has taken place,' Joe agreed.

'To my sorrow, his wife is influencing him in the wrong direction,' Steve added.

The discussion then shifted back to 'the Project'.

'Do you happen to know what kinds of materials they are working on?' Steve asked his guest.[13]

Joe hesitated but then said that most of the things he knew were common knowledge and had already been published, and when Nelson pressed him for details he seemed reluctant to be more specific. He said that he was worried that his communist background – he had been a Party member since 1938 – would count against him and cause him to come under scrutiny. When Nelson pressed him on whether any experiments had actually been completed, Joe stated that he did not know.

Nelson then tried an old trick, saying that he already knew something of the work Joe was engaged in, and at this the other man succumbed.

He again lowered his voice to a whisper. But he spoke slowly, so that Nelson – and the wiretappers – could take notes. Among the sentences they could make out was this: 'Separation method is preferably that of the magnetic spectrograph with electrical and magnetic focusing, or less preferably, that of the velocity sector ... sphere 5 centimeters in diameter with material ... deuterium ... this design is tentative and is being experimented upon.'[14] Joe added that there was already a separation plant under construction in Tennessee that would, he had been told, employ between 2,000 and 3,000 people.

The mention of deuterium suggested that Joe might also have

passed along information on an earlier occasion. Deuterium plays no role in electromagnetic separation but a seminar, held at Berkeley in July 1942, had discussed the 'Super' (US code for the bomb). In particular, the seminar had considered research by Eldred Nelson and Stanley Frankel on just what the dimensions of a U-235 atom bomb might be, and a plutonium bomb, which does need deuterium in its manufacture.[15] This was classified information of the highest quality.

Just before they parted, Steve and Joe discussed the arrangements for future meetings. Joe said he had a sister in New York and her movements could be used as the 'cover' for subsequent assignations, 'where he could pass additional information to Steve'. In return Nelson advised Joe to tell his comrades, who were being sent to Tennessee and other remote sites, to burn their party books as a precaution, and never to put anything in writing.

Pash was told of this exchange within hours and flew to Washington the next day to brief General Groves in his office in rooms 5120 and 5121 in the War Department Building. Following the briefing, Nelson's premises were put under 24-hour surveillance, so they could find out who 'Joe' was. In fact the very next morning Nelson was followed as he walked to a corner drugstore and made a telephone call from a public booth to the Russian San Francisco consulate. He was overheard using the name 'Hugo' as he asked to speak to 'Ivanov'. The two men agreed to meet 'at the usual place' a few days later.

On the evening of 6 April Nelson was followed to the grounds of St Joseph's Hospital in San Francisco, where he met Pyotr Ivanov. Ivanov was known to be a GRU operative, under diplomatic cover, attached to the Soviet Consulate in San Francisco. (GRU was the overseas intelligence branch of the Soviet Army.)[16]

A few days after that another visitor to Nelson's house was observed. This was 'a burly figure in an ill-fitting suit', later identified as Vasily Zarubin, officially third secretary of the Soviet embassy in Washington but in fact the NKVD *rezident* there. The phone-tap inside Nelson's home picked up the Russian counting out bills.

NELSON: Jesus, you count money just like a banker.

UNKNOWN MAN: Well, after all, I told [*deleted*], I used to pay out at Russia.

And then, with the counting done, for one hour and more, the two men talked about the Soviet espionage set-up in the United States.[17]

Two months later, in June, the unremitting surveillance of Lomanitz finally paid off. A G-2 agent following him witnessed him pose with three friends for a picture taken by a commercial photographer near one of the entranceways to the Berkeley campus. Enterprisingly, as soon as the four friends had departed, the agent approached the cameraman and bought the negative of the picture he had just taken. Apart from Lomanitz, the other faces in the photograph were identified as David Bohm, Max Friedman and Joseph Weinberg. All four were physicists at Berkeley, and three of them (all but Friedman) were students of Oppenheimer. It didn't take G-2 long to realise they had found 'Joe'.

Joseph Weinberg was indeed involved in the bomb project, working on magnetic field calculations, focusing the cyclotron beam with iron washers. He was another member of the Young Communist League, the [Berkeley] Campus Committee to Fight Conscription, and the Committee for Peace Mobilization, all branded 'communist Fronts' by the FBI.[18]

So some atomic secrets did pass between Weinberg and Nelson. But what of Oppenheimer's role? Pash, for one, did not have to be convinced that Oppenheimer was a security threat. He warned Washington that 'Oppie might have agreed to work on the bomb just so he could give it to the Russians.' He suggested the possibility of 'easing' Oppenheimer out of his job and replacing him.[19]

But of course this neglects the exchange between Nelson and Weinberg, that 'Nelson & Co.' now made 'the professor' feel uncomfortable, that a large change had come over him, and that his wife was influencing him in the 'wrong' direction. From their communist perspective this must surely have meant that Oppenheimer was now *not* as securely in their camp as he had been before. In fact, in 1999, in *The Haunted Wood*, their book on Soviet espionage during the Stalin Era, Allen Weinstein and Alexander Vassiliev were given access to certain Soviet files that confirm Oppenheimer was not a source for Russian intelligence.

General Groves was a stickler for security but he would not listen to Pash's entreaties to have Oppenheimer replaced. 'Lansdale [the general's deputy] stated that General Groves claims flatly that Oppenheimer is irreplaceable and that if anything happened to Oppenheimer, the project would be set back at least six months.' Such a delay could be catastrophic, he said, because recent intelligence showed that the Germans were laying new high-tension wires, 'leading the army to conclude that the Nazis might be building their own Calutrons [mass spectrometers used for separating uranium isotopes]'. The source of this intelligence has never been confirmed, and was clearly mistaken, but on the strength of it Groves now put a swift end to the agonising over Oppenheimer, ordering the district engineer to issue a security clearance, 'without delay, irrespective of the information you have concerning Mr. Oppenheimer. He is absolutely essential to the project.'[20]

One consequence of this was that on 7 May 1943 J. Edgar Hoover, head of the FBI, sent to President Roosevelt and his aide Harry Hopkins a memo outlining 'hard evidence' – first obtained in March that year (the second Nelson meeting) – of Soviet espionage against the Manhattan Project.

Nor was that the end of the matter. On 12 August, FBI agents were watching as Bohm, Friedman and Lomanitz arrived at Weinberg's apartment, where the others present were Steve Nelson and his assistant for Communist Party matters, Bernadette Doyle.[21] By now the surveillance of Lomanitz and the others had become formalised into an extensive operation codenamed 'CINRAD' – Communist Infiltration of the Radiation Laboratory. Eventually this amassed files on more than 300 Communist Party members in the Berkeley area.

The seriousness with which this operation was taken may be judged from the fact that it was discussed at the highest possible level. A few days later, on 17 August, Groves presented an account of progress on the Manhattan Project to the US government's Top Policy Group, but then went on to discuss what he referred to as his 'California trouble'.[22] On the same day he sent Stimson a note, attached to which was a draft memorandum intended for the president: 'It is essential that action be taken to remove the influence of FAECT from the Rad Lab.' (FAECT was the Federation of Architects, Engineers, Chemists and Technicians, which the Bureau was convinced was a communist front organisation burrowing away inside the Berkeley Lab.) Stimson passed the memo to Roosevelt two weeks later together with his own note: 'Unless this can be at once stopped, I think the situation very alarming.'

In that same month, August, moreover, Oppenheimer made what was described as a 'casual' remark to Lyall Johnson, an ex-FBI agent, now one of G-2's men in the Bay Area, to the effect that if, he, Johnson was concerned about security, 'George Eltenton was someone who was worth watching.'[23] Eltenton was a British-born chemist who had worked in Russia but was then employed by Shell and living in Berkeley where he was a member of the Communist Party. Johnson immediately phoned Pash who himself questioned Oppenheimer. Some months earlier, Oppenheimer now said, he had been approached by 'intermediaries' who were in touch with an 'unidentified official' at the Soviet consulate. One of these individuals had talked about passing along information regarding the operations at Berkeley. Oppenheimer said he had refused to collaborate but he did admit that there had been other occasions when colleagues had been approached and they had come to him for advice. He was reluctant to provide names but did concede that two of the three men approached were now at Los Alamos and the third was destined for Oak Ridge, a vast site in Tennessee where uranium enrichment was carried out.

This was clearly worrying and, pressed by Pash, Oppenheimer identified Eltenton as one of the intermediaries, though he had used another individual whom Oppenheimer again refused to identify. Somewhat dissatisfied by Oppenheimer's not entirely co-operative attitude, Pash transcribed the interview and sent it to Groves, recommending Eltenton be put under surveillance. This happened to coincide with an anonymous letter – written in Russian – that had been sent to Hoover by an out-of-sorts intelligence officer. The letter appeared to have been sent from near to the Soviet embassy in Washington and identified Vasily Zarubin, Leonid Kvasnikov and Gregory Kheifets as three of Russia's top spies in the United States, two of the three names being already known to the Bureau and the army. On the strength of this, King was given more than 100 reinforcements in the Bureau's San Francisco office.

The reinforcements paid off. On 3 September Weinberg was followed by a roundabout route to a post office where he mailed a thick envelope. Intercepted, this was found to contain a manuscript article headed 'The Communist Party and the Professions' and a note addressed to 'Dear A', asking him not to communicate with Weinberg 'during this period' and requesting him also to pass on the message to 'S' and 'B' 'without mentioning my name'. From previous surveillance, it was known that 'A' was Al Flanigan, a Berkeley graduate student and suspected Communist Party member who lived near Chevalier and was also a friend of Nelson. Pash suspected that 'S' and 'B' were Nelson and Bernadette Doyle.

Groves had decided to keep Weinberg on campus in case he led them to other sources but their other colleagues were now dismissed or refused clearance to go to Los Alamos.

In yet another attempt to get more names out of Oppenheimer, a meeting was arranged with John Lansdale, Groves's deputy. He tried a different tack, saying how much he (and by implication the army) admired and respected what the scientists were doing. Then he made an admission regarding the Russians. 'They know, we know they know, about Tennessee, about Los Alamos, and Chicago.' And, he added, 'It is essential that we know the channels of communication.'[24]

Lansdale got nowhere.

Despite the fact that Groves had by now 'cleared' Oppenheimer to go to Los Alamos, other doubts must have still lingered in his mind because the general himself travelled to Los Alamos and met alone with Oppenheimer in his office there. At this meeting, Groves ordered Oppenheimer to reveal the name of the intermediary who had approached his scientific colleagues. Oppenheimer promptly identified Haakon Chevalier.[25]

Moving on, Groves now ordered Oppenheimer to name the three scientists Chevalier had approached. Oppenheimer replied that, in fact, he had exaggerated with Pash and only one individual had been approached – and that was his own brother, Frank, also a physicist. What had happened, Oppenheimer said, was that Chevalier approached Frank 'about the possibility of either passing secrets on the bomb project to the Russians or persuading his brother to'.[26] Frank had come to Robert for advice and Oppenheimer had said they should both refuse.

Groves discussed this encounter with his aides. He himself thought Oppenheimer too valuable to doubt but it appears he was convinced Frank was involved in some way, and his aides did subsequently circulate a story that 'Haakon Chevalier had tried to recruit Robert Oppenheimer's brother to spy for the Soviet Union.'[27] Had Robert been telling the truth when he said to Groves that he had exaggerated with Pash, that three individuals and not one had been approached by Chevalier? At the least he was not being consistent.[28]

Alongside this, in January 1943, the Russians had approached the Lend-Lease Administration in Washington for 10 kilograms of uranium metal, and 100 kilograms of uranium oxide and uranium nitrate (see pp. 240–1 for a fuller discussion). General Groves authorised this request since a denial would have shown the importance he attached to the substance and might have drawn attention to the matter around Washington. A later request for more uranium was turned down though an order for 1,000 grams of heavy water, made later in the year, was also agreed to. Of course the timing of this otherwise opaque move shows that the Russians were signalling that they knew exactly what significance uranium and heavy water had.

Such developments were hardly welcome. For General Groves, the 'crown jewel' of atomic secrets, as he always maintained, was the very existence of the project itself. That must be concealed at all costs, and was one of the chief reasons why he insisted upon compartmentalisation.

But, a close reading of what was known to the American security services in early 1943, when Hoover sent the president his memo of 'hard evidence', showed – at the least – that:

Local communists in San Francisco knew that an 'important weapon' was being developed and that the Rad Lab in Berkeley had a role in it;
Soviet sympathisers were already attached to the research effort;
There were known or suspected communists working in the Rad Lab at Berkeley;
People were being paid for information, by the Soviets; a paymaster from Washington was in San Francisco.
The Soviets had been told that, in addition to the Rad Lab in Berkeley, there was a separate location in Tennessee, which was employing between 2,000 and 3,000 people, and a laboratory at the University of Chicago.
Sometime after March 1943 the 'project' was being relocated to a remote site, and that Oppenheimer, a senior physicist, would be going;
The project involved 'separation methods', 'deuterium' and 'velocity selection';
The code for the project was known as the 'Super'.
In January 1943 the Russians had approached the Americans for a supply of uranium metal, other uranium compounds and, later, heavy water.
Possibly, the Russians had been told the contents of a 1942 seminar that had discussed the dimensions of a U-235 bomb and a plutonium bomb (i.e. two types of bomb were being researched/developed).

Any physicist worth his salt knew that fission pointed to a bomb, using a chain reaction in uranium. Many also knew that heavy water – deuterium oxide – was vital as a moderator in the creation of a chain reaction. On top of that, free publication of research into nuclear physics had been suspended by American and British scientists. It would – or should – have been clear to General Groves, and others, in 1943, that Russia had learned, one way or another, that America had an atom bomb project, and that the 'crown jewel' of secrets was no longer secret. It should also have been clear that the Russians were aware the project was beginning to expand into an industrial-scale effort – in other words, it was more than just experimentation. (Groves allowed the Russians to acquire the uranium, he says in a note in the National Archive, because it would help 'keep Russia in the war' and because he hoped the Americans might track the metal to where it was used. That didn't happen.) As Joseph Albright phrased it, 'Groves didn't come close to realizing the extent to which his secret factories were penetrated by Soviet intelligence.' And this despite the fact that more than 100 cases of wartime atomic espionage were known about as well as 200 acts of sabotage.

It is not known if the Americans shared this information with the British, but there is no mention of it in any of the official accounts of the bomb's development, or in the memoirs of any of the individuals involved. In any case we can't be certain that Groves, Lansdale, Hoover and the rest had assimilated what they had. What we can say is that Churchill was just as much a stickler for security and secrecy as Groves. Had the prime minister been told by the Americans in early 1943 that Russia had a good idea that the Allies were working on a bomb, might he have changed his mind – and his tactics – accordingly? According to Martin Sherwin, Churchill wasn't told about all this until 7 March 1945.[29] Did Groves and Churchill always have unrealistic expectations of what secrecy could achieve? It is curious – to put it no more strongly – that if Groves felt that Russia was the main enemy right from the start, he took so little action to prevent them from gaining access to the results of Allied science.

12

The Secret Agenda of General Groves

In fact, Groves's role in the Manhattan Project needs a thorough reassessment. To an extent, it began with Martin Sherwin who in his 1988 book, *A World Destroyed*, concluded that 'nothing in the official history of the Atomic Energy Commission, in the files of the Manhattan Project, or in any other relevant source available, including Groves's own book, offers convincing support for his belief in the value of compartmentalisation – indeed, quite the contrary.' Moreover, 'Though originally conceived mainly as a defense against German espionage, compartmentalisation also came to be used to restrict discussion of the implications of the development of atomic bombs. Thus, when Groves learned in the spring of 1943 that a series of weekly colloquiums had been organized at Los Alamos, he sought to have them cancelled.' On this occasion, the discussion went ahead but in a very proscribed format. As the official history of the Manhattan Project put it, Oppenheimer agreed 'to avoid matters that, *whatever their importance in other ways* [my italics], were of little scientific interest'.[1]

But the reassessment needs to go much further than this. General Groves had been appointed to command the Manhattan Project on 17 September 1942. According to Ronald Clark, in his history of fission, published in 1980, Groves said only two weeks after his appointment that he was quite clear 'that Russia was our enemy and that this project was conducted on that basis'.[2] Groves doesn't mention this in his own memoir. In fact, what he says is that the basic requirement of the project was twofold: 'to provide our armed forces with a weapon that would end the war and to do it before our enemies could use it against us', adding later that 'the Axis Powers could very easily soon be in a position to produce either plutonium or U-235 or both. There was no evidence to indicate that they were not striving to do so; therefore we had to assume that they were.'[3]

This, of course, was not true. The British had more than one piece of convincing evidence that the Axis powers were not striving to build a bomb.

In a note in the National Archives in Washington, in the box with working drafts for the general's wartime memoir, the priorities are given a crucial different wording:

> Never once was any definite country named to me as the one against which our major security effort should be devoted. Initially it seemed logical to direct it towards the axis powers with particular emphasis on Germany. She was the only enemy with the capacity to take advantage of any information she might gain from us ... We did not feel that information secured by Japan would reach Germany accurately or promptly and we suspected that the Italian-German intelligence channels were also not smooth ... I learned within a week or two after my assignment to the project that the only known espionage was that conducted by the Russians using American communist sympathisers against the Berkeley lab. Our security aims were soon well established. They were three-fold: first, to keep the Germans from learning anything about our efforts or scientific or technical advancements; next, in so far as we were able, to keep the Russians from learning of our discoveries and the details of our designs and processes; and finally to do all we could to ensure a complete surprise when the bomb was first used in combat.[4]

Groves says he did not make 'any appreciable effort' during the war to secure information on atomic developments in Japan.[5] And, as we shall see later, despite what he said he never took the Russians as seriously as he might have done.

All of which confirms that, at that stage, only Germany could have comprised a real threat, but by then the intelligence the British were receiving showed that Germany had no bomb project. From what we now know, it seems unlikely that Groves was told this, then. If he had been, would he have believed it? Would he have welcomed the fact that the plum job he had just been given was now redundant? Did he, together with Bush and maybe FDR, assume from the word go that a bomb would give America an unrivalled position of power (or coercion) in the post-war world? Was this the real start of the Cold War? Stimson was not of that view. His assistant secretary James McCloy had many talks with Stimson about the nature of the German threat, but it was always a *German* threat.

Groves was a belligerent man. Was it his personal view that the Manhattan Project would be conducted on the basis that Russia – and not Germany – was the principal enemy? Or did it fit in with what Bush, Conant, Stimson and the president thought, even then? Again Stimson, for one, did not begin to doubt whether Russia could be trusted as a post-war partner until the autumn of 1944.[6]

Groves's most recent biographer says the general concerned himself with the domestic intelligence situation for the first year of his command and didn't really think about foreign intelligence until the latter half of 1943. A note in the archive dates it to 18 December 1943. But we also know that he set up an outfit on 30 March 1943 to monitor (and where possible suppress) media reports to do with nuclear matters, so he must have been aware of a report that had appeared in the London *Daily Telegraph*, in October 1941, written by its Moscow correspondent. This reported an appeal by several well-known Soviet physicists, led by Pyotr Kapitsa, in which the scientists had hinted that they knew about the Allied bomb effort and felt they should be part of 'new methods of warfare to be used against Germany'. (This is discussed in more detail on pp. 231–2.)

Groves may have been taking a realistic, realpolitik line, that a bomb was going to be built one way or another, come what may. He says in his memoirs that it is 'not extraordinarily difficult for anyone who will apply himself to learning them [the intricacies of nuclear energy] to understand the basic principles of atomic physics'.[7] But he wasn't very realistic – or very worried – about the early signs from Russia. As we shall see later, he felt that although it wasn't too difficult, the Russians weren't up to it.

In fact, the first time Groves made his statement about Russia being the obvious target of an atomic bomb was on 12 April 1954, when he gave evidence at the court hearing of Robert Oppenheimer, when the Los Alamos director's security clearance was revoked on account of the fact that he hadn't been entirely open and honest about his communist sympathies in the run-up to war and his appointment at Los Alamos. The security hearing was a rum business that need not concern us but what matters here is that we have only Groves's word that he actually identified Russia as the main threat in 1942, when he said it plainly and publicly for the first time nearly twelve years later. By then Russia had the bomb, the Korean conflict had come and gone and the Cold War was in full spate. While no one has ever queried Groves's statement, as we shall see he never seems to have been truly troubled by the risk that Russia would get the bomb and this does rather undermine whether he did really regard Russia as enemy number one from the start.

In fact, and more likely, given the wording in his memoirs that we shall come to, Groves was no doubt influenced by the predicament many scientists felt they were in at the beginning of the Manhattan Project and also highlighted by Martin Sherwin. It was four years since fission had been discovered in Germany and the American physicists could not be confident that the US was closing the lead that the Germans were assumed to have. 'A feeling of desperate urgency grew with every passing month, and with it one of hostility toward a precaution that caused delays.' The scientists were driven by a profound sense of danger and believed that nothing – not even security precautions – should be allowed to interfere with their work.[8] Until very late in the war, Sherwin confirms, American progress on the bomb was measured against Germany's 'presumed head start'. An exchange between Isidore Rabi and Oppenheimer at the end of February 1943 confirmed that a concern about Nazi Germany was the main worry of the senior scientists.[9]

At times, Bush and Conant were more frustrated by their ignorance of German progress than by their own scientific problems. In such circumstances they considered some 'imaginative' schemes for gathering intelligence on the status of the German programme. For example, in June 1942 – again the very month the British were receiving the Rosbaud report about the Speer meeting in Berlin – Bush and Conant considered a plan to send physicists to Switzerland 'with diplomatic status' for the purpose of contacting neutrals who attended German scientific meetings. Sixteen months later, in October 1943, a plan was conceived to examine relevant German literature and economic data to see what projects might be under way. This was of course much the same as the British had been doing since 1940. The more you think about it, the more extraordinary the situation becomes in retrospect. On one side of the Atlantic the British intelligence services knew there was no German bomb. On the other, the scientists were kept in ignorance of this fact and fretted for months unnecessarily.[10]

In some ways, perhaps it was not so surprising. Under General Groves's policy of compartmentalisation, 'he effectively suppressed organized discussion of post war atomic energy policy *during* the war'.[11]

Groves and his team had good contacts with the FBI. They had collaborated, as we saw in the last chapter, on the investigation of communists in San Francisco and elsewhere. It therefore comes as something of a surprise that Groves was apparently never informed by the Bureau about their investigation of Peter Debye.

Debye, to remind ourselves, was the Dutch director of the Kaiser Wilhelm Institute for Physics in Berlin. His reputation was such that he was invited to the first meeting of the Uranverein in the spring of 1939 but didn't attend.[12] Then, on 16 September that year, only two weeks into the war, he received a letter from Ernst Telschow, general secretary of the Kaiser Wilhelm Gesellschaft, the Kaiser Wilhelm Society (KWG), informing him that the institute was from then on to be employed 'for military technological ends and activities relating to the wartime economy'.[13] This clearly meant uranium research devoted to the manufacture of a bomb. Debye was also told that such sensitive work could not be left in the charge of a foreigner and so he was given the choice of either becoming a German citizen or resigning. Debye wasn't drawn to either option and so he decided to accept a long-standing offer to give a series of lectures at Cornell University in America in an attempt to stave off a final decision. He set sail for the US, arriving at the end of January 1940.

After the war, Debye was to become a controversial figure: Dutch journalists alleged that he was a Nazi collaborator, though as we have seen he played a prominent role in helping Lise Meitner escape; other accounts had it that he was one of Paul Rosbaud's main contacts.[14] Neither version has good supporting evidence, but from our point of view it is more important to consider what, exactly, Debye was doing in America. For example, was he a German spy?

This seems unlikely in view of the fact that, as he said after the war, Rosbaud had several meetings with Debye in Berlin in the early weeks of 1940, before the Dutchman left for New York. Very possibly, as we shall see, Rosbaud, who had got to know Debye well during Lise Meitner's escape, was briefing him on what to tell the Allies. But in Britain there was alarm when the news crossed the Atlantic, which took some weeks. (See Oliphant's complaint to Cockcroft, p. 86.)

It took the FBI four years to decide Debye wasn't a spy, by which time it was too late to make much difference on either side of the ocean. More intriguing is what he told the Americans about German plans.

He spoke early on to William Laurence, when the latter was preparing his article about the threat of fission (see above, p. 82), when he said that a large part of the Kaiser Wilhelm Institute was being turned over to research on uranium. Laurence had taken this as confirmation of his suspicion that Germany was working on an atomic bomb.

But Laurence wasn't the only person Debye spoke to. The FBI report on him says that, within two weeks of his ship docking in New York Harbor, he met with Warren Weaver. Weaver was director of the Natural Sciences Division of the Rockefeller Foundation, who had masterminded foundation funding for research in (among other places) Germany throughout the 1930s and so he knew many

European scientists well, including Bohr and Debye. In the war Weaver ran the applied mathematics section of the Office of Scientific Research. When they met, in February 1940, Weaver wrote up an account of what Debye had to say.

> The army has made this move [at the KWIP] because of their hope (which Debye considered quite misplaced) that a group of German physicists working feverishly with Debye's excellent high tension equipment will be able to devise some method of tapping atomic or subatomic energies in a practical way; or will hit upon some atomic disintegration process that will furnish Germany with a completely irresistible offensive weapon. That this is indeed the army's hope and the plan is supposed to be a great secret, and Debye himself is not supposed to know this. Nor is anyone supposed to know the German physicists who are entering into this scheme, although Debye has already told us who they are. Debye says that these German physicists very definitely have their tongues in their cheeks. With Debye they consider it altogether improbable that they will be able to accomplish any of the purposes the army has in mind; but, in the meantime, they will have a splendid opportunity to carry on some fundamental research in nuclear physics. On the whole Debye is inclined to consider the situation a good joke on the German army. He says that those in authority are so completely stupid that they will never be able to find out whether the German physicists are or are not doing what they are supposed to do.[15]

Several observations arise from this memorandum. Was Debye to be trusted, or was he spreading disinformation? Had he been *sent* to America with this in mind? Weaver knew him well, so would have been in a position to make an informed assessment. But of course Debye's dismissive comments about the German bomb programme and the physicists' motives in really wanting to carry out basic research under the guise of wartime urgency fitted very well with what the British were learning from their sources.

We do not know to what extent this intelligence was shared, if at all and, as we have seen, the Americans got it into their heads that the German bomb project was a full two years *ahead* of the Allied effort. Given what Debye had to say, how did this wild overestimate occur? Groves seems never to have learned of it.

In fact, in his memoirs, written after the war, Groves wrote that 'it had begun to seem possible to us in 1943 that the Germans could have progressed to the point where they might be able to use atomic bombs against us, or more likely against England. Although this possibility seemed extremely remote to me, a number of senior scientists in the project disagreed.'[16] Those scientists included Hans Bethe and Edward Teller who had, apparently, heard on the scientific underground about the Uranverein and Heisenberg and von Weizsäcker's involvement. In a memo to Oppenheimer in August 1943, they noted that 'Recent reports, both through the newspapers and through secret service, have given indications that the Germans may be in possession of a powerful new weapon that is expected to be ready between November and January.'[17]

Though lacking in detail, this was seemingly new information. Back in March 1942 Vannevar Bush had told the president that 'I have no indication of the status of the enemy program, and have taken no definite steps toward finding out.'[18] The Americans, as we have seen, did not take any such steps on any kind of scale until the autumn of 1943.

Things had begun to move in summer that year when, in June, Arthur Compton wrote a memo entitled 'Situation in Germany' which was based on his discussions with several German refugees and a review of German physics journals. He considered the work of Bothe, Hahn and Debye, and reported that Bothe had done work over the previous two years on neutron diffusion related to chain reactions in the form of a reactor or a bomb. The fact that his results had been published indicated, said Compton, 'no close censorship over such material' and the same applied to Hahn's work on the radioactivity of caesium, which had even been checked by a Finnish scientist. Despite this, the memo went on to state that it was a 'reasonable guess' that Heisenberg and Weizsäcker in particular were involved in chain reaction studies as a source of power, 'using heavy water and uranium enriched in 235 content' though here too Compton concluded, 'The openness of the discussion of neutron diffusion, however, makes it appear possible that this aspect of the chain reaction is not being developed.'[19]

The Americans sought indirect evidence where they could, such as information on earthquakes 'or anything that looked like earthquakes', the increased radioactivity of selected rivers (none was found), large industrial facilities being constructed (none were identified), emissions of xenon gas, a by-product of fission (none was found), and they were also interested in Vemork though, on the face of it, as we saw earlier, the plant's production seemed inadequate for a full-scale bomb programme.

Despite this, the Americans were not so sanguine as the British. They were not so strapped for resources, and were influenced by Oppenheimer who argued that nuclear weapons involved such a new set of techniques that it was always possible the Germans would come up with a much simpler way of doing things than the Allies had.

These remarkable differences in atomic intelligence assessments continued. In August 1943, Wallace Akers, the ICI scientist who had taken over as director of Tube Alloys (see chapter 10), had a meeting with Bush and later sent an account to his deputy, Michael Perrin:

> We got to talk about possible German progress and he told me he did not think that American intelligence had picked up much of real value, but that we must, as soon as possible, put together what we had both got to see what it amounted to. He said that he was nervous because there seemed good evidence that the German high command was surprisingly confident about the outcome of the war, in spite of the present unfavourable military position.
>
> I replied that ... it was clear that the Swedes, who are better placed than any other nation to know what is going on in Germany, clearly had no knowledge of any impending success with a 'secret weapon' or they would not have terminated the military transit agreement so abruptly. [After its annexation of Norway in 1940, Germany had concluded an agreement with the Swedish government to transport theoretically wounded soldiers across Sweden in theoretically unarmed trains. In practice

> soldiers on leave also used the trains and this was, again theoretically, a contravention of Sweden's neutral status. As the war started to go against Germany, the agreement was terminated in August 1943.] Bush told me he felt happier after our talk.[20]

The Americans were never to lose this fear of German progress entirely. Chadwick wrote a note from the Pentagon in September 1943: 'They suspect that Germany is well advanced and estimate they are nearer realisation of a weapon, in some form or other, than the US. They have no certain information and would welcome any that we can give them.'[21]

At much the same time, Akers wrote to W. L. Webster, the individual in charge of the Tube Alloys office in Washington, saying that the Germans 'had not yet reached the stage of erecting a boiler [nuclear reactor]', and concluding that 'the use of fission was not imminent'.[22]

Another American memo was prepared in September 1943 by Philip Morrison, a physicist at the Met Lab in Chicago, which laid out several ways in which the progress of a German bomb might be monitored. This included a literature survey, wider than just physics and taking in electronics, chemical engineering, geology and medicine, assessing for example if there had been any developments in, say, therapy for fluorine burns or radiological problems. Morrison also recommended an economic survey of crucial raw materials (uranium, boron, deuterium), plant construction and, finally, links between established scientists and engineers, indicating when the project might be converted from a research activity to actual production.[23] It is not clear that this led to any new information or policy initiative.

More intelligence was received in 1943 about another visit to Switzerland by a German nuclear physicist. This was an account by Paul Scherrer of a meeting with Klaus Clusius who, as we have seen, was well integrated into the German Uranverein and present at the June 1942 Harnack Haus meeting with Speer and the generals. Clusius told Scherrer that he had abandoned work on uranium at Munich, that the separation of uranium isotopes by diffusion had been given up as 'hopeless' and that 'nothing of practical value had yet emerged'.[24]

Paul Rosbaud reported to the British that, before the end of 1943, because of the Allied bombing of Berlin, Werner Heisenberg had moved much of the Kaiser Wilhelm Institute out of the city to Hechingen, a small town outside Stuttgart. General Groves referred to this report in his memoirs; he didn't name Rosbaud but described him as 'one of Britain's most reliable agents in Berlin', so it is all the more surprising that the general didn't pay more attention to Rosbaud's other reports, one of which said that Heisenberg had received information that the latest research in Germany was in full agreement with theory but that they were 'still eight to twelve years' away from a bomb.[25] Almost simultaneously, the newly formed Office of Strategic Services (OSS, predecessor of the CIA) in Berne reported that Paul Scherrer had learned from Heisenberg that he was living near Hechingen, near the Black Forest. And, at about the same time, American postal censorship had intercepted a letter from an American prisoner of war in which he mentioned a research laboratory where he was working. The letter was postmarked 'Hechingen'. As Samuel Goudsmit remarked, the 'excellent maps and aerial photographs' of Hechingen showed, however, no building of any kind that would be needed if a bomb project was going ahead there.[26] To be frank, the evidence was everywhere building on Rosbaud's 1942 report that Germany didn't have a serious bomb project.

Despite this, still later American intelligence continued to be no less alarmist. One account, by Robert Furman, Groves's aid on foreign intelligence, written in March 1944 as part of the Foreign Intelligence Service's *Report on Enemy Activities*, concluded that the comments by German scientists that their bomb programme was not progressing were a 'pretence' and that the delay in publishing articles was part of the 'enemy security policy' – in other words, a deception – and that the absence of any articles in the literature on plutonium only showed that it was taking place in secret.[27] This would appear to show that the British had by then shown the Americans at least some of their intelligence from the Norsk nucleus (the reports by Suess, Wirtz, Jensen) but had not come clean about the high quality and reliability of the Rosbaud reports, which reinforced the arguments of the German scientists.

The most alarmist report of all was produced in the same month, March 1944, by Karl Cohen, a chemist who worked at Columbia University on isotope separation under Harold Urey. In his *Report on German Literature on Isotope Separation* for General Groves, he argued that Germany was pursuing a deliberate 'planned partial publication program', pointing to the absence of articles on reactors, chain reactions and the separation of uranium hexafluoride, which showed, he said, that the Germans wanted 'to deceive us about the extent and progress of his programme, and so cause us to relax the pressure on our own ... The German publication program is thus a blind for diligent work.' Cohen then went on to give it as his opinion that the German programme had been given the go-ahead 'at full speed' before Stalingrad, as it was becoming clear that the war would not be over so quickly as originally thought – that is in spring 1942 (shortly before it was in fact abrogated). As a result of this reasoning Cohen anticipated that the Germans would have a completed bomb plant by the autumn of 1943 and a bomb based on enriched U-235 by the spring of 1945.[28]

At the same time, US reconnaissance aircraft were flying low over Germany, on sorties so dangerous that many pilots were in favour of aborting them, searching for traces of radioactive krypton and any huge and mysterious building plants that would indicate an industrial-sized bomb project (though the pilots of course did not know this). Nothing was found.

All this activity took place almost two years after the British had reached the conclusion that the Germans were not working on a bomb.[29] And Groves repeated his apparent worries more than once. 'Unless and until we had positive knowledge to the contrary, we had to assume that the most competent German scientists and engineers were working on an atomic program with the full support of their government and with the full capacity of German industry at their disposal. Any other assumption would have been unsound and dangerous.'[30] And: 'I could not help but believe that the Germans, with their extremely competent group of first-class scientists, would have progressed at a rapid rate and could be expected to be well ahead of us.'[31]

These differences in the assessment of German progress were remarkable. But despite the general's repeated assertions about the quality of German physicists there is one powerful piece of evidence that suggests that he was being less than candid – much less.

As part of the end game in the atomic intelligence war, Groves sent a small team of specialists, codenamed Alsos (Greek for 'grove'), behind the Allied troops, first in Italy in the autumn of 1943 (where they found out very little of a definite nature), and then behind the invading forces in Normandy in summer 1944, through northern France and into Germany, where they first found Joliot-Curie in Paris, who told them he didn't think the Germans had got very far in the development of atomic weapons. Only when the American Alsos team arrived in Strasbourg did Groves concede publicly that Germany did not have a bomb programme, and when of course he could attribute the discovery of this 'dramatic breakthrough' to American intelligence.

The Alsos team found that many documents in the Strasbourg department where Von Weizsäcker had worked were in fact unclassified, not even written in code, and that letterheads had such wording as 'The Production of Energy from Uranium', confirming that the work was not secret. They also found that Weizsäcker – universally assumed to be one of the central players in the programme – was not even involved full-time in nuclear physics. The documents showed that he had lectured in Lisbon, Madrid, Paris, Helsinki and Copenhagen during wartime so the Germans were not at all worried by what might happen to him on his travels. Samuel Goudsmit, the physicist in charge of Alsos, was also surprised about the large amount of pure physics that was done during the war. 'There could hardly have been any time left-over for war research.'

This all accorded exactly with what Peter Debye had told Warren Weaver (see above, p. 159) and emphasised the 'pitiful smallness', as Goudsmit put it, of the whole enterprise, 'all on the scale of a rather poor university and not of a serious atomic energy project'. They did come across one newspaper clipping from the *New York Times* describing some of the research being done at Columbia (this was the Laurence article described above, p. 82), but that was all.[32]

They also found out that the Germans had never thought of using fast neutrons, but instead believed they had to moderate them as in a pile or reactor. As Groves put it, 'In effect they thought that they would have to drop a whole reactor, and to achieve a reasonable weight they would need [an] enormous amount of U-235.'[33] This shows once more that the Germans did not invariably draw the same conclusions from existing evidence as Allied scientists did.[34]

And then, as the Nazi state collapsed, the Alsos group tracked down the German scientists – Heisenberg, Hahn, Weizsäcker, Wirtz and the others, one by one – and interned them.[35]

While Alsos was clearly a useful and even necessary exercise, its director Samuel Goudsmit lamented in his memoirs the fact that they never found 'a Mata Hari with a physics degree'. But of course that is exactly what Rosbaud was, a physicist with a higher degree and Mata Hari-style access. The Alsos exercise proved in 1944 and 1945 only that the British (Rosbaud) intelligence from the summer of 1942 had been correct all along and in every detail, down to the fact that many German scientists had spent the war years 'bickering' and pretending that their science might be of value to the war effort when it was in fact a ploy to gain funding for their (basic) 'purely academic' research and keep them from being drafted (exactly as Debye had told Warren Weaver after he arrived in America in early 1940).[36] As Margaret Gowing put it in her official history of the British midwife role in the atomic bomb's development: 'The correctness of their 1943 assessment and their success in thereafter building upon it "an almost uncannily accurate picture of the German effort" were remarkable.'[37]

To which we can now add R. V. Jones's note to Ronald Clark that 'until summer 1944 we [British intelligence] were producing really all the information available to Western Allies'.[38]

This, too, takes some digesting. Groves always claimed that only after the Alsos mission could the Allies be certain that the Germans had no bomb.[39] Yet the British intelligence services had actually been first aware of Speer's decision since the middle of 1942. And as Hinsley put it in his official history of British intelligence in the

Second World War, it had become the '*settled* [my italics] opinion' of the British scientists attached to the Directorate of Tube Alloys that, since the summer of 1943, 'the danger that they [the Germans] would acquire an atomic weapon before they were defeated could be discounted'.[40] In fact, the joint Anglo-US Report to Anderson and Groves, presented on 28 November 1944, explained specifically that:

> From the end of 1942 or beginning of 1943 the conclusion seems to have been drawn that large-scale plants could not be built in time to produce a weapon for use in the war and the priority attached [by the Germans] to T.A. [Tube Alloys] was lowered in favour of work on other 'secret weapons' such as V.1 and V.2. Research, nevertheless, has continued and from the end of 1943 has even increased in intensity. There is no positive evidence that this is due to a successful solution of the various technical problems or is associated with the final design or construction of large-scale plants, though the new plants in the Bisingen area [in southern Germany, away from the heavily bombed Berlin] might be connected with T.A. It may well be due to a renewal of official interest through the acquisition of information in Germany about the scale of the T.A. work that is being carried out in the USA. [Including the news that Bohr had escaped to London and America.] It must certainly be assumed, and there is some supporting evidence, that the Germans know of the American programme in general terms if not in considerable detail. We believe that this increase in T.A. research activity in Germany is the psychological result of their knowledge of the great progress that has been made by the Allies, and is not due to any real hope that they can build plants themselves to be of use in the war.[41]

The report – which appears to contain information, possibly from Rosbaud but never declassified – also argued that 'because of the very unusual nature of the problems that must everywhere be faced in the early stages of research on the T.A. project', British intelligence had been satisfied in its conclusion 'that work on the project in Germany had not reached production scale and was not being regarded as of direct military importance during the war'.[42]

It then went on to say, however, that since the beginning of 1942 'summary reports of the information in the UK and the conclusions drawn from it were sent to the American T. A. Authorities and, at the end of 1943, very close liaison was established ... as a result of an exchange of cables between the Chancellor of the Exchequer and General Groves. Since this close liaison was established, the information obtained by independent British and American investigations has been made fully available to all parties interested.'[43]

This was an important development, if somewhat late in the day, and General Groves subsequently always tried to downplay the differences between the British and American intelligence assessments. In 1967, Professor J. J. Ermenc, a historian of engineering at Dartmouth College, at Hanover in New Hampshire, interviewed a number of survivors involved in the production of the atom bomb, on both sides, including Werner Heisenberg, Paul Harteck and General Groves. Here is part of the exchange with Groves:

ERMENC: The intelligence was never quite up to date. You really didn't know precisely what they [the German physicists] were doing?

GROVES: You couldn't tell.

ERMENC: Do you think the British knew more?

GROVES: No, they didn't know any more. They knew just what we did.

ERMENC: It was said after World War I that the British knew about Ohain's jet the following day.

GROVES: We didn't know and British intelligence didn't know. The first time we were really definite about what the Germans were doing was when we captured the supply of uranium that had been in Belgium and the Germans had seized and put in the salt mines.

Groves's account here is clearly at variance – and by some distance – with the true picture. Groves was utterly dogmatic in his denials that the British knew more than he did, denials that were so extreme as to amount to falsehoods that have seriously distorted our understanding of this episode.

Because the British did not share the *quality* of their intelligence – only the summaries – with their principal ally, the Americans retained doubts about German capabilities.[44] Still further light is thrown on American thought processes by a bizarre plan that was seriously entertained and, perhaps more than anything, calls General Groves's judgement and paranoia into question.

This was a plot for Werner Heisenberg to be either kidnapped or murdered when he gave a lecture in Switzerland. Groves said Heisenberg, when captured, 'was worth more to us than ten divisions of Germans', who would be of 'enormous value' to the Russians.[45] The man deputised to carry out the assassination was Morris 'Moe' Berg, well known as a multi-talented Princeton graduate, gifted in several languages (his Japanese was good enough to be used in propaganda broadcasts), who gained fame as a career baseball player, starring for four professional teams (the Brooklyn Robins – the 'Dodgers', the Chicago White Sox, the Cleveland Indians and the Washington Senators). The idea of kidnapping or even killing Heisenberg had first occurred to Victor Weisskopf and Hans Bethe in the autumn of 1942 and the idea caught Groves's imagination and thereafter seems never to have left it.

When it was discovered, via the OSS office in Bern, that Heisenberg was scheduled to give a lecture in Zurich on 18 December 1944, Berg, using OSS influence, flew to Switzerland via London and sat in the audience, among professors and graduate students, with a pistol in his pocket (for Heisenberg) and a cyanide tablet (for himself), 'just in case'. His German, as it turned out, wasn't quite up to the lecture, which was in fact a very abstract theoretical seminar. Not being able to make head or tail of the words, he didn't shoot Heisenberg then, nor at a later dinner arranged by Paul Scherrer, the director of the Physics Institute and an invaluable Allied source, codenamed 'Flute' by the OSS. (According to one account, it was the OSS's idea to have Scherrer invite Heisenberg to Zurich in the first place.)

But what on earth were the Americans, and Groves, thinking of? As several physicists who worked on the Manhattan Project have observed, putting Heisenberg 'out of action' in December 1944 made no sense. Rudolf Peierls was just one who dryly commented, 'If there was a German project that had any chance of success before the war ended, it would have progressed by this time to the stage where the technological problems would have been central and individual scientists would not have been vital.' (Just as Groves knew – because the physicist himself said so – that Bohr was not essential to the Allied effort.) On top of this, Peierls found it 'wholly incredible that it should have entered anyone's head that Heisenberg could, or would, talk about an atom bomb in an open academic seminar in Switzerland'.[46]

Then there is the very fact of Heisenberg *being* in Switzerland at all. In 1943, the British had drawn the common-sense conclusion that Otto Hahn would never have been allowed out to lecture in Stockholm had there been a viable German nuclear project, because the risk of kidnap or murder (exactly what Groves had in mind for Heisenberg) would have been too great. So why did Groves, security maniac that he was, not draw the same conclusion regarding Heisenberg? According to other accounts, Weizsäcker (whose wife was Swiss) was also at the meeting.[47] This makes it all the more incredible that Groves and his team didn't question why the Germans were allowing so many leading 'bomb makers' out of the country.[48] Groves himself told Major Robert Furman, his chief aide on foreign intelligence matters, that Furman 'was not to travel beyond the reach of American protection in the European Theatre of Operations – he knew too much *to risk even the hazards of neutral Switzerland* [my italics]'.[49] Did Groves not realise that if *he* were taking such precautions, and about Switzerland, the Germans – if they were so well advanced in an atomic project that it could decide the war – would too? In August 1944 Groves criticised the British intelligence services for their 'typical lack of common sense'. He did not always show common sense himself.[50]

In his memoirs published after the war, Groves wrote that Alsos had clearly justified itself in demonstrating definitively that Germany had no bomb. Why did Groves not set more store by the British conclusions, reached two years before those of Alsos?

There was more than one reason. In the first place, his Anglophobia cannot be ruled out. This, it must be said, was no ordinary Anglophobia (if there is such a thing) but bordered at times on the pathological, meriting an entry all to itself in the index to one of his biographers' books. It was of long standing, says Robert Norris, and inherited from Groves's father, who regarded the English as 'incorrigibly selfish'. The general told Eisenhower, when he was president, that it was wrong for Americans to think they owed the British anything. He thought the British were lax on security and he claimed, wrongly, that he was responsible for bringing Bohr out of Denmark (who actually escaped on a British plane organised by Lindemann and Chadwick, as we have seen). He thought New York City was 'too English' and not patriotic enough. He deprecated the 'extreme friendliness' that he felt was shown to the British by the *New York Times* and the *Herald Tribune*. He was deeply distrustful of Europe as a whole and its intellectual culture. To him Cambridge (both of them) and Göttingen were 'alien places'.[51] He was beside himself with fury when Norman Ramsey, the head of the Delivery Group in Army Ordnance, favoured the British Lancaster bomber to actually deliver the first atomic bomb. 'It was beyond comprehension that Ramsey could consider using a British plane to deliver an American atomic bomb ... The new Boeing B-29 Superfortress would carry the atomic bomb.'[52] In the summer of 1943, when Conant warned that it was 'quite conceivable' that circumstances might enable the Germans to develop a weapon that was a concentration of radioactive solids, Groves took comfort in the fact that, as he put it, 'If any attack is made by the Germans, it seems extremely unlikely that it will occur in the United States but in Great Britain.'[53] It almost sounds as if he wouldn't have minded.

Most important, in his memoirs, although he credited the British with being the 'midwife' to the bomb, and Churchill as 'probably the best friend the Manhattan Project ever had', their actual contribution he described as 'preliminary ... helpful but not vital. Their work at Los Alamos was of high quality but their numbers were too weak to enable them to play a major role.' On another occasion he described the British contribution as 'puny'.[54]

This is inaccurate to the point of distortion: six of the twenty-four

British members were group leaders, including leadership of the complex implosion hydrodynamics group. Groves himself wanted Geoffrey Taylor, Britain's foremost explosives expert, so badly that, as Chadwick wrote in a letter to a colleague, 'anything short of kidnapping would be justified'. Peierls's use of punch cards helped solve the calculations defining implosion; when it was mooted that Peierls return to London in early 1945 to take part in a discussion of Britain's post-war atomic energy policy, he was not allowed to leave Los Alamos because 'his absence might very well delay the date on which the weapon would be ready'; Frisch devised a method to avoid the need to test the fission bomb, and James Tuck (who had warned Jones about the upcoming 'BIG BANG') had developed a successful form of X-ray photography that helped understand the implosion process; Philip Moon spotted discrepancies between experiment and theory that increased the efficiency of the bombs 'manyfold'; Ernest Titterton 'had proved himself to be such a master of electronics that he was entrusted with the task of generating the electronic pulse that would fire the plutonium bomb'; William Penney's knowledge of blast waves made him 'indispensable' at Los Alamos and earned him a seat on the observation plane over Nagasaki on 9 August; Hans Bethe praised Fuchs as 'perhaps the most hard-working member of our entire division', who 'contributed greatly to the success of the Los Alamos project'; Peierls, William Penney and Francis Simon received the Presidential Medal of Merit after the war, the highest medal a civilian can receive in the US, and Cockcroft received the Medal of Freedom. (Groves received the Legion of Merit and Britain's Companion of the Order of the Bath.) Fuchs worked with John von Neumann, one of the founding figures of computing, on the hydrogen bomb and the pair filed a joint patent application for the radiation implosion principle, using 'unsurpassable' mathematics (Fuchs told at least some of this to the Soviets).[55]

These are more than narrow nationalistic gripes: they call into question Groves's motives and judgement in ways that would come to matter, as we shall see. An undated letter from Chadwick to Anderson says that the general was always ready to concede in private 'the importance of our early contribution to the project' but that 'Whether or not Groves would be anxious to admit this in public seems more doubtful.' The general ran the two-billion-dollar 'state-within-a-state' without the merest hint of financial corruption, but his nationalism was – as it turned out – his Achilles heel.[56]

The Americans were in general more wary of German intentions than the British, which was surprising given that the British were closer to Germany – much closer – and therefore, as even Groves acknowledged, much more at risk. But in the case of both chemical weapons and biological weapons, as well as atomic weapons, the British concluded that there was little threat from that direction. This was due partly to decrypts of Enigma intercepts, which showed: (a) that the Germans believed that the Allies had superior (chemical and biological) weapons and so they didn't want to provoke reprisals; (b) that intelligence from Italy showed that when the Italians had given up developing gas weapons, the Germans had too; (c) that the Germans had instructed their retreating forces to dump Italian gas canisters in the sea at least 6 miles off the coast; (d) that in the Enigma traffic there was no mention of gas or biological warfare other than the preparation of countermeasures; (e) intercepts of communications between the Japanese and the Germans, to the effect that Speer had told the Japanese that Germany did not intend to use gas or biological weapons; and (f) the complete absence of decrypted instructions for troops to be prepared for gas warfare, such as being inoculated.

The Americans were unconvinced by these not inconsiderable arguments, and in January 1944 the US authorities stepped up their own biological warfare research. Is this evidence of common sense?

But the British assessment of the threat from gas and biological warfare underwent no change in the six months before D-Day and, as with atomic weaponry, proved extremely accurate. To repeat: R. V. Jones, in a letter to Ronald Clark, on 4 June 1960, when the latter was preparing his book on the atomic bomb, said, 'Until summer 1944 we were producing nearly all the information available to western Allies.'[57]

Why the Americans in general, and Groves in particular, failed to heed these accurate reports is not clear, even now. In 1944, fairly late on in the war, according to his memoirs, Groves remarked that Hitler would obviously have made use of his top scientists. His staff therefore started scouring 'all the present and back issues of the German physics journals' to ascertain the whereabouts of leading scientists, work that Fuchs and Peierls had been doing repeatedly since 1941. Was Groves ignorant of the considerable British achievement in these areas, or did he wilfully discount it?[58] Barton Bernstein, professor of history at Stanford University, notes that Groves was no buffoon, with great practical intelligence, who had 'a deep sense of his own historical importance' but 'often made himself unduly important'. 'He was neither reflective nor, it appears, self-critical.'[59] One of his biographers concurs, saying that Groves 'tried to take the credit for *everything*', that his personal papers are replete with 'braggadocio', and that he was a 'victim of his own ego'.*

In Groves's case, however, there was another issue. In view of the importance of this matter, it is worth quoting both Groves's biographers who make the same observation. Robert Norris puts it this way:

> As he did with all information, Groves carefully controlled the facts about the Germans' lack of progress in their bomb program. It is difficult to tell precisely when Groves became absolutely confident in his own mind that there was no threat, and to whom he may have confided this information. *He feared perhaps that if it were known that the Germans had not advanced very far on their bomb, it might cause some scientists to stop working on the American project* [my italics]. After all, this had been a key inspiration for many of them to enlist in the project in the first place.

* At the same time, he put down others mercilessly. In a 174-page memoir in the Groves file in the National Archives, these are some of his verdicts: Pegram 'did not have any backbone', Oliphant 'often spread false rumours', Simon and Peierls 'had a superior attitude ... the British refused always to admit that they were not superior beings', Harry Hopkins was 'a typical give-awayer', 'everyone knew that Winant [the US ambassador in London] was not very bright', Averell Harriman, the US ambassador to Moscow, was 'always a weak sister', Urey was 'at heart a coward', and *The New World*, the official history of the Manhattan Project, was 'completely unreliable and cannot be trusted'.

For William Lawren, Groves:

> had carefully avoided revealing both the absence of an all-out German bomb project . . . even to those on the level of Compton, Lawrence and Oppenheimer [but see below regarding Oppenheimer]. Instinctively, Groves probably felt that as a motivating factor the threat of a German atomic bomb was too good to lose. As far as many of the scientists were concerned, Germany was the principal enemy and the threat of a German atomic bomb was the raison d'être of the entire American nuclear effort . . . with Germany out of the picture, [many of the physicists] might simply resign. This was a risk Groves could not afford to take . . . it might deal a fatal blow to the entire fission project.[60]

These are most revealing words. Groves's biographers are both implying – more than implying – that the general knew that there was no German bomb and yet concealed it from the scientists for some time, so that they would keep motivated. In fact, there is one important – compelling – piece of evidence that suggests that Groves, and others, accepted (despite what they said publicly) that Germany had no bomb as early as May 1943. And it was written down.

In the autumn of 1942, after the Manhattan Project was formally launched and Groves appointed to run it, Bush and Stimson decided that it needed in effect an executive board to oversee developments, 'a small group of officers to consider strategic uses and tactics' with respect to the bomb.[61] Thus was born the Military Policy Committee (MPC), with Bush as chairman, Conant as his alternative, Rear Admiral Reynolds Purnell as the navy representative and General Wilhelm Styer for the army. In theory Groves reported to these men as a CEO does to a company board in the world of business, though in practice the members of the MPC often merely rubber-stamped initiatives he had already set in motion. He himself admitted as much. By the summer of 1943 'the principal function of the Military Policy Committee was to approve my actions and proposals'.

One of the most pressing questions the committee faced was where to drop the first bomb. The obvious choice, given what had gone before, was somewhere in Germany. As Sean Malloy phrases it, 'The spectre of a Nazi A-bomb had been the driving force in the American atomic project from the outset.'[62]

Stimson, the American secretary of war, wholeheartedly supported the overall Allied strategy throughout the war and that strategy gave priority to the defeat of Nazi Germany. Stimson thought this was the bomb's first purpose. His assistant secretary John McCloy later recalled that he had 'many conversations with him in regard to the menace of a possible German development of an atomic weapon', usually when they were 'off-duty', relaxing at Stimson's Washington estate. In January 1943 Vannevar Bush even contemplated a plan whereby the Allies would use atomic weapons against suspected nuclear sites in Germany in a pre-emptive strike. Groves said he had no problem dropping the bomb on Germany 'because Mr. Roosevelt told me to be ready to do it'.[63]

With all this background pointing to an attack on Germany, on 5 May 1943, in its first and only formal discussion of the matter, the MPC came up with a total surprise: the minutes of the meeting show that it selected *Japan* as the site for the first use of the atomic bomb. This too takes some digesting. According to Malloy, Bush and the others were apparently swayed by concerns that the Germans could use knowledge gained from an American bomb to further their own efforts. That is to say, they were not aware of British intelligence and were particularly worried about what would happen should a nuclear 'dud' land on German soil. 'Japan, on the other hand, was not believed to have an active nuclear weapons programme.'[64]

This was an obviously momentous decision and raises the equally important question as to why the MPC could ignore the German threat so early on. Is it really the case that a worry about the bomb being a 'dud' was a clinching argument for not dropping a bomb on Hitler and opting in favour of Japan? As Frisch and Peierls had pointed out as early as February 1940, the chief point of having a bomb was to *deter* Hitler from using his, if he had one. Such a bomb had to work, yes, but if a 'dud' had landed on Japanese soil, say, the Germans would still have found out, Japan and Germany being Allies, and they could even have obtained the hardware itself.[65] In any case, as we have seen, most of the Americans in the know thought that Germany was ahead.

This detailed chronology underlines the fact that this 'dud' argument doesn't stand up.

An alternative explanation is connected with the way the war was changing. In November 1942 the British forces led by General Bernard Montgomery had routed Rommel's army at the battle of El Alamein in Egypt. At much the same time the Americans landed a massive force in Morocco and Algeria, ready to join up with the British, in obvious preparation for an invasion somewhere in southern Europe. May 1943 was just two months after the Battle of Stalingrad had been concluded in Russia's favour, forcing the Germans to begin their retreat across Eastern Europe. The lineaments of a pincer movement were emerging.

Arguably, though, the Russians were now doing more to defeat the Nazis than were their Allies.[66] The winning of Stalingrad marked the crucial appearance on the world stage of Russia as a superpower. Between February 1943, when Stalingrad concluded, and May 1943, when the MPC settled its target priorities, the balance of the war, and of world realpolitik, changed for all time. No one could foresee precisely how the war would end, not then, but clearly Stimson, Bush, the MPC – and Groves – now accepted, whatever they said publicly, that Germany was not a nuclear threat. This fitted neatly, as we have seen, with what British intelligence concluded at much the same time.

Later – much later, in 1967 – Groves said that one reason for the change of target was because casualties were mounting and he felt the bomb would end the war in the Far East more quickly. But if so this view never reached Stimson or Chadwick, for example, for months.[67]

Keeping the change of target secret was by no means easy, for the choice of Japan changed the *type* of target and that determined the *design* of the bomb. In theory, at least, the altered design should have made it obvious to the scientists and in particular the engineers at Los Alamos that the target had changed. The original design had been for a deep-water explosion in a German harbour. Frisch and Peierls had suggested this in their very first paper because they thought that detonation in a deep-water harbour would minimise civilian casualties. But this was now changed to a bomb exploded *above* a military target surrounded by 'light housing' structures (which is how Hiroshima, for example, was configured).

Conant was one of those who had originally considered 'underwater' use as well as an 'airburst' weapon but in the MPC's May 1943 meeting 'the general view appeared to be that its best point of use would be on a Japanese fleet concentrated in the Harbour of Truk' (a massive naval base on an atoll in the Caroline Islands in the middle of the Pacific Ocean).[68]

The Ordnance Division's staff, shortly after the opening of Los Alamos in 1943, had set about designing weapons for use in harbours. They considered a nuclear depth charge and even an atomic torpedo.[69] At the same time, others were working on designs for a bomb to be dropped on land targets. At the end of 1943 Oppenheimer became so concerned about the proliferation of designs being pursued – or he said that he did – that he suggested the more 'technically challenging' designs for underwater weapons be 'temporarily postponed'. Or, was he – alone of all the scientists at Los Alamos – privy to the MPC decision to opt for an airburst weapon over targets in Japan?

Airburst weapons had the disadvantage that they would kill far more civilians and spread far more radiation. But they had the undoubted advantage that their effect would be far more spectacular, especially against the lightly built, wood-frame houses common in Japan.

A final element in the design equation was that, in order for the bomb to be delivered, and for the bomb crew doing the delivering to escape, the device needed to be dropped from a very great height – say 30,000ft – so that the aircraft would be several miles away by the time of the explosion.* In turn that meant that only very large targets were suitable. Cloud cover could easily obscure smaller targets, like harbours and fleet installations, and the bomb might even miss such a target. All of which pointed to entire cities being the object of attack.

And all of which confirmed Japan as the location for the explosion of the first atomic weapon. It had nothing to do with the risk of 'dud' bombs.

But this was not the only set of non-sequiturs in General Groves's behaviour. In his memoirs, he wrote, 'As plans for the invasion of Europe began to take form, we considered very seriously indeed the possibility that the Germans might lay down some kind of radioactive barrier along the invasion routes. We could not calculate with any certainty the likelihood of their doing this, for we were *truly in the dark* [my italics] then about their progress in atomic development.'[70]

But, as we have just seen, Groves and the more senior Americans were not at all in the dark about German progress in atomic development. Long before plans for the invasion of Europe took their final form, the Allies were well aware that there was no German plan for a bomb – that is why the MPC felt able to select Japan as the target.

More even than that, Groves's sentiments do not sit well with the fact that British intelligence had confided that there was *no* threat of radioactive material being used in the invasion, because they knew from intelligence received at that time that Hitler felt that the Allies had better poison gas resources and use by the Germans would have provoked retaliation. British intelligence insisted that the same reasoning applied to radioactive materials.

In fact, Washington had raised this matter several times, in the summer of 1942 and again in August and September 1943.[71] On the second of these occasions, says Hinsley in his official history of British wartime intelligence, US experts had suggested, 'not on the

* Hiroshima was bombed from 31,600ft, and Nagasaki from 29,000ft.

basis of any evidence but in the course of a feasibility study', that the Germans might, over the next year, use heavy water to produce 'radio-active solids' in bombs, and to do so on a scale sufficient to make it necessary to evacuate parts of London.

This was in theory very worrying but in September 1943 the British discounted 'any imminent danger of this weapon being used against us. There was *no evidence* [my italics] that Germany had the requisite power machine, and if she had, she would be much more likely to use it to separate plutonium for the development of a weapon that would be much more effective than fission products.'[72] Compare this reasoning with General Groves's above (see p. 170). Who was showing common sense now?

Washington returned to the subject again in December 1943, after reports had appeared in the press in neutral countries (Sweden, Switzerland) about the existence of a German secret weapon and this had suggested to the Americans that the enemy 'might be planning to use radioactive powder in the V-weapons'. Once again the British dismissed this speculation, adding pointedly that any authorisation of widespread precautionary measures might, by alerting the enemy, 'increase the difficulty of obtaining reliable intelligence about Germany's actual TA work'. Using the same argument, it recommended there was no need for the development of simple detectors – for there was no evidence of any large-scale work on radioactive substances, as there would need to be to have an effect on the war. Here again, the sheer absence of any mention of radioactive substances in the Enigma traffic was instructive. For once, the absence of evidence really did mean it was evidence of absence.[73]

Groves responded: 'We agree that the use of TA weapons is unlikely. The indirect and negative evidence developed by your agencies to date is in support of this conclusion. But we also feel that as long as definite possibilities exist that question the correctness of this opinion in its entirety or in part, we cannot afford to accept it as a final conclusion.'[74]

This sounds reasonable. But throughout the war the British, to their credit, concentrated on *evidence*, whereas, more than once, as we have seen, the Americans responded to what was rumour or speculation, however worrying or believable. This difference in approach does not seem to have registered much with General Groves, but the series of alarmist reports, which turned out to be a significant collective *overestimate* of German capabilities by the Americans, was played down by him after the war. Even Jeffrey Richelson, normally so measured, gives only two lines to this momentous disagreement: '[T]he Strassburg material [discovered in the last week of November 1944] tended to support the conclusion that British intelligence had reached in 1943, that a German bomb was not a serious threat.'[75] Tended?

However, being so much more strapped for cash than their primary ally was, the British needed finer judgements at all times, so as not to waste precious resources. On top of which, the intelligence they were receiving from Enigma and elsewhere was in such abundance, and of such quality, that they could attach much more significance to *negative* evidence than, perhaps, Groves could.

As Hinsley (and others) make plain, the intelligence reports – about the V-1 and V-2 rockets, for example, or about the new submarines the Germans were developing, or the new four-engined bombers, the Me 264 and the He 277, or the new 'rocket-propelled' Do 335 – were received in such copious amounts that the absence of intelligence on atomic matters was starkly revealing. The Enigma traffic disclosed not only the technical developments of these new weapons, but aspects of crew-training, production numbers, organisational matters – none of which were mentioned in connection with atomic devices.

One significant long-term consequence of the American *overestimate* of German capabilities may well have been to make them *underestimate* Russian abilities later on, to dramatic effect.

Many of the differences between the British and American approaches might have been avoided if the British had been more open with their American colleagues, not just about the intelligence they were receiving, but about the quality and status of that intelligence. But, as Hinsley put it, 'On account of the need for preserving secrecy about the Allied interest in the [TA] work, the Allied intelligence-gathering agencies could not be briefed in detail, only instructed generally to search for information on new weapons, unusual new materials, and new scientific activity.' This is underlined by Michael Goodman in his official history of *The Joint Intelligence Committee*: 'For much of the war, Britain had refrained from liaising with the US on the German atomic programme.'[76] This must count as a major – a strategic, an egregious – error. And, from Hinsley's wording, is this itself a code of sorts? Could Hinsley have been using spy-speak and *apologising* for the fact that such vital information was not shared, when it could have had such important consequences?

We may say then that there were three important moments during the Second World War when the course of events could have been changed in such a way that, taken together, an atomic arms race would have been avoided.

1. If, in mid-summer 1942, the British had shared with the Americans the strong status of the Paul Rosbaud intelligence, about Speer's meeting with Heisenberg, and all the other approaches made by Suess, Jensen and Wirtz, and since the Manhattan Project had not formally begun then, with expenditure relatively confined, there might have been no need at all for the Project itself.
2. If, in early 1943, the Americans had shared with the British the fact of Soviet penetration of the Manhattan Project, then there might have been two corollaries: more attention might have focussed on Fuchs, as an ex-communist; and Churchill might not have been so implacably against Niels Bohr's ideas for dialogue with the Russians (see chapter 22).
3. If it had become known in May 1943 that the MPC had chosen Japan as the primary target of the bomb, any number of scientists at Los Alamos might have baulked at this 're-alignment' of priorities and refused to continue working on the bomb, especially if they had also concluded

> that Germany was therefore no longer a threat. Secretary of war Stimson was in particular most exercised by the compromised morality of this choice (he thought a naval base should be the target), and there is no reason to doubt that many scientists would have been too, had they known about it.[77] Alsos was part necessity but part conspiracy and cover story, to keep the scientists at Los Alamos 'on track'.

The fact is that, by the time Bohr and Fuchs arrived in the United States, at the end of 1943, the status of the Manhattan Project had changed. General Groves, Bush, Conant and Stimson all knew that Germany did not have the bomb, and had no prospect of acquiring one during the war. The Military Policy Committee had selected Japan as the primary target without anyone really knowing what the state of the war would be by the time a bomb was ready. The underlying truth is that Japan was a convenient target. Bush and Groves in particular (but also Churchill and Lindemann) were the first to grasp that, as they thought, a bomb – if it could be developed in secret – would give America and the West a significant advantage over Soviet Russia once the war was over and its ally had become its adversary.

Because important intelligence was not shared – in both directions across the Atlantic – the world was maladministered into a new, unprecedented era of danger.

13

The Bohr Scare

Bohr grasped this instinctively. This was clear as soon as he arrived in London after his dramatic escape from Denmark in October 1943.

Bohr was in fact required to make two escapes, equally fraught. He and his wife had escaped Copenhagen in the dead of night, being secretly ferried across the Øresund, the narrow stretch of water separating Denmark from Sweden. Once in neutral Stockholm, however, it was soon clear to Bohr that the Nazis there still had him in their sights and there were fears that an attempt would be made on his life. He moved around but an armed guard was always on hand, just in case.

His sons narrowly escaped capture when they made the run a few days later from Denmark to Sweden. One of them, Erik, had a baby daughter and she was smuggled into Sweden in the shopping basket of the wife of a Swedish embassy official.

Soon after the family had safely assembled in Stockholm, Bohr received a telegram from London, from Lord Cherwell, formerly Dr Frederick Lindemann and now Prime Minister Churchill's personal advisor on scientific matters. It was in effect an official invitation to come to England. Bohr understood that it would be impractical for Mrs Bohr and the rest of the family to go with him but replied in the affirmative, asking only that his eldest son, Aage, then aged nineteen, be allowed to travel to London also, as his assistant.

On Wednesday 6 October 1943, a British Mosquito bomber landed at the Stockholm airport. This was days after Allied forces had gained a toehold on the foot of Italy, when the inhabitants of Naples, sensing their approach, rose up against the German occupiers. The Mosquito bomber carried no arms, so as not to contravene Swedish neutrality.

It was made of light wood so it would be difficult to pick up on radar, and was similar to aircraft used in commercial flights between Sweden and the UK. Bohr was then flown, alone, from Stockholm to Scotland. Bohr was a big man with an exceptionally large head. He travelled in the empty bomb bay of the aircraft, together with an oxygen tank by which to breathe, since they would be flying high to avoid enemy planes. Unfortunately, the headset Bohr was given by the pilot was too small to fit over his ears and so he never heard the pilot's instruction as to when to put on his oxygen mask. He passed out.

Not hearing from his distinguished passenger throughout the entire flight, the pilot was fearful that Bohr had died, and on landing in Scotland he hurried round to the bomb bay. In fact, Bohr had quickly revived as the plane descended into Scottish airspace and there were no after effects – he was very fortunate. He was flown on conventionally to London, where he was met by James Chadwick.[1]

On Bohr's first evening in Britain, a highly secret dinner was arranged in his honour at the Savoy Hotel in the Strand. It was a large party – many people wanted to meet the renowned scientist (who became known throughout Whitehall as 'the Great Dane'). Apart from Chadwick, the most senior of those present was Sir John Anderson. Also there that night was Lord Cherwell, officially professor of experimental philosophy (as physics was known) at Oxford and Churchill's personal advisor on scientific matters as they affected the course of the war. Others present were Stewart Menzies, head of MI6 (SIS), Wallace Akers, Michael Perrin, Eric Welsh, the London end of the Norsk nucleus, and Charles Frank, a theoretical physicist who had worked in Berlin before the war, partly on the lookout for developments in physics that might be weapons-related. Finally, there was R. V. Jones, the man who ran Britain's outfit that assessed enemy sciences for their weapons potential, and who was also responsible for bringing Britain's own science developments forward so that they could be used in the war.[2]

Bohr, they found, was a mix 'of companionability, wisdom and an endearing incoherence'. He concentrated so much on what he was saying, or trying to say, that his pipe was always going out and he was constantly forced to cadge matches. Despite this, that night Bohr forcefully expressed the view that if the bomb could be built, it was foolish to assume that the underlying know-how could be kept secret. Before long the government of any industrially advanced country would find out how it was done. This meant that a terrible arms race was inevitable. Wouldn't it be better, Bohr said, for the Allies – and that included the Soviet Union – to share the idea and use the opportunity to create a new era of harmony and trust? As a world-class nuclear physicist, and with unrivalled experience of Russian physics and physicists, Bohr was in a unique position. It seems that Anderson and even Lindemann that night saw the beginnings of a new approach to world affairs. They both thought it was worth trying to have Bohr talk to the prime minister.[3]

One other thing discussed that night was the visit Bohr had received in Copenhagen in September 1941 from his friend Werner Heisenberg. By the time he arrived in London he had formed a view that, at that time (1941), 'the Germans had concluded that the Bomb project was impractical'. He was, however, by 1943 much less certain. The ambiguity he evinced during his journey from Copenhagen to Los Alamos, as we shall see, would play an important role in subsequent events.

On the day after the dinner, Bohr was shown the headquarters offices of 'Tube Alloys' in Old Queen Street, near Birdcage Walk and St James's Park. He was given a ration book and used it to go shopping for clothes. He spent some weeks in Britain visiting the various sites – Liverpool, Birmingham, Oxford, Cambridge – where early work was going on.

By then, over eighteen months, Peierls and Frisch had worked out just how big the U-235 bombs needed to be; Francis Simon, at Oxford, had done much the same for a gaseous diffusion plant, estimating manpower and costs, which underlined just how massive any bomb project would have to be. At Cambridge, the French physicists von Halban and Kowarski showed that a chain reaction with uranium and heavy water was feasible, which in theory could produce the new element 94, while others confirmed that fission did indeed occur in element 94. Along the way, Bohr was filled in on what was being done in the US and he soon grasped fully that the bomb would now be built. 'Only technical – not fundamental – problems stood in the way.'[4]

He also grasped that his escape had occurred at a critical time. The fall out between the principal Allies had recently ended and a score of British scientists were in the process of being transferred to America for a concerted effort to build a bomb.

He went to see the chancellor, John Anderson, again and asked him what he thought the possession of a definitive weapon could do to the relations between the possessor and other states. Would any nation, Russia most of all, stand by without taking any action? Was the bomb bound to start an arms race? And was such a bomb not capable of destroying civilisation?

Bohr knew at first hand what it was like trying to deal with the Russians. He had tried to rescue more than one scientist in Russia and been involved in attempts to promote scientific co-operation. So he was under no illusions.[5] Even so, he wondered aloud whether this sudden advance to a new level of scientific achievement would not offer an unprecedented opportunity to introduce a new level of international co-operation? Such an extraordinary advance, Bohr argued, surely opened up a unique opportunity, a transcendental advance, pointing to a degree of co-operation unthinkable and unattainable in the past. The terrible threat of obliteration, now coming in to view, might well be a key to unprecedented co-operation.

Anderson mulled this over. The idea would stay with him.

Although Bohr's safe arrival in London was good news, both as far as his own personal survival was concerned and for the contribution he might now make to the bomb project, it was accompanied by a scare. News of his escape had leaked out and his presence in Britain was broadcast on the radio and published in the newspapers. As we saw in chapter 1, the reports said that Bohr was carrying plans for a new invention involving atomic explosions that would be of great significance for the war effort.

The British had known since the summer of 1942 that the Germans had no bomb. But there was obvious concern that Bohr's presence now among the Allies would cause the Germans to think anew about their stalled project. If the Allies were embarked on a nuclear project, perhaps they should too.

This helps explain an extraordinary report that was published on 26 December 1943 (Boxing Day in Britain), the day the *Scharnhorst* was sunk and a short while after Bohr (and Fuchs) had crossed the Atlantic. The article appeared in the London *Sunday Express*, headlined 'THE SECRET WEAPON MAY NOT COME OFF'. The text of the article consisted of a long and more or less accurate feature by Kai Siegbahn, the son of Lise Meitner's reluctant host in Stockholm, Manne Siegbahn, and a colleague of Njål Hole, who we know was a British source in Sweden. Siegbahn outlined the basics of nuclear energy before giving an account of pre-war research. When it came to a bomb, however, he summed up by saying: 'Despite all the secretiveness about researching the uranium problem, I venture to say that the uranium bomb is still non-existent, except as a research objective ... It is rather difficult to say if it is possible at all to construct such a bomb, but for the present it seems as if an essential link is missing for making the uranium bomb a reality.'

In an editorial in the same issue, the *Sunday Express* went further than Siegbahn's conclusion, reassuring its readers that: 'It may therefore be a source of consolation to know that the able Swedish atoms-scientists believe that the Germans have not succeeded in creating atom explosives.'[6] The *Express* also felt it necessary to explain how the information had come its way. It said that in neutral Sweden, scientists like Siegbahn had close contacts with German scientists. The Germans occupied Norway and had recently arrested Norwegian professors and students. These Norwegians, it was implied, had exaggerated the threat and the Germans had told as much to the Swedes.

Nonetheless, there were puzzling aspects of the article, especially in view of the tight security surrounding atomic research. Although the overt thrust of the piece was to reassure the British, the subtext was that the Allies – like the Germans – had failed to make a bomb, that they had faced the same difficulties as, according to Siegbahn, the Germans had faced. The argument was all the more powerful, at least in theory, because it wasn't written by a partisan British scientist, but someone from a neutral country using 'leaked' information. It was, again in theory, as objective as could be.

It was of course an attempt at a deception. What the British knew from Rosbaud and others was put into the mouths of Norwegians. The appearance of the article, and its thrust, can be explained by the fact that *Express* newspapers were owned by William Maxwell Aitken, Lord Beaverbrook, a good friend of the prime minister and therefore familiar with the history of the atomic bomb project.

It was in fact a classic piece of what we would now call disinformation. Rumours about Hitler's 'secret weapons' were widespread, and the Freshman assault on Rjukan and General Groves's bombing of Vemork the month before, had once again raised the profile of a potential weapon. But in reality the *Express* article was aimed as much at the Gestapo and the Abwehr (German military intelligence) as it was at the British public. By then, and for some time before, the Germans were meant to draw the conclusion that heavy water was essential for atomic research and that the Allies would 'do anything' to halt its manufacture. But now, in a further twist to the argument, an article written by a distinguished physicist in a neutral country strongly implied that the British were still in the early research stages and without much hope of 'making the uranium bomb a reality'.

The article was an SIS plant. The idea behind it was conceived by MI5, in particular its Double-Cross Committee (the XX Committee), chaired by John Masterman, and designed to make the Germans think that Allied nuclear research was in much the same state as (they knew) German research was.[7] In saying that the Norwegian/German experiments on heavy water were not proving very productive, the idea was to suggest to the Abwehr that the same was true in Britain. The Allies had faced the same problems as they did, with the same disappointing results.

This was probably the most ambitious act of press deception concerning atomic intelligence throughout the war.

14

A Vital Clue Vanishes

When Bohr reached Los Alamos, after he had left London and passed through New York, he was received ecstatically – 'like a monarch returning from exile', according to one observer. As the physicist Victor Weisskopf put it:

> In Los Alamos we were working on something that is perhaps the most questionable, the most problematic thing a scientist can be faced with. At that time, physics, our beloved science, was pushed into the most cruel part of reality and we had to live it through. We were, most of us at least, young and somewhat inexperienced in human affairs, I would say. But suddenly, in the midst of it, Bohr appeared in Los Alamos. It was the first time we became aware of the sense in all these terrible things, because Bohr right away participated not only in the work, but in our discussions. Every great and deep difficulty bears in itself its own solution ... This we learned from him.

Bohr had arrived at Los Alamos in the evening and not until the following morning did the magnificence of the location and the chaos of the work going on fully reveal themselves. Los Alamos had originally been a boys' school, built on a high mesa, the Pajarito Plateau, the weathered and flattened tabletop of a long-extinct volcano, 7,200ft above sea level.[1] The scale of building activity – trucks galore, wiring, the din of hammering – belied the remoteness of the site. Santa Fe was the nearest outpost of civilisation, 20 miles away with only a rutted, winding track connecting the laboratory with trains and planes and decent roads. Los Alamos was, in effect, a small city 'of unacknowledged existence', consisting of about 6,000 persons. Correspondence was addressed only to PO Box 1663. (While the scientists' chief preoccupation professionally was fission, as Gerard DeGroot has put it, personally their focus was fusion: 200 babies were born in Los Alamos during the war.)[2]

To combat rumours about what was going on, which inevitably occurred (that poison gas was being manufactured there, or whiskey, or spaceships), scientists were sent to the bars in Santa Fe to spread rumours about a secret electric rocket.

The laboratory was where more of the thinking about atomic and nuclear physics went on, but it was also the place where the purifying of plutonium would be carried out, the plutonium itself being produced at a vast site at Hanford, in Washington State. Los Alamos was where the actual bombs would be assembled.

On that first morning Bohr was given a badge, reading 'Nicholas Baker', and Aage had one with 'James Baker', and they were shown a copy of the *Los Alamos Primer*, a secret booklet prepared to bring recruits up to speed on the theory of bomb-making. By the same token, Fermi became 'Mr. Farmer', Eugene Wigner was 'Mr. Wagner' and Ernest Lawrence was 'E. Lawson'. Other security measure included 'burn boxes' in every office, in which coded documents were incinerated after they were read. The émigré scientists were also admonished not to speak German.

However, as Bohr moved around the dusty landscape, or stopped by the workrooms and laboratories, shouts of recognition and delight could be heard. Though many of those present hadn't see him since before the war, four long years earlier, there was no mistaking the domed head, the heavy-set features, the large bones. Frisch, Weisskopf and Hans Bethe, among others, rushed to renew their acquaintance with the man many soon called 'Uncle Nick'.[3]

At Los Alamos, Bohr was astounded by the size of the project and the amount of progress that had been made since he had last seen so many colleagues. He was able to help with one or two technical problems but his main contribution, at least at first, was to act as a sort of elder statesman for the younger scientists (he was fifty-eight), many of whom were troubled by what they were in the middle of.[4]

And yet, it has to be said that as the years have passed the role of Niels Bohr at that time in Los Alamos has, if anything, grown more and not less ambiguous.

The moral worries that many of the scientists at Los Alamos felt they were confronting were thrown into stark relief by something that Bohr had brought with him from Denmark. This was a drawing, a notorious sketch as it turned out, because he said it had been given to him by Werner Heisenberg. And that is what the fuss has all been about – what, exactly, were Heisenberg's motives in contriving a meeting with Bohr.

Heisenberg's visit to Bohr is one of those iconic episodes in the atom bomb story, well known but still mysterious and, even now, nearly eighty years after it took place, not fully understood or settled in all its details – what motivated it, what exactly was said by the two men, *and what was meant.*

It is agreed that the meeting formed one item in an itinerary whose main purpose had been, according to Heisenberg, a conference at the German Cultural Institute in Copenhagen on astrophysics, attended by five German physicists, Heisenberg and Carl Weizsäcker among them. Paul Rose, in his book on Heisenberg and the Nazi atomic bomb project, dismisses this conference as a 'transparent propaganda exercise', a sort of disguise for the real reason for the visit, which was for the Germans to assess what Bohr knew about Allied plans.

We must remember that, in the autumn of 1941, the United States was still a non-belligerent, there was a US embassy in Copenhagen and, in theory at least, American scientists were free to visit Denmark. It was – again in theory – not inconceivable that Bohr had received such visits, though had they occurred the occupying Nazi administration would surely have taken note of them and informed Heisenberg and Weizsäcker.

The idea for the meeting with Bohr, as mentioned earlier, appears to have been sparked by a report published at the beginning of September 1941 in a Swedish newspaper, *Stockholms Tidningen*, to the effect that: 'In the United States scientific experiments are being made on a new bomb, according to a report from London. The material used in the bomb is uranium, and if the energy contained in this element were released, explosions of heretofore-undreamt-of power could be achieved ... All structures within a range of 150 kilometres would be demolished.'

Weizsäcker had learned about the report from his father, Ernst, who was a senior figure in the German Foreign Office and, says Rose, at the time of their visit to Copenhagen, 'The possibility of an American bomb must have been very much on their minds.'[5]

After the war, Heisenberg claimed that the whole exchange brought about a misunderstanding on Bohr's part and that his real motive in the meeting was to get across to the other man – without exactly saying so (because that would have been disloyal and treasonous) – that a German bomb was not on the cards, at least not so soon that it could affect the course of the war. The discussion must have become fairly technical because during their deliberations Heisenberg produced a drawing of a possible design for a reactor bomb. (In a reactor, controlled fission produces heat, not explosions, but can, under certain circumstances, produce plutonium, which *then* can be explosive.) Almost as much mystery surrounds this drawing as the meeting itself. Though Bohr took it to Los Alamos, it has since gone missing and Paul Rose's attempts to trace it in six separate archives proved fruitless. Apparently the drawing resembled 'a crude box shape with rods sticking out at the top'.[6]

Not surprisingly, at Groves's behest Oppenheimer called a special meeting to consider this drawing, shortly after Bohr's arrival, on New Year's Eve 1943. Besides Bohr and his son, Aage, and Oppenheimer himself, the meeting was attended by Hans Bethe, Edward Teller, Victor Weisskopf, Robert Bacher (who had worked with Hans Bethe at Cornell), Robert Serber, 'a lean, gentle, Philadelphian' (from the University of Illinois) and others.

By all accounts, the assembled physicists found it difficult to take the drawing – no more than a sketch – seriously. Weisskopf, says Rose, 'thought that Bohr's anti-German prejudice had deceived his better judgment, while Bethe was bewildered by the whole thing'. As

Bethe recalled later, 'A little drawing ... As far as we could see, the drawing represented a nuclear power reactor with control rods. But we had the preconceived notion that it was supposed to represent an atom bomb. Are the Germans crazy? Do they want to drop a nuclear reactor on London?'

There is no written record of Oppenheimer's meeting and, as stated above, the drawing itself has gone missing too. But that doesn't end the matter. Far from it. Following the New Year's Eve meeting, Oppenheimer asked Bethe and Teller to prepare a report on the possibility of an atomic weapon based on the drawing.[7] Their report on the Heisenberg/Bohr device, produced in a day or so, concluded that the explosion resulting from the machine in the drawing would liberate energies that would 'probably' be smaller than an equal mass of TNT.[8]

Oppenheimer immediately passed on Bethe and Teller's report to Groves with a crisp covering note.

> The calculations referred to and described in the accompanying memorandum were carried out by Bethe and Teller but the fundamental physics was quite fully discussed [at the New Year's Eve meeting] and the results and methods have been understood and agreed to by Baker [Bohr]. No complete assurance can be given that with a new idea or a new arrangement, something along these lines might not work. It is, however, true that many of us have given thought to the matter in the past, and that neither then nor now has any possibility suggested itself that had the least promise. The purpose of the enclosed memorandum is to give you a formal assurance, together with the reasons therefore, that the arrangement suggested to you by Baker [Bohr] would be a quite useless military weapon.[9]

Questions proliferate. Why would Heisenberg take the trouble to travel all the way to Copenhagen from Berlin to give Bohr a diagram of a 'bomb' that wouldn't work? Was he really trying to show, via the drawing, that the Germans were on the wrong track, barking up the wrong tree? But this is not what Heisenberg tried to claim after the war and, as it turns out, there is much more to it than that. Thanks to no fewer than three remarkable acts of scholarly sleuthing, not carried out until relatively recently, we now think we have a much fuller picture of what actually went on.

The first scholar-sleuth was Jeremy Bernstein, a multi-talented physicist who has worked at CERN, Oxford, the University of Islamabad and the Princeton Institute for Advanced Study. Bernstein revealed his fascinating piece of detective work at a seminar that took place at the Smithsonian Institution in Washington DC in March 2002, called in the wake of Michael Frayn's play *Copenhagen*, which sought dramatically to recreate the September 1941 meeting between Bohr and Heisenberg. The play sparked controversy in that it portrayed Heisenberg as a hero, someone who 'never killed anyone', whereas Bohr – or his theories – had. (Heisenberg to Bohr: 'You have actually done the deed we were tormenting ourselves about . . . because it wasn't a toy at all.') This stung the Niels Bohr Foundation in Denmark to release, earlier than had been intended, a number of letters Bohr had written to Heisenberg but never sent. These starkly contradicted the thrust of Frayn's play.

We shall come to Bohr's unsent letters in just a moment, but for now let us stick with Bernstein's detective work. It began in 1977, he tells us, when he did a series of interviews with the physicist Hans Bethe, who had first mentioned the Bohr–Heisenberg meeting, which at that point was news to Bernstein. It was Bethe who told Bernstein that Heisenberg had given Bohr a drawing and he, Bethe, had seen it at Los Alamos during the celebrated New Year's Eve meeting. When Bernstein had written up his interviews for a three-part *New Yorker* magazine profile, others had taken issue with the account, arguing that Heisenberg giving Bohr such a drawing would have been an act of treason on Heisenberg's part, and the German was known to be a staunch patriot (see above, p. 82), making such a gesture extremely unlikely. Since the drawing had gone missing, no one could tell in whose hand it had been made.

One of Bernstein's critics on this matter, Thomas Powers, a Pulitzer Prize-winning journalist, who was to write his own book on

Heisenberg, had the sensible idea of asking Aage Bohr, Niels's son, himself a Nobel Prize-winner, about the drawing. 'Aage said in no uncertain terms that the notion that Heisenberg gave Bohr a drawing in 1941 was pure fiction.'[10] He later added that if there had been a drawing in 1941 that his father had thought represented a plan for a German atomic bomb, he would certainly have used his contacts in the Danish underground to promptly transmit it to the Allies, and this he didn't do. At the same time Bethe repeated to Bernstein that 'there was absolutely no doubt in my mind' that he had seen a drawing at Los Alamos. But he did add that he couldn't say whether the drawing was by Heisenberg or made by Bohr, from memory.

Bernstein was stumped. He asked other physicists who had been in Los Alamos. Weisskopf replied that he had no recollection of a drawing. Rudolf Peierls had not seen a drawing but conjectured that Bohr had kept it from Aage and the rest of his family to protect them from knowing something treasonable. Not entirely satisfactory.

But then Bernstein thought of Robert Serber. Serber ended his career as professor of physics emeritus at Columbia University and had been one of those at the New Year's Eve meeting at Los Alamos. During Bernstein's subsequent conversations with Serber, Serber told him that Bohr had shown a drawing to General Groves during their two-day train journey from Chicago to Santa Fe. It is not clear whether it was a drawing he made on the train, or something he had with him, something he had brought from Denmark. The result was that by the time he arrived at Los Alamos on that occasion, Groves was alarmed. 'He thought Bohr was describing plans for a German nuclear weapon.'[11] This is why he asked Oppenheimer to convene the New Year's Eve meeting, the result of which we have already seen.

But that didn't end Bernstein's detective work. He asked Bethe to reconstruct the drawing he had seen on that New Year's Eve in 1943. On several attempts Bethe always drew what looked like the cross-section of a cylindrical container, inside which heavy water and uranium metal sheets were placed. From what we now know about the German plans, none of Heisenberg's reactor designs until 1943 resembled the design that Bethe and Teller examined. The designs of other German physicists did, but not those of Heisenberg. Indeed, in the autumn of 1941, the same time as he visited Copenhagen,

Heisenberg published a drawing showing a test reactor of *spherical* design.[12] Bernstein was stuck again.

He next turned to two secret exchanges that Bohr had had during the war with James Chadwick, which had been published in Andrew Brown's biography in 1997. The first of these took place in January–February 1943, when Chadwick had sent a top secret message written on microdots hidden in minute holes bored in some keys. Chadwick's first message invited Bohr to Britain to take part in 'scientific matters'. Bohr replied that circumstances prevented him taking up the offer for the moment, and added crucially that he did not for the time being see any practicable use for the 'recent discoveries in atomic physics'.[13]

A few months later, however, in the autumn of 1943, Bohr changed his mind and, in a second secret letter to Chadwick, referred to 'rumours going around the world, that large scale preparations are being made for the production of metallic uranium and heavy water to be used in atomic bombs'. He concluded this second letter by saying he was modifying his earlier statement 'as regards the impracticability of an immediate use of the discoveries in nuclear physics'.

What, Bernstein asked himself, had changed Bohr's mind? It seems that, during the course of 1943, Bohr had learned enough about the German programme to become seriously alarmed. What had he learned and how had he learned it?

Paul Rose suggests that one way was via Eric Welsh, who had been informed of German progress by Paul Rosbaud but this seems unlikely because Rosbaud was the man who had alerted the British to the fact that Germany was *not* pursuing a bomb. Instead, Bernstein's answer was far more plausible – the figure of J. Hans D. Jensen. Jensen was never part of the Uranverein, the Uranium Club, the central unit of the German bomb programme – he was judged as being too left wing (even communist) to be totally accepted. But he was an excellent physicist who would go on to win the Nobel Prize after the war, by which time he was professor of theoretical physics at Heidelberg.[14]

In the summer of 1942, as we have seen in chapter 9, Jensen visited Copenhagen, very possibly at Heisenberg's instigation, in an attempt to repair the damage that had been done by the 1941 visit. Nothing much was achieved on that score, though that was when Jensen told the Norwegians that Germany was not working towards a bomb. However, Jensen returned to Copenhagen in the summer of 1943 when he told Bohr about Heisenberg's latest ideas, given in a lecture on 5 May 1943 to the Aviation Academy. This lecture was part of a series in which, once again, it was made clear to all concerned that atomic energy would not play any military role in the near future, and certainly not in the war. But in his lecture, Heisenberg produced a diagram, Bernstein tells us, and it showed a reactor in a cylindrical container with a heavy water moderator and metallic uranium plates. Heisenberg's idea was that the uranium plates were to be immersed in the heavy water.

Bernstein argues that there is every reason to think that Jensen knew about this reactor – he was working on heavy water (as we know) and clearly in contact with Heisenberg. To clinch matters, after the war Jensen had dinner with members of the Berkeley Faculty at the home of Mal Ruderman (now an astrophysicist at Columbia) where he told them that he was indeed the one who had informed Bohr about the German programme and Heisenberg's lecture.

And so, Bernstein concluded, it was *this* occasion that prompted the drawing, and Bohr's change of mind about the seriousness of the German threat, which in turn prompted his second letter to Chadwick. The irony here, as Bernstein points out, is that it shows 'that at the time Bohr really did not understand what a nuclear weapon was. He also did not seem to understand this reactor very well.'[15]

Bernstein goes on:

> One can only wonder what Chadwick made of this communication. He surely knew that heavy water has nothing to do with atomic bombs except that it can be used in reactors that make plutonium. There is no heavy water in a [U-235] fission bomb and Bohr did not know about plutonium. Furthermore, on his arrival in Britain, after his escape toward the end of 1943, Bohr seemed to be sending mixed messages. Chadwick [whom we know had dismissed the relevance of heavy water in his MAUD report of autumn 1941] got the impression that what Bohr was saying was

> that Heisenberg was *not* working on atomic weapons at that juncture. This was true but Bohr was nonetheless presenting his information as if it did concern the design of such a weapon. This persisted until he got to Los Alamos where, as we have seen, he prompted Groves to call an emergency meeting. 'Bethe once told me that it was clear to him that Bohr, when he arrived at Los Alamos, was quite ignorant about nuclear weapons. In fact, Oppenheimer assigned Richard Feynman to Bohr to get him up to speed. He also did not understand the physics of this reactor very well. This is what Bethe and Teller sorted out.'[16]

Bernstein's detective work therefore produced a result that was plausible. The New Year's Eve meeting at Los Alamos considered Heisenberg's thinking not in 1941, as had always been thought, but in 1943. And what it showed was that the Germans had not made much progress towards a bomb. It is also worth pointing out that there is no suggestion that Bohr showed a drawing to the meeting in London at the Savoy Hotel attended by Chadwick and others. He would surely have done so, had he possessed one.

Now we turn to the second piece of detective work, this time by Paul Lawrence Rose, professor of European History and Mitrani professor of Jewish Studies at Pennsylvania State University until his death in 2014. Rose's starting point was a remark made at Farm Hall by Karl Wirtz, who was one of the Uranverein and worked in Berlin on a horizontal layered reactor design. Farm Hall, to remind ourselves, was a secret intelligence establishment in East Anglia in Britain, where around a dozen German nuclear physicists were held at the very end of the war, and their conversation secretly listened to, thanks to bugging devices installed in their quarters. During this eavesdropping, when the incarcerated German physicists were startled to learn about Hiroshima and were trying to draw up a memorandum recapitulating and justifying their wartime work, and trying to show they had not been left totally behind, Wirtz reminded his fellow prisoners that 'they should remember that there was a patent for the production of such a bomb at the Kaiser Wilhelm Institute for Physics, and that the patent had been taken out in 1941'.

A German bomb in 1941? The memory of this patent, says Rose, 'has been conveniently suppressed for the last fifty years'.[17] Nonetheless 'traces' of it exist in German archives, which he set about unearthing. In particular, he sifted through a comprehensive survey of the German uranium project prepared by the German Army Weapons Research Office in February 1942 (but dated August of the previous year), which includes a detailed discussion of reactor-bombs, and contains the following reference:

> P1. Patentanmeldung. Technische Energiegewinnung. Neutronenerzeugung und Herstellung neuer Elemente durch Spaltung von Uran oder verwandten schwerem Elementen 28.8.1941. 14S (Patent application, technical specifications. Neutron generation and production of new elements by fission of uranium or [its] heavy relatives.)

Rose searched other archives.[18] None of these turned up anything remotely suggestive.

Then, much as Bernstein had done with the drawing, Rose asked himself who would have prepared the P1 patent? He concluded that it would most likely have been registered in the name of the organisation and possibly of the administrators rather than the true inventors. Although much of the work on reactors was carried out at the Kaiser Wilhelm Institute for Physics, the actual unit in charge of the project in 1941 was the Nuclear Physics Section of the Weapons Research Office, whose administrative head was H. Basche and the scientific leader of which was Kurt Diebner (who had been present at the June 1942 meeting between Speer and the physicists). Following up this insight, Rose eventually came across the following report published by Diebner in 1956:

> T-45. K. Diebner, H. Basche u a [und andere, and others]
> Geheimpatent über Uranmaschine mit verschiedenen geometrischen Anordnungen von Uran und Bremssubstanz. (Secret

> patent about a uranium machine [reactor] with different geometrical arrangements from uranium and moderator substance.)[19]

Rose considered that it is reasonable to conclude that this T-45 reference in the Diebner papers is the same as P1 and if true it confirms that the Germans *had* developed a reactor, whose main purpose was the production of power but used heavy water and therefore that might – far down the line – produce plutonium, which is the only way heavy water is related to an atomic bomb.

It is plausible to suggest that it was this design, developed in August 1941, that Heisenberg discussed with Bohr in the following month. It was a reactor, not a bomb. Heisenberg may have known about plutonium at the time (Weizsäcker certainly did, see immediately below) but not Bohr. Which is why he didn't feel the need to pass on the details to the Allies.

The third sleuth was the German historian Rainer Karlsch who, in his book *Hitlers Bombe* (2005), and afterwards, discovered a number of documents in the Russian archives taken by Soviet forces at the end of the war from the Kaiser Wilhelm Institute for Physics in Berlin. Four in particular are of interest: an official report written by von Weizsäcker after a visit to Copenhagen in March 1941; a draft patent application written again by von Weizsäcker sometime in 1941; a revised patent application in November that year; and the text of a popular lecture given by Heisenberg in June 1942.

Von Weizsäcker's report of his March visit to Copenhagen is interesting because, as we shall see, after the war some of the Danish physicists accused the two Germans of being spies, an allegation that appears to be supported by the wording of his report, written at a time when Germany had not yet invaded the Soviet Union and a quick victory appeared likely. This is part of von Weizsäcker's report to the German army:

> The technical extraction of energy from uranium fission is not being worked on in Copenhagen. They knew that in America

> Fermi has started research into these questions in particular; however, no more news has arrived since the beginning of the war. Obviously, Professor Bohr does not know that we are working on these questions; of course, I encouraged him in this belief . . . The American journal *Physical Review* was complete in Copenhagen up to the January 15, 1941 issue. I have brought back photocopies of the most important papers. We arranged that the German Embassy will regularly photocopy [make photographs of] the issues for us.

So, it is clear: on that visit at least, von Weizsäcker behaved as a spy of sorts.[20] Taken together with the fact that elsewhere he regretted the original publication of Hahn and Strassmann's fission paper, and sought to inform himself of American progress, we see that Weizsäcker was among the more committed German physicists.

The second Russian document was von Weizsäcker's patent application written 'sometime' in 1941, which 'makes it crystal clear' that he did understand both the properties and the military applications of plutonium. 'The production of element 94 in practically useful amounts is best done with the "uranium machine" [nuclear reactor] . . . It is especially advantageous – and this is the main benefit of the invention – that the element 94 thereby produced can be easily separated from uranium chemically.' Von Weizsäcker also makes it clear that plutonium could be used in a bomb. 'With regard to energy per unit weight this explosive would be around ten million times greater than any other [existing explosive] and comparable only to uranium 235.'

On 3 November 1941 (so after the September meeting between Bohr and Heisenberg in Copenhagen), von Weizsäcker resubmitted his patent with the same title: 'Technical extraction of energy, production of neutrons, and manufacture of new elements by the fission of uranium or related heavier elements.' Rainer Karlsch says this submission differed in two significant ways. 'First the patent was now filed on behalf of the entire Kaiser Wilhelm Institute, instead of just von Weizsäcker. Second, every mention of nuclear explosives or bombs had been removed.'[21]

This change could be put down to one of several reasons, Karlsch adds. It could reflect the change of fortune in the war: in November 1941 a quick German victory no longer appeared as certain as it had done earlier in the year. But 'another possible explanation is that von Weizsäcker and his colleagues had a change of heart – perhaps their initial enthusiasm for the military applications of nuclear fission had cooled. This would support Heisenberg's and von Weizsäcker's post-war claims that they had visited Bohr in September 1941 because they were ambivalent about working on nuclear weapons.' A third reason might have been that they didn't want to promise something they couldn't deliver.

These various patents also show that there were, as Karlsch says, two competing groups working on nuclear reactors in Germany – one under Diebner in Gottow, near Berlin, and one directed by Heisenberg in Leipzig and Berlin.

The fourth Russian document, which gives the first verbatim account of Heisenberg's popular lecture, the one where Speer was present, does make it clear that the physicist knew the Americans were working on a bomb, one reason being that 'American radio provided extensive reports', and that 'one day' Germany would be able to follow the path revealed by von Weizsäcker (plutonium) and that if the 'war with America lasted for several years' a bomb might 'play a decisive role'.

The wording is clear in one regard and ambiguous in another. It is clear that the Germans did understand the explosive potential of both U-235 and element 94 (plutonium), but it is not clear whether Heisenberg (or von Weizsäcker) were deliberately slowing down their own research efforts, or whether they genuinely thought that a bomb was a long shot in the current war. In support of this is the evidence – contained in several documents – that the German nuclear physicists could not believe that the Anglo-Americans could overtake them. Whatever the real reason, the result was the same: there was not enough enthusiasm to embark on an industrial-scale effort towards a bomb.[22]

Before summing up, let us turn to Bohr's unsent letters. There are between a dozen and a score of these, in various stages of completeness. There is also one from Heisenberg that he sent *to* Bohr, and one is written by Bohr's wife, Margrethe, also unsent.

Six things of importance emerge from these letters, apart from the fact that Bohr never sent them.

1. He insists that he remembers the encounter very well.
2. He remembers the atmosphere of the meeting as very hostile on Heisenberg's part.
3. Germany would win the war and that this would be 'a good thing'.
4. He remembers that Heisenberg told him that he was actively involved in working on the preparation of an atomic bomb which, he was convinced, would decide the war if it went on long enough.
5. He refers four times to Jensen's visit, saying that he emphasised that he was trying to increase the production of heavy water in Norway but that he and other German physicists were working only on the industrial – not military – applications of atomic energy. While Bohr and his colleagues felt that Jensen was telling the truth, so far as he himself was concerned, they were unsure as to how much he actually knew about an atomic weapon.
6. He wonders what authority Heisenberg had for disclosing what he disclosed.[23]

Certain observations are in order.

The first is that, as Aage Bohr points out, his father made no attempt to alert the Allies about the 1941 encounter (when the US was still a non-belligerent). We may conclude, therefore, that, whatever was said, Bohr saw no imminent threat.

Second, Heisenberg's statement, as described by Bohr, that he and his colleagues were actively working on an atomic bomb and that he thought nuclear weapons would decide the war 'if it went on long enough', is much more ambiguous. Bohr later said that his impression of the 1941 meeting was that the Germans were not near to making a bomb (partly perhaps because he himself at that point didn't think that a weapon was practicable). Which is one reason – perhaps the main one – why he made no attempt then to report the meeting to the Allies. This would fit with what Rose found out about the reactor patent: it was essentially a preliminary, exploratory academic experiment.

Bohr's behaviour and moral rectitude, in strong contrast to Heisenberg's, are unimpeachable. But we do need to scrutinise this part of his memory. (The unsent letters were written between 1957 and 1962.)

In 1941, Heisenberg and Weizsäcker felt – and had good reason for feeling – that the war would soon be over, in Germany's favour. The Blitz was in full spate, Buckingham Palace had been hit, the German U-boats in the Atlantic were increasingly successful in sinking British ships, the Wehrmacht were approaching Moscow, General Rommel had just taken over the Afrika Korps, Austrian Jews were being deported to Poland and Charles Lindbergh was urging the United States to conclude a neutrality pact with Hitler. We know from the contents of the infamous speech Heisenberg gave to the German Cultural Institute in Copenhagen, on the same visit, that he stressed how important it was that Germany should win the war, that although it was 'sad' that Denmark, Norway, Belgium and Holland had been occupied, 'as regards the countries in Eastern Europe it was a good development because these countries were not able to govern themselves'. He thought that perhaps democracy wasn't energetic enough to stand up to dictatorship.[24]

Therefore, Rose (among others) asks pertinently why Heisenberg would need to emphasise the bomb and the fact that it was possible but far off. In 1941, so far as any nationalistic Germans (like Heisenberg) could see, the country had no need of atomic weapons. There are two possible answers.

One is that he was indeed subtly trying to tell Bohr that Germany was a long way from producing any weapons, and he was delivering this information in such a way that it wasn't, strictly speaking, treason. This would also answer Bohr's other point: Heisenberg had no authority to say what he was saying but he was speaking in such a way that, should his words somehow get back to the authorities in Berlin, they would be 'deniable', as we would say today. On this account, his speech to the German Cultural Institute was in effect camouflage, designed to be part of the record, showing him publicly as a patriot, and deflecting attention from what went on in his meeting with Bohr.

The other possible explanation, favoured by Rose, is much more blood-curdling and much more cynical. What Rose thinks Heisenberg meant was that Germany would build a bomb even after a German *victory* in the war, and use it to subdue – or threaten to subdue – other nations as part of a *Pax Nazica*.[25]

Such ideas may or may not have got across to Bohr. From our point of view, however, the main point is that Heisenberg worried and angered Bohr, but Bohr saw no reason to act. Whatever technical discussion Heisenberg and Bohr shared on that occasion, we now know, courtesy of Bethe, that Bohr was out of the loop, did not fully grasp nuclear weapon technology, thought atomic bombs not feasible and thought it unlikely, even impossible, that the Reich was on its way to a bomb any time soon. He did not realise that a reactor could lead to plutonium, as Heisenberg and Weizsäcker did. So Bohr did nothing.

Now, fast-forward to 1942 and 1943, and the visits of Jensen. As we have already seen in chapter 9, Jensen was one of several German physicists who, in the summer of 1942, tried to alert British intelligence (via the Norwegian underground) to the fact that the German A-bomb programme had been aborted by Speer in June of that year. So the historian of today knows something that Bohr didn't know at the time. Jensen, when he told Bohr in 1943 that Heisenberg was working on a nuclear reactor that could, given time, and by implication, produce plutonium, he was also, as an expert on heavy water, telling Bohr, again in a roundabout way, that even that was a long way off.

But Bohr, it must be said, and with due regard to his unimpeachable reputation, appears to have got the wrong end of the stick. Not because he was incapable of understanding but because, living in occupied Denmark, he did not have access to the work going on in both Germany and in Britain and America. This is why, when Bohr escaped to Britain later in 1943, Chadwick was confused by the mixed messages Bohr brought with him. As Bethe recalled, when Bohr arrived at Los Alamos, he didn't understand nuclear weaponry and had not thought things through to the fissionability of plutonium. One of Oppenheimer's biographers observes that 'Bohr had much to learn and little to teach about the physics of the atom bomb.'[26]

In considering the fallout from all these various recent developments and discoveries, six things stand out.

The Germans *had* developed a reactor in 1941, were aware of the relationship between a reactor, heavy water and the creation of plutonium, and that plutonium would be fissionable.

Bohr was not up to speed about plutonium at the time, which is why he didn't feel the need to pass on the details to the Allies.

During the course of 1941 the initial enthusiasm Weizsäcker and Heisenberg had had for the military applications of fission had cooled, whether they were ambivalent about such work, as they later said, or because they felt Germany was doing so well in the war that she didn't need such a weapon, or because they didn't want to promise something they couldn't deliver.

They had some knowledge that the Americans in general, and Fermi in particular, were working towards a sustainable chain reaction.

The drawing that Jensen showed Bohr in 1943, or more likely the drawing Bohr himself made as a result of their deliberations, was for a reactor, not a bomb: the design of the reactor was different to that discussed in 1941, but it was still a reactor, nothing more.

Bohr – sequestered as he was in occupied Denmark – was, to an extent, confused about the way developments were going. He had heard rumours to the effect that large-scale preparations were being made for atomic bombs, and had been visited by Jensen with Heisenberg's latest ideas. We now know from the Karlsch evidence that Heisenberg was aware, in June 1942, when he gave his lecture that Speer attended, that American radio had mentioned work on an atomic bomb in the United States. Jensen was at that meeting, so he knew what Heisenberg knew. Jensen was possibly the source of the 'rumours' that Bohr told Chadwick were 'going round the world'. Given that he felt Heisenberg had been so hostile in 1941, Bohr was quite likely intimidated by Jensen's 1943 visit, even though all he was being shown was a reactor.

Now we must turn to certain events that took place in London between Bohr's flight from Denmark and his arrival in Los Alamos. In particular, there was the secret dinner held at the Savoy Hotel the very day he arrived in the capital (p. 185 above). Among the fellow guests that night were no fewer than four intelligence chiefs – Sir Stewart Menzies, Commander Eric Welsh, R. V. Jones and his deputy Charles Frank. Jones had the ear of Churchill and so did Frederick Lindemann – it was an extremely high-powered group.

No doubt the intelligence chiefs were there partly to find out at first hand from Bohr what he knew about German plans. At the same time it is inconceivable that, in such a gathering, Bohr wasn't told about the British evidence that Germany was not working towards a bomb. We can be pretty certain of this because, as several witnesses have testified (Jones, Anderson, Lindemann), one of the main topics of discussion that evening, if not *the* main topic, was Bohr's disquisition on Russia and the need to share atomic information with them, in order to create a new form of international cooperation post-war. Indeed, this was the occasion when, for most of them, such an idea was first mooted. Would this have happened if a German threat still hung over the dinner table? If Bohr had a drawing given to him by Heisenberg or Jensen, would he not have produced it? Would it not have dominated the conversation?

Bohr stayed in London from early October until early December, travelling to the main sites of research – Liverpool, Birmingham, Oxford and Cambridge. Overlapping with Bohr's travels, in November, General Groves's aides Richard Tolman and Robert

Furman arrived in London and this was when, officially, there was the first widespread exchange of nuclear intelligence. A series of handwritten pages, in pencil, in the Groves papers in the Washington National Archive, appears to comprise Tolman's itinerary for his time in London. This shows that on Wednesday 17 November he met with Akers, Perrin, Welsh and Jones, much the same individuals as Bohr met at the Savoy dinner, when the full weight of British intelligence was made known to the Americans.[27] We shall see shortly that there are good grounds for doubting this official story but even so this casts a new light on Bohr's arrival at Los Alamos.

When Bohr arrived on the Pajarito Plateau his mind was in turmoil. He had escaped from Denmark worried about German progress, feeling, as he said, that Jensen wasn't completely in the swim in Germany, so had been unwilling – or unable – to tell Bohr the whole truth about their plans. In London Bohr had been overwhelmed by the progress the Allies had made, so much so that for him, given what he had also been told by the British intelligence services, the entire nature of the project had changed and he now saw relations with Russia as the major problem going forward. For Bohr this was a fundamental change of perspective. He had left Denmark worried about what Heisenberg was doing. Now he was faced with the Allies' accomplishments.

He surely discussed all this with General Groves during their two-day train journey between Chicago and Lamy. It was also on this journey, we know, that Groves told Bohr what could and could not be discussed at Site Y. And we know from others on the 'Reservation', as the Russians called it, that discussion of political issues and use of the bomb ('non-scientific matters') was off-limits. The significance of this will become clear.

According to Groves, Bohr now agreed with the British intelligence assessment, that there was no serious German bomb. The general, we also know, according to the official version, had been informed by Tolman and Furman about the British intelligence material just a few days beforehand and, as was his way (discussed several times above), was disinclined to believe it. Groves, who couldn't bring himself to accept the British intelligence assessment because it might demotivate his scientists, and damage his own importance in the war, chose to regard the drawing as fresh evidence from, as it were, 'the horse's mouth', which would keep his scientists on their toes.

To maintain secrecy, Groves adopted a policy of committing as little as possible to paper. So there is no record, in his archive or in Bohr's, of what transpired on the train journey. All that we know is that Groves outlined to Bohr what could and could not be spoken about. But consider the situation from Groves's point of view. Bohr was, with Einstein, the most senior figure in world physics, certainly on the Allied side, with the status of an elder statesman. He had just come from London where he had met leading physicists but, more importantly, a raft of intelligence chiefs who gave him evidence – as they had, virtually simultaneously, given the same material to General Groves's aides – showing that there was no German bomb. Yet Bohr arrived in Los Alamos with a drawing that he and/or Groves thought might be evidence *for* a German weapon. We know that, as both Groves's biographers say, the general feared that if the Los Alamos scientists knew there was no German bomb they might be crucially demotivated. As we shall see in Part Three of this book, after his visit to Los Alamos, Bohr's entire behaviour was concerned with Russia. That behaviour shows that he accepted there was no German bomb.

The suspicion surely exists, therefore, that on the train journey Groves somehow prevailed on Bohr to downplay his concerns about Russia, and to go along with the idea that the drawing Bethe thought made no sense was in fact some kind of German prototype, in order to keep his men motivated. Given the difference between the Bohr in London at the end of November 1943 and the Bohr in Los Alamos on New Year's Eve, a few weeks later, something like this must have happened.

Bohr must have shown the drawing to Groves on the train, or even made it himself, and Groves was worried enough – or said he was worried enough – to prevail on Oppenheimer to hold the special New Year's Eve meeting. That meeting soon quashed any idea that the reactor was a bomb, except insofar as it might eventually produce plutonium, which Bohr didn't know about at the time, according to Bethe. But it achieved its purpose in keeping the scientists' minds focused on Germany.

According to Thomas Powers, the sheer scale of Los Alamos frightened Bohr into changing his mind about the German bomb, but this can't have been true given Bohr's elaborate views about Russia, outlined so recently in London, and his subsequent behaviour, which we shall come to in Part Three. He was no longer worried about a German bomb but even so he may well have brought along the sketch to see what his more informed colleagues would make of it. And Rose reports that Serber told him that when, at the New Year's Eve meeting, Oppenheimer referred to the drawing as Heisenberg's, Bohr did not contradict him. Why? Even if he was confused, Bohr would surely have said so. He was that kind of man. Could it have been that Jensen had brought the drawing, or a set of ideas, as a basis for a drawing, to Copenhagen and Bohr assumed the idea was either *by* Heisenberg, or a faithful reproduction of what Heisenberg then thought? On any grounds it would be good to have an insight into Heisenberg's latest thinking.

The central point, however, is that the missing drawing was yet more evidence that the Germans had no bomb project. They were still experimenting on reactors for power. This in fact only added to the copious and convincing evidence that, whatever Groves might say, there was no serious German bomb programme at the time Bohr reached Los Alamos.

Hardly less important, this intelligence information was kept from the scientists working on the Manhattan Project. Had the assembled scientists at Los Alamos known this, what would they have done? We can't be certain but we can be sure that it would have affected both the course – and the impact – of Bohr's later arguments that we shall follow in subsequent chapters.

A final factor at this time concerned James Chadwick. On the day after Tolman and Furman met Akers, Welsh, Jones et al. in London, Tolman's pencilled diary shows that he met Chadwick. No doubt some of the previous day's talk was repeated (Chadwick thought that Heisenberg was by far the most dangerous German physicist, 'because of his brainpower'). Chadwick, as well as being the most senior British physicist in the atom bomb project, was also one of the most down-to-earth and realistic. Shortly after his meeting with Tolman, he left for America. He spent the next few weeks travelling across the US 'acquiring a virtually complete picture of the Manhattan Project'. What especially impressed him was the Oak Ridge site in Tennessee. Despite the fact that this secret city did not exist officially, it measured 16 miles by 7 miles, contained dwellings for 13,000 people and by then was dominated by a gigantic factory, itself more than a mile long, designed to house Lawrence's electromagnetic separation plants. The gaseous diffusion plant occupied a 5,000-acre site in one corner of the reservation (Francis Simon had originally conceived the British site at 40 acres). Seventeen cafeterias served 40,000 meals a day, the plant had its own cattle and chicken ranches, 800 buses carried the workers and the specially built hospital had 337 beds.

All this brought home to Chadwick two things of importance. The sheer scale of the site was so staggering that he wondered why he had ever thought it could have been carried through in Britain; and, second, that a place like Oak Ridge was impossible to camouflage from aerial reconnaissance. He concluded that unless the RAF discovered huge new industrial sites being built in Germany, the threat of a Nazi atomic bomb was non-existent.[28]

By the end of 1943 hardly anyone in the swim at senior levels on the Allied side thought Germany had a bomb programme. That included Leslie Groves but not his scientists at Los Alamos.

American Institute of Physics

Rudolf Peierls (front row, left) in Leipizig 1931, with Werner Heisenberg (front row, right). In the back row: second from left, George Plazcek; second from right, Victor Weisskopf.

Pyotr Kapitsa (right), the Russian nuclear physicist (and later a Nobel laureate), in Cambridge, dressed in traditional morning suit as James Chadwick's best man, with a dented top hat. A Fellow of Trinity College and of the Royal Society, he was forbidden from returning from Russia to Britain in 1934.

Werner Heisenberg. A Nobel Prize-winner in 1932 at the age of thirty-one, he was one of the most brilliant and charismatic scientists and one of the founders of modern physics. He was also highly controversial and it is still not clear whether, during the war, he worked towards a German bomb, or did his best to deceive Hitler about the chances one could be produced before the fighting ended.

Getty Images

The Seventh Solvay Conference of Atomic Physicists, Brussels, 1933, devoted to the atomic nucleus. Note the three women in the front row: second from left, Irène Joliot-Curie; fifth from left, Marie Curie; second from right, Lise Meitner, with Chadwick at the end. On Irène Joliot-Curie's left is Neils Bohr, and next to him Ioffe. Standing: Frédéric Joliot-Curie is third from left, with Heisenberg next; Enrico Fermi is seventh from the left, Peter Debye tenth; next to him is Klaus Fuchs's professor at Bristol Nevill Mott and three beyond him is Walther Bothe; next to him is Patrick Blackett. John Cockcroft is sixth from the right, and Peierls fourth.

Getty Images

Niels Bohr. Prime Minister Winston Churchill thought he was a Soviet spy meddling in politics. As well as being one of the top two physicists of all time, he was probably the most profound thinker on atomic strategies.

Getty Images

Enrico Fermi, the 'pope' of nuclear physics. He escaped from Fascist Italy in 1938, after he and his Jewish wife had been to Norway to collect the Nobel Prize.

Pyotr Kapitsa in Moscow after he had been refused permission to return to Britain. His Cambridge lab was transferred to Moscow 'lock, stock and barrel'. He tried to alert the British and Americans to the fact that Russia had some idea what atomic research was going on in secret.

Getty Images

Otto Hahn and Lise Meitner at the institute set up in their name in 1959. Her escape from Nazi Germany was more fraught than any of the other physicists. But in exile in Sweden she was first to realise that Hahn had discovered nuclear fission, bringing the prospect of an atomic bomb much nearer.

Bridgeman Art Library

Paul Rosbaud. His extraordinary anglophilia stemmed from his treatment by British soldiers as a PoW at the end of the First World War. He became Britain's number one atomic spy and was the first to reveal that Germany did not have a bomb.

Irène and Frédéric Joliot-Curie at the Laboratoire de Chemie Nucléaire in Paris. Their insistence on publishing their results about nuclear fission in April 1939 alerted the Germans, the British and the Russians to the very real possibility that nuclear weapons were feasible. Many émigré physicists wanted the French to keep their results secret and were distraught at the open publication.

Alamy

James Chadwick, left, and General Leslie Groves, commanding officer of the Manhattan Project. Groves 'could have won almost any unpopularity contest', and was as much an Anglophobe as Rosbaud was Anglophile. However, Chadwick, the most senior British physicist in the Manhattan Project, was one of the few scientists the general respected.

Arthur Compton of the University of Chicago. Despite being a senior figure in the American bomb programme, he was deliberately kept in the dark by General Groves in regard to Germany's non-progress towards a bomb and so was one of those who worried unnecessarily.

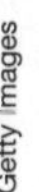

Peter Debye, a Dutch Nobel Prize-winning physicist who, on the outbreak of war, was director of the Kaiser Wilhelm Institute in Berlin. He left Germany rather than resign or convert his nationality, and moved to the United States to lecture, where he informed the authorities that Germany did not have a bomb, and that German physicists were using the bomb project to avoid military service. He was not believed.

Hans A. Bethe left Germany in 1933, for England, where Rudolf Peierls served under him, moving to America two years later. Bethe became head of the theoretical division at Los Alamos and was one of those who recommended the assassination of Werner Heisenberg. He remembered Bohr bringing to Los Alamos a drawing (which went missing) of German nuclear ideas in 1943–44.

Getty Images

Hans. D. Jensen, a specialist in heavy water from Hamburg. He attended the fateful meeting at which Speer killed off the German bomb project and visited Bohr to explain there was no threat from Germany. He also informed the Norwegian underground of the same message.

J. Robert Oppenheimer and General Groves at the Trinity test site in the Alamogordo desert, in July 1945. Oppenheimer's 'sweat-stained pork-pie hat' was a famous icon at Los Alamos.

Getty Images

Klaus Fuchs. A brilliant mathematical physicist and a dedicated communist, much influenced by Henri Barbusse in France. 'His principal achievement was learning how to live without people rather than learning to live with them.' He told his Russian controller that he 'avoided falling in love deeply'. His betrayal of nuclear secrets advanced the advent of a Russian bomb by as much as two years.

Getty Images

A missing link. Louise Bransten, a wealthy socialite who was close to certain Russian operatives in San Francisco. The FBI never spotted her links to Elizabeth Bentley, a Soviet spy with a large network of contacts, one of whom would have led to Harry Gold – and Fuchs.

Alamy

Harry Gold under arrest for espionage. He had great respect for Fuchs. They had in common a love of music, chess and their solitary life of secrecy.

Alamy

Los Alamos from the air. Santa Fe, 17 miles away, was the nearest outpost of civilisation. Rumours abounded that spaceships, poison or whiskey were being manufactured on the base, so scientists were sent into town to spread stories that a new form of rocket was being developed.

British physicists who, in 1946, received the American Medal of Freedom for their part in the Manhattan Project. From the left: William Penney, Otto Frisch, Rudolf Peierls, John Cockcroft.

Hiroshima in ruins. The change of target from Germany to Japan was kept secret from the scientists labouring on the bomb, because it was feared they would be demotivated and might refuse to continue working.

15

'Don't Bother Me with Your Scruples': The Loss of Innocence

We can now add to this an account by Josef Rotblat about an encounter that took place in Los Alamos a few weeks after Bohr arrived there, in spring 1944. Rotblat was a Polish physicist who had been on a scholarship, studying with James Chadwick in Liverpool in 1939 when war broke out, and had been unable to return home.[1]

Documents show that he arrived in America on 16 February 1944 to join the Manhattan Project. According to an account he wrote after the war, he had at first stayed in Los Alamos with his old Liverpool professor and his wife, in the Chadwicks' house on the site, before moving into the building where single men were housed, known as the Big House. While he was still staying with the Chadwicks, they had given a dinner party in March where one of the guests was General Groves. During the dinner, Groves had used words to the effect that 'the real reason for the bomb was to subdue the Russkies'. 'Until then,' Rotblat added, 'I had thought that our work was to prevent a Nazi victory, and now I was told that the weapon we were preparing was intended for use against the people who were making extreme sacrifices for that very aim.' On hearing Groves's words, he says, he felt 'a deep sense of betrayal'.[2]

Putting the chronology of chapters 8, 9 and 12 together in the way that we have, we begin to see that the purpose of the atom bomb underwent a fundamental change during 1943. And there are other details that support such an idea. In 1985, Rotblat reported that he had come across a letter that Oppenheimer had sent to Enrico Fermi on 25 May 1943 (three weeks after the MPC had selected Japan as the target for the first bomb), in which he had discussed the development of radioactive poison – specifically, radioactive strontium – which he thought might be developed in secret, to be used in foods, but the technology should be given the go-ahead only if it would 'kill at least half a million people'. This surprisingly bloodthirsty idea coming from Oppenheimer suggests he was no longer thinking of a bomb to serve only as a deterrent. And Robert Wilson, another physicist at Los Alamos, noted that when scientists suggested to Oppenheimer that Soviet physicists should be on site, just as the British were – since they were all Allies – or at least informed in advance before the bomb was actually produced, 'he closed down all discussion'.

In fact, any discussion of the political aspects of nuclear weaponry and its use was discouraged at Los Alamos, and was further hindered by Groves's policy of compartmentalisation, not to mention his lack of doubt throughout that the bombs would be used.[3]

Yet another tantalising insight is provided by David Hawkins, Oppenheimer's assistant. Hawkins was in Oppenheimer's office on 31 December 1943, when Groves turned up. It was the day after the general had arrived with Bohr, and Hawkins heard Groves tell Oppenheimer that a report had been received 'from a German source' to the effect that the Nazis in fact had no bomb programme. According to Hawkins, 'Oppenheimer said nothing.' Groves apparently went on to say that they weren't sure they could trust the source, but Hawkins added that 'the truth of the report made no difference – things had gone too far at Los Alamos; we were committed to building a bomb regardless of German progress ... Oppenheimer only shrugged.'[4]

We have seen before that Groves was apt to discredit, or even lie about, British intelligence, and he seems to have done much the same here. We now know that the Americans had been shown the British intelligence file on 17 November, a little over a month earlier, and that Major R. R. Furman had been in London in December, to set up a Manhattan liaison office and initiate a joint Anglo-American intelligence effort. We don't know all the details they were shown or, more to the point, what Groves was aware of even before that date. But we do know that Tolman had met Welsh and Jones in London and we know that Furman reviewed several hundred RAF photographs of the Nazi-controlled mines at Joachimsthal. These showed that, over the years, a massive pile of uranium pitchblende had lain untouched. Franklin Matthias, head of construction at Hanford, had told Groves that if the Germans had a bomb project they would need an industrial-sized plant, much like Hanford, with 'lots of material going in and almost nothing coming out'. Inspecting the Joachimsthal photographs while in London, even Furman began to think that the Germans 'weren't actively interested in the bomb. In fact, our information led us to believe that their chances of having one were only about one in ten.' This wording, and arithmetic, suggests the British were *still* not sharing the full status of the intelligence they possessed.

Thomas Powers thinks that when Groves disparaged the British source he was referring to the Jensen evidence, which of course was the subject of the drawing Bohr had brought with him.[5] If it had been more than that – the Rosbaud report, alongside his other contributions (to confirm validity), and an analysis of the negative Enigma traffic, as discussed in chapter 9 – Groves must surely have been more convinced.

But Enigma was too important to the rest of the war effort for its existence to be shared and in any case Groves discounted the new evidence because had another agenda.

We shall return to that in just a moment, though first it is worth dwelling on why, in these circumstances, Rotblat would become the only physicist to leave the Manhattan Project. Even he did so only in December 1944, when the findings of the Alsos Commission finally confirmed, to Groves's public satisfaction, that Germany did not have a bomb.* What this also confirms, of course, is that the Los Alamos scientists did not have – were denied – access to the impressively accurate British intelligence that Groves was shown (and see below).

Rotblat wrote after the war that there were, he thought, three

* And when Rotblat learned that he was himself suspected of espionage and that there was an inch-thick security dossier on him, which speculated that because his family was in Poland, he might be susceptible to blackmail. He always said afterwards that the security services might have spotted Fuchs if they hadn't been so intent on nailing him. In *Copenhagen*, the author makes Heisenberg accuse the Allies of dropping the bomb 'on anyone who was in reach'.

reasons why other scientists did not take the line he did, and this helps us further understand what was going on in the Manhattan Project in those crucial days.

He thought that one reason scientists were 'committed' to the bomb was sheer scientific curiosity, the desire to see whether it would all work out as theory had predicted. 'When you see something that is technically sweet,' as Oppenheimer put it, 'you go ahead and do it and you argue about what to do about it only after you have had your technical success. That is the way it was with the atomic bomb.' 'Don't bother me with your conscientious scruples,' Fermi complained to colleagues at the time. 'After all, the thing's superb physics.'[6] Fermi also said later that it wouldn't make any difference whether the bomb went off or not 'because it would still have been a worth-while scientific experiment'. He may have been trying to defuse the tension in the days before the test of the plutonium bomb in the Alamogordo desert, as Groves put it, but Freeman Dyson, a British-American mathematical-physicist, also said something similar about his co-workers: 'Los Alamos had been for them a great lark. It left their innocence untouched.'[7] David Hawkins, who had been committed to a bomb regardless, nevertheless said he was disturbed by the 'manic, joyous, delirious' enthusiasm about the bomb, and warned that some of his colleagues had 'lost sight of the grave consequences' of doing their job.[8] This is a point underlined by Admiral William Halsey, mentioned in the Preface, to the effect that the scientists had a 'toy' that they wanted to try out, 'so they dropped it'.* A second reason, which came into force later, was that if it was used against Japan it would save lives (this argument, too, has since been discredited, as also outlined in the Preface). And a third reason, Rotblat argued, was that for a scientist to pull out of the

* These sentiments find an echo in the words of Professor Brian Taylor, generally regarded as the 'father' of Britain's H-bomb. In a BBC documentary, aired on 3 May 2017, which described itself as the 'inside story' of Britain's nuclear programme, Taylor – who referred in passing to parts of the weapon as 'Tom' and 'Dick' (but not 'Harry') – was asked how he felt after the bomb had been successfully exploded. He replied: 'For the scientists it was a triumph. I mean, it was just the feeling that, scientifically, we'd demonstrated this was the way to go. It was just scientific people looking at one another and [saying] "It looks as though we've got something good here".'

project, however late in the day, would have jeopardised his career. 'Scientists with a social conscience', he said, 'were in a minority. They left morality to others.'

Ray Monk, in his 2012 biography of Oppenheimer, adds another reason: loyalty to the leader. Despite the fact that more than one colleague found Oppenheimer a 'ridiculous intellectual snob' and a 'sham', the leader, he says, was for many others an inspirational figurehead, someone who, as the British physicist James Tuck put it, had created in Los Alamos 'the most exclusive club in the world'.[9] It was a place where Tuck 'found a spirit of Athens, of Plato, of an ideal republic'.

Rudolf Peierls later said he thought many of the scientists at Los Alamos were naïve, adding that it 'never occurred' to him that the bomb would actually be used on a city, rather than a strictly military target.

But the crucial point is that by the time Bohr and Fuchs arrived on American soil, the purpose of the bomb had changed. They knew it but many of their colleagues did not, and would not find out the truth for some time. It is extraordinary that such different levels of knowledge could coexist in such a close-knit community but it seems to have occurred at Los Alamos. The change of target was not then as obvious as it appears in retrospect.

The official American history of the Manhattan Project, *The New World*, is muted in its discussion of this. All it says is: 'The bomb project had begun as an effort to overcome a Nazi head start. As fears eased that German scientists would win the race, American thinking turned toward Japan.'[10] That is it. No elaboration or discussion. No mention of Russia.

It is an interesting question – though perhaps unanswerable at this distance – as to whether Hawkins's view was widely shared – that all the physicists at Los Alamos were 'committed' to building the bomb. No one else, other than Rotblat, pulled out but had they been privy to the full extent of British intelligence, and earlier, would they have carried on 'regardless', without knowing how the war with Japan might end? Both of Groves's biographers were convinced that Groves kept the British intelligence from the scientists because he feared they might indeed object to carrying on regardless, but the full extent – and consequence – of his deception has not been understood until now.

Christoph Laucht, in his book on Fuchs and Peierls, also confirms that the 'vast majority' of scientists at Los Alamos were kept in ignorance of the findings of the Alsos Commission, that Hitler had not come 'even close' to building a bomb, and he accuses Groves of having 'no interest in informing Peierls and his colleagues ... because this information might undermine the scientists' morale and thus jeopardize the continuation of work on the atomic programme. As a consequence, most Los Alamos scientists were not aware of British and American intelligence reports that revealed that Germany did not possess a working nuclear device and voiced, at first, virtually no moral or political concerns about the continuation of their work on nuclear weapons.' He goes on to say, 'By suppressing debates on moral and political issues among the scientists through enforcing a strict regime of scheduling and time management, J. Robert Oppenheimer also contributed significantly to the project's progress.'[11]

There certainly seems to have been different levels of awareness among the scientists. In James Chadwick's papers at Churchill College, Cambridge, in the files of his Los Alamos correspondence, there are some documents about the effect of bombs on different locations, the height at which aircraft should fly for safety when dropping bombs of various weights, and a collection of photographs, dated between February 1944 and March 1945, showing *German* targets before and after raids.[12] Hawkins, Peierls and Rotblat appear to have had varying levels of knowledge or understanding of the bomb's intended target. That Rotblat could have felt 'betrayed' in March 1944 shows that, even though he was staying with the Chadwicks, in their own home, he had a very different understanding compared to his host and long-time colleague. Or, given the German photographs in Chadwick's archive, was he too in the dark about Japan at that time?

Towards the end of his life Chadwick told an interviewer that, in 1941, 'I realized then that a nuclear bomb was not only possible – it was inevitable.' He recognised, he said, that 'scientists everywhere' would sooner or later find out how to make nuclear bombs, they would be built and that 'some country would put them into action'.[13]

After the war, in his memoirs, Groves put it this way:

> When we first began to develop atomic energy, the United States was in no way committed to employ atomic weapons against any other power. With the activation of the Manhattan Project, however, the situation began to change. Our work was extremely costly, both in money and in its interference with the rest of the war effort. As time went on, and as we poured more and more money and effort into the project, the government became increasingly committed to the ultimate use of the bomb, and while it has often been said that we undertook development of this terrible weapon so that Hitler would not get it first, the fact remains that the original decision to make the project an all-out effort was based upon using it to end the war . . . The first serious mention of the possibility that the atomic bomb might not be used came after V-E Day, when Under Secretary of War [Robert P.] Patterson asked me whether the surrender of Europe might not alter our plans for dropping the bomb on Japan. I said I could see no reason why the decision taken by President Roosevelt when he approved the tremendous effort involved in the Manhattan Project should be changed for that reaon.[14]

Is it simply naïve to think that the Manhattan Project could have been abrogated so far in, with all the effort and money that had been spent, so much of it without Congressional approval? Vannevar Bush, much senior to Hawkins, regarded the prospect of America's unilateral possession of the bomb as a way to 'guarantee peace in the world'. It was one of the underlying reasons behind the falling out between the Allies: Bush, Groves, Stimson and even FDR for a while thought America could and should have a nuclear monopoly.

Was a specific decision ever made? Later in the war, Stimson insisted that it was always President Roosevelt's intention to use the bomb when it became available: 'on no other ground could the war-time expenditure of so much time and money be justified'.[15] Can it be said that the atomic bombs were dropped on Japan partly for book-keeping reasons? It is worth repeating that *The New World*, volume one of the history of the United States Atomic Energy Commission, written under the auspices of the Commission's Historical Advisory Committee, whose members included Arthur Compton and Glenn Seaborg, both Nobel Prize-winners, devoted only two sentences in its 766 pages to the change of targets.[16]

Was it blindingly obvious to those in positions of power that, once a bomb became a certainty in a technical sense, it was going to be built anyway, because if one 'Ally' could do it, the other could too? The quote of John Moore-Brabazon, mentioned earlier, about the 'unprecedented political clout' an atomic weapon would give the *solitary* nation that possessed it, shows the lure and the danger of such thinking.

Against this background, and thinking, is the following chronology suggestive in an important way?

June 1942: Rosbaud learns crucial intelligence about Speer's abrogation of the German bomb; Roosevelt gives the Manhattan Project the green light.

23 August, 1942–2 February 1943: The Battle of Stalingrad, won by Russia, which now emerges as a world power.

13 January 1943 (with the end of Stalingrad in sight): The US aborts co-operation with Britain and Canada.

5 May 1943: The Military Policy Committee selects Japan as the target for the first atomic bomb.

7 May 1943: J. Edgar Hoover informs President Roosevelt about the 'hard evidence' that Russia has infiltrated the Manhattan Project, which had been known about for some time.

Or did the world drift into the nuclear age without any real understanding of the physics by the politicians concerned and without real thought of where events were leading? Those meagre two sentences in *The New World* suggest that this official document avoids an awkward discussion. Chadwick was host at the dinner where Groves surprised and disturbed Rotblat with his remarks about Russia. By then, as we have seen, in his heart of hearts Chadwick knew that Germany didn't have a bomb. Having visited the vast expanse of Oak

Ridge he knew that any equivalent German project would have been spotted by the RAF's aerial reconnaissance.

We can never get away from the fact that the British knew, before anything at all had been spent on the Manhattan Project, in the summer of 1942, that the Germans did not have a bomb that would play any role in the war. That does not mean they would never have developed the requisite technology, of course, but – if the Allies did win the war – then German science could have been constrained for the foreseeable future, as indeed it has been.

Nor can we get away from the fact that the MPC, of which Groves was the dominant member, selected Japan as the target on 5 May 1943. This was fully four months before Alsos was created to follow the Allied invasion of Italy, ostensibly looking for a German bomb. Moreover, as we have seen, when Chadwick visited the Pentagon in September 1943 he was told by his hosts that they suspected 'Germany is well advanced and estimate they are nearer realisation of a weapon, in some form or other, than the US' (p. 162).

Something here doesn't add up. If the US Army top brass thought the Germans were ahead in September 1943, why did Groves feel able to target Japan four months *earlier*, in May of that year, when the whole point of the bomb was conceived as a deterrent to Hitler? Even before that, in January 1943, Bush and Conant were willing to sideline the British, even though they (said they) believed Germany was then as much as a year ahead and knew that turning their backs on the British might slow down the development of the American bomb by six months. 'But Conant and the others took the view that though a delay would occur it would not be fatal or a serious hindrance. The postwar monopoly was a – if not the – major consideration.' For Samuel Goudsmit, on the other hand, the thought of German superiority drove some people 'almost to panic' and they sent their families into the countryside.

None of this adds up and, given Groves's stated policy of committing as little as possible to paper, the suspicion surely exists that Groves, Bush and Conant, at the very least, knew far more about the lack of progress by the German scientists than they ever let on. Nothing else can explain the curious chronology of events during

1943. Only if they knew for certain that Germany had no viable bomb programme could they have taken the decisions that they did take. As we have seen, American intelligence was wildly inaccurate in its assessments of German progress, forcibly exaggerating their capabilities. None of this squares with the order of events in 1943. The official British history of wartime intelligence says that the British did not pass on any information of real value to the Americans until the end of 1943. That is hard to swallow.

So far as the other enemy, Japan, was concerned, Groves admitted that the Allies were not worried that she could develop a bomb. That left Russia.

But Groves, as we shall see in Part Four, never seems to have been as worried by Russia as he might have been. So it is pertinent at this point to ask: if the Manhattan Project had not gone ahead, would the Russians have mounted their own nuclear effort? Would Chadwick's argument – that some nation like the Russians would build a bomb and use it – come about? More specifically, would Russia have been the threat that Groves, Churchill and others were convinced of?

16

The Claws of the Bear

If the purpose of the bomb was changing, and Russia was now the main object of attention, what the Soviets actually knew about atomic physics was clearly important.

During his time in Britain, Bohr naturally spent much time with Chadwick, as the most senior physicist attached to the project, and someone he had known since their time at the Cavendish Laboratory in the 1920s. And they certainly discussed the progress of Russian physics. We can be sure of this because, in February 1941, Chadwick had approached Francis Simon, at Oxford's Clarendon Laboratory, asking him to send a particular article in a Russian journal, but one published in German. Chadwick had studied in Berlin and had been interned there throughout the First World War, so spoke good German. Simon copied the journal for Chadwick. It was a paper entitled, '*Uber die spontane teilung von Uran*' ('On the spontaneous division/splitting by uranium'), by Konstantin Petrzhak and Georgii Flerov, from the Institute for Radium Research of the Academy of Sciences of the USSR, the Physicaltechnical Institute, Leningrad.

We don't know what provoked Chadwick to consult this paper, which was published on 14 June 1940, or even to know about it, though the Tube Alloys Directorate did try to keep up with foreign publications on nuclear matters throughout the war. But what is especially interesting is that Bohr was also familiar with this article because a copy of it had been sent to him in Denmark by the man who was probably the most influential physicist in Russia. Because of the importance of this coincidence, we now need to step back in time.

On 5 December 1940, Abram Fedorovich Ioffe, a 60-year-old Ukrainian who had been director of the Leningrad Physicaltechnical Institute since 1923, wrote to Bohr, enclosing the same Petrzhak and Flerov paper. Ioffe was the doyen of Russian physicists, the individual who had done more than anyone else to push the pursuit of nuclear research in the Soviet Union. In his letter to Bohr, however, he wrote that 'Unfortunately, we have almost no news about scientific results outside our own country.'[1]

Western scientists were in general ill-informed about Russian physics – and Russian science in general – because their results were published mainly in Russian journals with limited availability in the West and, of course, were written in the Russian language, though there was at least one Russian journal published in German. But Russian physicists, as Bohr knew, took pride in keeping up with their Western colleagues, and did read Western journals assiduously, even though it sometimes took weeks for them to arrive.

In his letter to Bohr, Ioffe seems to have been hinting that he, along with colleagues, had noticed that Western scientists were publishing less and less, so far as fission was concerned. Behind his comment was the unspoken but pregnant thought that there were political and military reasons for this silence. Bohr was aware of this.

In his reply to Ioffe, which was dated 23 December 1940, Bohr didn't remark on this silence (Denmark was occupied, his letter might be read by censors). But Bohr did write that 'it is exceedingly interesting that the experiments of Petrzhak and Flerov actually seem to confirm our expectations [Bohr and Wheeler's paper, p. 63 above] ... It is most desirable that these important experiments will soon be further extended.' Without going into details – for the moment – this exchange confirms Bohr's place at the centre of world physics and his crucial understanding of that community.

A brief account of Russian physics up to this exchange of letters at the end of 1940 and up to Bohr's arrival in Los Alamos will tell us what Bohr knew and how it shaped his views. This has only really become generally available since the Russian archives were opened up following the fall of the Soviet Union in 1989 and are thanks mainly to the imaginative scholarship of David Holloway, director of the Center for International Security and Arms Control at Stanford University. It confirms both Bohr's view about the high quality of Russian physics prior to the Second World War, and its integration with the subject elsewhere.

Ioffe was Jewish. He was born in the small Ukrainian town of Romny in 1880 and attended St Petersburg Technological Institute, from which he graduated in 1902. On graduating, he went to Munich to work with Wilhelm Röntgen, the discoverer of X-rays and where he received his doctorate in 1905 for a study of conductivity in crystals.[2] As a Jew he found advancement difficult at first but began to make his mark after he formed a friendship with the Viennese physicist Paul Ehrenfest (also Jewish), who lived in St Petersburg from 1907 to 1912 and was largely responsible for bringing modern theoretical physics to Russia. Ioffe became a professor at the Physicaltechnical Institute in 1913 and two years later the Academy of Sciences awarded him a prize for a study of the magnetic field of cathode rays.[3]

Throughout, he maintained his international links, visiting Germany every year until the First World War. As early as 1916 he had organised an influential seminar on the new physics, which was attended by well-known Russian physicists, including Pyotr Kapitsa and Nikolai Semenov, future Nobel Prize-winners. In 1921 he set off on a six-month visit to Western Europe, buying scientific journals, equipment and books, spending most of his time in Germany and Britain. For part of his trip to Britain he was accompanied by Kapitsa, and took him to Cambridge where he persuaded Ernest Rutherford to take him into his laboratory. In 1926 Ioffe was offered a professorship at Berkeley but preferred to stay at home.

It is Ioffe, Holloway says, who deserves most credit for the continued growth of Soviet physics – his institute was known variously as the 'cradle', the 'maternal nest' and the 'forge'. Until the clampdown of 1933, he still spent part of each year travelling.

Other physicists were affected by the clampdown no less than Ioffe, in particular Kapitsa. Kapitsa had enjoyed his time in Cambridge enormously, becoming a fellow of Trinity College, a fellow of the Royal Society and James Chadwick's best man. He was in Cambridge when the exciting discovery of the neutron was made by Chadwick. But in 1934, at the end of his regular annual summer return to Russia, where he attended the hundredth-anniversary celebrations of the birth of Mendeleev, he was prevented from leaving. All efforts by Rutherford, Bohr and Paul Rosbaud to intervene on his behalf failed. Eventually a special institute was created for him, and his laboratory in Cambridge was transferred – lock, stock and barrel – to Moscow. His place at the Cavendish was taken by John Cockcroft.

One other influential figure to be added to this group was Vladimir Vernadskii, a mineralogist intensely interested in radioactivity. In a lecture to the General Assembly of the Academy of Sciences as early as 1910 he had argued that steam and electricity had changed the structure of human societies. 'And now in the phenomena of radioactivity new sources of atomic energy are opening up before us.'[4] With the help of Vitali Khlopin he set up the Radium Institute in 1922, working towards the realisation of atomic energy.

In 1933, again on Ioffe's initiative, an international conference on nuclear research was held in Moscow, attended by some familiar names: Frédéric Joliot-Curie, Paul Dirac, Franco Rasetti – a colleague of Enrico Fermi – and Victor Weisskopf, then Wolfgang Pauli's assistant in Zurich, and who got to know Vernadskii well. The conference helped to stimulate nuclear research in Russia and among the young Soviet physicists who attended were several who were to play leading roles later in the Russian bomb project, including Iulii Khariton, Lev Artsimovich and Aleksandr Leipunskii, who had studied in Cambridge.[5]

Above all there was Igor Kurchatov who may have been influenced by his friend V. D. Sinel'nikov, who had also studied for two years in Cambridge and married an Englishwoman he met there. Kurchatov, Holloway says, was known as 'the general', because 'he liked to take the initiative and to issue commands'. Like Rutherford, he was also known for his swearing at experiments. He and his colleagues published seventeen papers on artificial radioactivity, expanding our understanding of what the Joliot-Curies had discovered.[6]

In the late 1930s Kurchatov ran a neutron seminar which, without his knowing it, would provide preparation for a bomb when war came. At that time, Russian physicists found it increasingly hard to travel abroad but several foreign scientists visited the various institutes, including Bohr, John Cockcroft, Paul Dirac, Victor Weisskopf (who spent eight months in Russia at one time), Friedrich Houtermans and Fritz Lange, both Germans. The Ukrainian Physicotechnical Institute published a journal of Soviet physics in German. In his book on the great purge, *The Accused*, Alexander Weissberg wrote that, until 1935, the Ukrainian institute was an 'oasis of freedom in the desert of Stalinist despotism'.[7]

Many in the institute were arrested during the purge and, obviously enough, the clampdown had some effect on the amount of physics that took place. But it did not seem to affect the quality. Rudolf Peierls knew the Soviet physics community well in the 1930s (remember he had a Russian wife whom he had met at a conference of physicists in Odessa) and said that when it came to the choice of research problems, 'he did not have the impression in those days that in the way science operated there was any real difference between the Soviet Union and other countries'. Victor Weisskopf also had a high regard for Soviet physics: they 'did not lag behind in their understanding of nuclear structure'.[8]

Moreover, when physics was transformed again at the end of 1938, with the discovery of fission, one of the first papers produced in response to the breakthrough was brought out by Iakov Frenkel, a theoretical physicist at Ioffe's institute. He reported his findings to Kurchatov's nuclear seminar on 10 April 1939, and the paper was published soon after in a Soviet journal. Partly as a result, Kurchatov directed Lev Rusinov and Georgii Flerov to investigate whether Bohr was right about slow neutrons and U-235. They concluded he *was* right and published their results – again in a Russian journal – on 16 June 1939. As Holloway points out, in the wake of fission Russian physicists were asking the same questions as their Western colleagues, and were closely attuned to what was being done in the West. 'But their research made little impact outside the Soviet Union.'[9]

Except on Bohr. He, as we have seen, had been sent news from Ioffe of the work by Flerov and Petrzhak. This had arisen from the most important Soviet theoretical work being done in the wake of fission, by Iulii Khariton (who had a Cambridge PhD) and Iakov Zel'dovich, at the Institute of Chemical Physics in Leningrad, under Kurchatov's supervision, on the conditions under which a chain reaction could take place. Khariton and Zel'dovich reworked experiments done by Szilard, Joliot-Curie, Fermi and others, their aim being to observe how the flow of neutrons from a sphere of uranium changed when neutron sources with different energy spectra were placed inside it. Petrzhak and Flerov devised a highly sensitive ionisation chamber to register the fission episodes. They began their study in early 1940 and found, much to their surprise, that the ionisation chamber continued to click – that is, it registered fission – even *after* they had removed their neutron source. They soon realised that what they were observing was *spontaneous* fission – 'fission without bombardment by neutrons'. This had been predicted theoretically by both Frenkel and by Bohr and Wheeler but the Russians had provided the first experimental proof of the phenomenon. In order to eliminate any other source of neutron bombardment, Kurchatov insisted that Flerov and Petrzhak repeat the experiment underground (actually in the Dinamo metro station in Moscow) so they could be certain the fission was not being caused by cosmic rays. (A parallel discovery was made in Liverpool and tested in the same manner, by repeating the experiment underground.)[10]

Ioffe's letter to Bohr was sent in the crucial window of time between the Ribbentrop–Molotov Non-Aggression Pact of 23 August 1939, and the Nazi invasion of the Soviet Union on 22 June 1941. Bohr was therefore well aware of all the main developments – and capabilities – of Soviet physics to that point. And, to an extent, so were other Western physicists, because Kurchatov, conscious of the importance of the results, sent a short cable to the American journal *Physical Review*, which published it on 1 July 1940. The results received widespread attention in the Soviet Union where, says Holloway, '[They were] taken to show that Kurchatov and his colleagues were now working on the same level as the leading researchers in the West.'[11]

Bohr's (and Peierls and Weisskopf's) respect for Russian physics was not misplaced but of course it was not easy or practicable for Bohr – or anyone else in the Allied countries – to follow closely what was happening in Russian physics after the Nazi invasion of the Soviet Union in June 1941, not least because many scientists were taken off the problems they were investigating and tasked with more immediate weapons-related work designed to repel the German advances. But even before that the Soviet Pact with Germany discouraged Western physicists from contacting their Russian counterparts. Here too we can see that Bohr, partly thanks to his eminence and partly to his nationality, was different, and in a privileged position.

Despite their scientific excellence, the pattern of behaviour of Russian physicists was different from their counterparts in Britain, the United States or Germany in one crucial respect. As Holloway tells the story, there is no evidence that Russian physicists tried to alert their government to the possibly belligerent implications of nuclear fission. They continued to publish freely and no effort was made, either by the government or by the scientists themselves, to restrict the release of information. One reason for this, as Holloway also observes, was that expressions of alarm at the prospect of a German bomb would not have been 'compatible' with the Nazi–Soviet Pact of August 1939 or the Friendship Pact that followed it in September.[12]

In fact, no action was taken until May 1940 and only then, it seems, because of an article in the *New York Times*. This was the story, already described, by William Laurence, which ran on the front page on Sunday 5 May 1940, headlined: 'VAST POWER SOURCE IN ATOMIC ENERGY OPENED BY SCIENCE' (see p. 82). As we have already seen, in America there was no official reaction. But there *was* a reaction of another kind, although the author of the article, William Laurence, never knew about it.

One of the teachers on the history faculty of Yale University, just up the coast from New York, at New Haven, Connecticut, was George Vernadsky – none other than the son of Vladimir Vernadskii, the man who had put radioactivity on the map in Russia. He immediately sent his father the *New York Times* article.

Vernadskii was then staying at a sanatorium in Uzkoe, the southwest region of Moscow, together with Khlopin. They both found the *New York Times* report very exciting and immediately began to consider whether the Soviet Union had enough uranium deposits to make nuclear energy a viable proposition. They wrote to the Geological and Geographical Section of the Academy of Sciences, who quickly set up a 'troika', or a Uranium Commission, to consider the problem – Vernadskii, Khlopin and Aleksandr Fersman, from Murmansk. Not long after, in July, Vernadskii and Khlopin wrote to Nikolai Bulganin, deputy premier and chairman of the government's Council on the Chemical and Metallurgical Industries, drawing his attention to the discovery of fission and the great quantities of energy that could be released, and making the point that both the Americans and the Germans appeared to be working on a bomb. This was the first attempt by Soviet scientists to alert their government.

Throughout the second half of 1940 there was growing awareness among Russian physicists of the possible military significance of nuclear fission, and the scope of research widened. At a meeting of the Uranium Commission on 17 May 1941, the sweep of research increased further, ranging from fluorescent tests for detecting uranium to methods of isotope separation, and chain reaction calculations.

And then disaster struck. In the following month, on 22 June, Hitler invaded Russia without warning. Stalin was taken totally by surprise. By November that year Germany had control of territory where 45 per cent of the Russian people lived, and where 60 per cent of its coal, iron and steel were produced.

Scientists, no less than anyone else, were mobilised. Some were drafted, institutes were evacuated from Moscow and Leningrad and reorganised to work on matters more closely related to the hostilities – radar, armour, rocket artillery, the demagnetising of ships. Kurchatov dropped his work on fission and his laboratory was disbanded. He himself worked on demagnetising. The Uranium Commission ceased to function. Vernadskii was evacuated to Borovoe, a lakeside resort in Kazakhstan, 2,500 miles to the east. In the middle of July he recorded in his diary the fear that Germany might use gas or 'uranium energy' in the fighting but in a letter to his son in New York, written on the day before he was evacuated, he wrote, 'I am profoundly glad that we are now indissolubly linked with the Anglo-Saxon democracies. It is precisely here that our historic place is.'[13]

He was not alone in this feeling. On 12 October 1941 Pyotr Kapitsa and several other well-known Soviet physicists issued what David Burke has called 'a veiled appeal for Soviet inclusion in the atomic bomb programme as a full member of the Grand Alliance'. A day later, the London *Daily Telegraph* published a sympathetic article under the headline 'SOVIET CALL TO WORLD SCIENCE', which reported on an appeal issued by '20 world-famous Russian scientists', and called on 'scientists of the world to concentrate on discovering new methods of warfare to be used against Germany'. Quoting Kapitsa personally, the *Telegraph* article discussed the possibility of the future development of an atomic bomb by the Allies, although it was sceptical that such a weapon could be ready for use in the current war:

> A hint of a new explosive was given by Prof. Kapitsa, who for many years worked on atomic problems with Lord Rutherford at Cambridge ... 'In principle we can foresee another increase of the demolition power of explosives by about 100 per cent ... New possibilities are also opening ... Thus, for instance, the use of subatomic energy. I believe, however, that the chief difficulties in the use of this energy are today still so great that the probability of an atomic bomb being used in this war, unless it lasts a very long time, is small. It is worthwhile mentioning, however, that theoretical calculations prove that if the modern powerful bomb that can destroy today a whole district is replaced by an atomic bomb, it could really destroy a large capital with a population of several millions.'

The crucial elements here are, first, the very fact of the appeal by Russian scientists, hinting that they knew something of what was going on among their Allies and, second, the mention of 'theoretical calculations', which clearly indicate that they themselves had been working towards a weapon, if only in theory.[14]

Added to what we already know – including the various intelligence leaks – the *Telegraph* article offered an opportunity for Britain and the United States to share what they had. It was a warning that went unheeded and contributed to the serious *underestimate* of Russian capabilities later on.

Vernadskii and Kapitsa were senior scientists. Their more junior colleague, Georgii Flerov, then twenty-eight, born in Rostov-on-Don, in the south of Russia, but a graduate of Ioffe's institute, had joined up at the outbreak of war and been sent to work on dive bombers at Ioshkar-Ola ('Red City'), 500 miles to the east of Moscow, to which the air force academy had been evacuated. He was much less sanguine than his more senior colleagues, was dogged by a sense of urgency they didn't share and he could not put nuclear research out of his mind. He contacted Ioffe and persuaded him to let him address a seminar in Kazan, 75 miles away, where, in mid-December 1941, he spoke to a number of physicists, including Ioffe himself and Kapitsa. Flerov told them that he had observed that scientists in the West had stopped publishing their work on fission and this could mean only one thing: that both sides were working on a bomb. His arguments had some effect on those present but Flerov returned to his air force base without, as he thought, having persuaded Ioffe and the others that nuclear research should be resumed. But he still wouldn't give up. He now wrote to Kurchatov a long letter covering thirteen pages of a school notebook.[15] He argued that a slow-neutron chain reaction in natural uranium was impossible but that a fast-neutron chain reaction could produce an explosion equivalent to 100,000 tons of TNT, and was therefore 'worth the time and effort'.[16]

Kurchatov didn't respond to Flerov's letter but the young man was even now not daunted. At the beginning of 1942 his unit was moved to Voronezh, 500 kilometres south of Moscow, where the university had been evacuated but the library was still there, and intact. Flerov used the opportunity to consult the American journals that, despite the war, had found their way to the library. He expected to find a reaction to his own work, with Petrzhak, on spontaneous fission (the work Ioffe had drawn to Bohr's attention). But Flerov found nothing. Not only that, he found no articles on nuclear fission – none at all (and remember that more than 100 articles had been published between Hahn and Strassmann's discovery of fission at Christmas 1938 and the outbreak of war in September 1939). Even more to the point, there were no articles by leading nuclear scientists on any other subjects either. They had all gone very, very quiet.

For Flerov this was a convincing confirmation of his argument and a major example, as he put it, of 'the dogs that did not bark'. It was clear to him that nuclear research in the United States and Britain was being carried out in secret. More worrying, such reasoning also applied to German scientists. Germany, he knew, had significant supplies of uranium ore, a heavy water plant that they controlled in Norway, and pre-eminent experts on separating isotopes.

Now on a crusade, Flerov wrote to Sergei Kaftanov, the State Defence Committee's 'plenipotentiary' for science. He drew the minister's attention to the absence of publications on fission in the international journals. 'This silence is not the result of an absence of research ... In a word the seal of silence has been imposed, and this is the best proof of the vigorous work that is now going on abroad.'[17] He floated the idea that the powers-that-be simply ask the Americans and British – now their Allies – what they were doing. The earlier Kapitsa appeal, in the *Daily Telegraph*, had produced nothing.

Despite the fact that he followed his letter with *five* telegrams, he received no reply from Kaftanov. A lesser man might have given up but, still undaunted, his next move was the only one left to make. In April 1942 he wrote to Stalin.

In his letter he called for a high-level meeting of all the leading nuclear physicists in the Soviet Union. He said he hoped that Stalin would himself attend, to gauge properly the threat that he, Flerov, was outlining. This was a risky manoeuvre, and in fact Stalin didn't reply to him either and no such meeting ever took place. But, as it turned out, the dating of his letter, April 1942, was all important.

On 21 September 1941, Anatoli Gorskii (codename Vadim), the NKVD *rezident* in London, transmitted to Moscow details about a secret meeting that had been held nine days earlier to discuss the MAUD report. Among the comments he made about the meeting, Gorskii said that the British thought it was possible that an atomic bomb could be produced in two years. Contracts had been issued for the production of uranium hexafluoride and the construction of a pilot factory 'for producing uranium bombs'.

This information, we now know, had come from one or more meetings of the Defence Services Panel of the Cabinet Scientific Advisory Committee held to discuss the MAUD report. Just over a week afterwards, Gorskii sent further information to Moscow Centre about the Scientific Advisory Committee's report to the War Cabinet. So, at this point, the Soviet government knew that Britain had decided to build an atomic bomb, that British scientists thought it would take between two and five years, and that the UK had decided to build a gaseous diffusion plant in North America.

The individual who made this information available to Vadim was John Cairncross, the 'Fifth Man' of the notorious 'Cambridge Five', who had been recruited as a Soviet agent by Guy Burgess while he was still an undergraduate at Cambridge in the 1930s. On graduation, in classic style, Cairncross had entered the Foreign Office (coming first in the Civil Service Exam), was subsequently transferred to the Treasury and in time became private secretary to Lord Hankey, minister without portfolio in the War Cabinet and chairman of the Cabinet Scientific Committee. Hankey was also chairman of the Defence Services Committee, giving Cairncross a ringside seat to watch Britain's atomic ambitions.[18]

This basic information about the MAUD Committee was not provided by Klaus Fuchs but, in fact, by then Fuchs had also begun to provide intelligence after the Nazi–Soviet Pact had collapsed with the German invasion of Russia on 22 June 1941.Before Fuchs went to the US at the end of 1943, he had some six meetings in Britain with his Soviet control, Ursula Kuczynski, codename Sonja, revealing details mainly of the gaseous diffusion aspect of isotope separation, though he also told her about a pilot plant in Anglesey, North Wales to test the process, and that the US and UK* were collaborating.[19]

David Holloway says that the information Gorskii supplied to Moscow (and by implication that of Fuchs) had no immediate effect. 'It arrived in Moscow less than a month before the great panic in the middle of October when most of the Soviet government was evacuated to Kuibyshev, 1,000 kilometres to the east, and tens of thousands fled the city. The British decision to build an atom bomb that would not be ready for two to five years, doubtless seemed a great deal less urgent than the effort to stop the Germans seizing Moscow in the next few weeks.'[20]

In fact it was not until March 1942 that the Soviet leadership responded to the information coming from Cairncross and Fuchs. An added factor was that, in February 1942, a German officer had been captured in a raid on the Ukrainian town of Taganrog, and he had in his possession an exercise book that, in the opinion of experts, contained notes indicating that work was progressing in Germany on an atomic weapon. In March, Beria, who was convinced by the German material as much as anything else, sent a memorandum to Stalin and the State Defence Committee urging that steps be taken

* In their book, *The Crown Jewels: The British Secrets at the Heart of the KGB Archives*, published in 1998, Nigel West and Oleg Tsarev say that there were two other atom spies in Britain, codenamed 'K' and Moor. K began to offer his services, using Vladimir Barkovsky, a young engineer who specialised in scientific matters in London, in December 1942 and, from what they say, much of the material he offered was similar to that offered by Fuchs. Furthermore, like Fuchs, K was an ideologically committed communist, scrupulous about money, who refused to be paid what he considered excessive sums for his services. K, they say, was 'something of an adventurer' who filched what was kept in the safes of colleagues. K provided enormous numbers of genuine documents in particular on the construction of uranium piles. But they do not say that K *was* Klaus Fuchs. Nor do they say what became of K or Moor later on. Nigel West told the author that he suspects that K was Engelbert Broda, an Austrian chemist and physicist, who worked on nuclear fission at the Cavendish Laboratory from 1941. His codename in the KGB files was 'Eric' and MI5 suspected it was he who recruited Alan Nunn May. West said he had no candidates for Moor.

to evaluate the information. The delay was partly due, as we have seen, to the German incursions into Russia, but also to the fact that Beria and others were suspicious that the intelligence being provided was in fact a feint, an attempt by the British and Americans to get the Soviets to spend time and money on a technology that could never work.

To address this, Beria proposed two courses of action. One was to create an authoritative committee, which could organise research. The other was to acquaint selected scientists with the intelligence that was being received. The memo by Beria showed that not only was he informed about British plans, but that in September 1941 the NKVD had begun to receive intelligence about American nuclear research, 'but not of the same quality'.[21]

The crucial element, as with so much in this story, lay in the timing. Beria's memorandum was written in March 1942, a month before Flerov's letter to Stalin. Flerov couldn't know it, after all the rebuffs, but he was now pushing at an open door.

There were still difficulties to be overcome, however. In fact, there now followed a bewildering time in which leading scientists were consulted about the feasibility of the bomb without being told about the intelligence reports that had been received. There must still have been a residue of doubt in the Soviet hierarchy as to whether or not the intelligence reports were genuine or a 'plant', though of course, in keeping the two approaches separate, it was a useful way for the Russian leadership to 'test' whether and to what extent they corroborated each other.

Furthermore, in 1942 Russia was still in mortal danger. The Red Army had succeeded in defending Moscow but then Stalin had launched a poorly coordinated offensive and the *Wehrmacht* had regained the initiative. In the summer the Germans pushed east towards Stalingrad, breaking through in August to the Volga. Russia was again threatened and on 28 July Stalin issued his notorious Order No. 227 – 'Not a step backward!'[22]

It was against this desperate background that the physicists were at last consulted about the atom bomb project. Khlopin, Kaftanov and Leipunskii were all brought in. The general reaction was that the Soviet Union should still not attempt to build a bomb. Such a project was too expensive and too risky, given the threatening situation the country faced.

But then attitudes had begun to change and this may have been due in part to Flerov's letter being sent so close to Beria's memorandum.[23] Whatever the reason, in September or October 1942, Molotov, the pince-nez-wearing old Bolshevik, decided to show the intelligence material to Mikhail Pervukhin, deputy premier and people's commissar of the Chemical Industry, and to ask his advice on what should be done with the reports. Pervukhin recommended that the material be shown to qualified physicists and *their* advice sought. Molotov was still not convinced.

Events moved tantalisingly slowly. With the senior Soviet authorities still holding back the intelligence reports, a meeting was arranged between Pervukhin and Kurchatov and his colleagues, but not until January 1943. At this meeting, Kurchatov confirmed that a chain reaction in U-235 could release enormous amounts of energy, that it was possible the Germans were trying to develop such a weapon, and that he and his colleagues were worried by the secrecy that had undoubtedly descended on nuclear research, both in Germany and elsewhere. Pervukhin asked for a memo on the subject, which was quickly prepared by Kurchatov and the others, and the deputy premier passed it to Molotov.[24] Molotov talked to both Kapitsa and Ioffe. Kapitsa said he thought the atom bomb was a weapon for the next war, so distant did its prospects appear, and Ioffe was also doubtful. Molotov needed more enthusiasm than they were able to offer and therefore he appointed Kurchatov, the youngest and least well-known physicist, to run the project, under the supervision of Pervukhin and Kaftanov.

Two things may be said about this decision. One is that when Kurchatov went to Moscow for his meeting with Pervukhin, the Battle of Stalingrad was entering its final phase. The Red Army was closing in on the Germans, who would surrender on 2 February. This was a major turning point in the fortunes of both sides in the war, when confidence in an Allied victory grew and, no less important, it marked the beginning of the Soviet Union's emergence as a world power.[25]

The second thing was that in February 1943 Kurchatov told Molotov that he was not sure that a bomb could be built or, if it could, how long it would take. And he added that 'there was a great deal that was unclear to him'. And that was when Molotov at last decided to show Kurchatov the intelligence materials that had been received. As Molotov said later, 'Kurchatov sat in my office in the Kremlin for several days studying those materials. Some time after the Battle of Stalingrad in 1943.'[26] This, one might say, was when the claws of the Russian bear began to sharpen.

Kurchatov was now wearing a beard and had been nicknamed *Boroda* by his Russian colleagues. He was following an ancient Roman practice, he said, and, although it made him look like an Orthodox priest, he vowed not to shave it off until 'Fritz' was beaten.

He was hugely impressed by what he read in the intelligence files. He compiled an extensive fourteen-page memorandum for his immediate superior, Pervukhin, which he wrote by hand, for reasons of secrecy, and he said the files had 'huge, inestimable significance for our state and science ... On one hand, the material has shown the seriousness and intensity of research in Britain on the uranium problem; on the other it has made it possible to obtain very important guidelines for our research, to bypass many very labour-intensive phases of working out the problem and to learn about new scientific and technical ways of solving it.'[27]

He said there were some overlaps between Soviet and British work but that the British had also broken new ground. The most interesting thing, he said, was the research that proved that a chain reaction was possible in a mixture of uranium and heavy water and he noted that the intelligence material indicated that an isotope of element 94, with a mass number of 239, might be used instead of U-235 in a bomb.[28] This suggests that information about the 1942 seminar, hinted at in Steve Nelson's bugged conversation, had indeed reached Moscow (p. 145).

Kurchatov said that the work appeared to have been carried out by major scientists and 'he had the impression' that it was genuine. This was important, he added, because, owing to the lack of a technical base, Soviet scientists were not in a position to check some of the British results.

Two other things came out of the intelligence reports. In the first place, the Soviets received – via Fuchs – details about German progress in nuclear research. Fuchs, as we have seen, had become involved in 1941 and 1942 in making assessments of German achievements. Before he left for the United States at the end of 1943, the details he had passed to Moscow had 'confirmed', in the words of Fuchs's KGB controller in London, 'that, first, the corresponding research in Hitler's Germany had reached a dead-end; second, that the USA and Britain were already building industrial facilities to make atomic bombs'.[29] This is a most important insight for our story.

The second thing to come out of the intelligence material was that it persuaded Kurchatov that the plutonium path to the atomic bomb was the most promising one. The material directed him back to the most recent papers published on the subject in *Physical Review*, which gave him what he called 'a new direction'.[30] This new direction, it is now generally conceded, 'would accelerate the Soviet program by a full two years'.[31]

It also led him to a series of scientific questions to which he needed answers, answers that could only be provided from America, because, in the wake of Barbarossa (the code-name for the invasion of Russia)and Stalingrad, Russia's damaged cyclotrons would not be up and running again until the summer of 1944. Kurchatov therefore provided a list of the laboratories in America 'where the corresponding work might have been done' (the Radiation Lab at Berkeley stood at the top of the list). Pervukhin sent Kurchatov's list to Gaik Ovakimian, deputy chief of the foreign department of the NKVD Chief Directorate for State Security, who was instructed to pass on Kurchatov's questions to agents abroad.[32]

On the strength of all this, Molotov introduced Kurchatov to Stalin. They appear to have got on. Kurchatov 'received every kind of support', Molotov later said, 'and we began to be guided by him. He organised a group and it turned out well.'

Kurchatov's laboratory was known as Laboratory 2, for secrecy, though Richard Rhodes tells the story that, when walking from the Moscow subway station to the lab one morning in 1944, the physicist Anatoli Alexandrov lost his way and stopped to ask a 'gang of neighbourhood children' for directions. 'It's over the fence where they're making the atom bomb', one of the children told him.[33] The lab was authorised – by the State Defence Committee, via the Academy of Sciences – on 12 April 1943.

Kurchatov had no illusions about how long the research and development would take and, indeed, it was not until the end of September 1944 that the cyclotron in Leningrad produced a deuteron beam.* An expedition to Central Asia was organised to look for uranium deposits. The metal was found but its mining had a low priority and Kurchatov calculated that it would take almost five years to amass the amount he needed for his pile. This was when the Soviet government sent a request to the Lend-Lease Administration in Washington DC for 10 kilograms of uranium metal and 100 kilograms of uranium oxide and uranium nitrate. General Groves approved the request, as we have seen, fearing that a refusal would alert the Russians to the American project or 'excite curiosity in Washington'.[34] (Delivery was not made until months later, however, and Groves made sure the metal was of poor quality. He also entertained hopes that if he could somehow track the shipment to Russia he might get some idea of Russian progress towards a bomb. That didn't happen.)[35]

But intelligence continued to come in. We now know that, at the end of 1942, Pyotr Ivanov, an official at the Soviet consulate in San Francisco, contacted George Eltenton, a British engineer who had worked at the Institute of Chemical Physics in Leningrad and was now living in the Bay Area, to ask if he could obtain information about the work of the Radiation Laboratory in Berkeley (which was top of Kurchatov's list). Eltenton in turn sought help from Haakon Chevalier, who was a close friend of J. Robert Oppenheimer. Chevalier got nowhere with Oppenheimer but Ivanov didn't give up.

* A deuteron is the nucleus of deuterium.

Over the course of the next few months several suspect scientists in the radiation laboratory were observed, or suspected, of passing information to the Soviet consulate. No charges were ever brought, since that would have drawn attention to the whole matter, so they were quietly relocated by the army to places where they could do no damage. Much the same was true of the Metallurgical Laboratory in Chicago and the isotope separation plant at Oak Ridge where, again, suspected offenders were quietly dismissed or relocated.

But the Americans didn't spot everyone, not by any means. The size of what they missed may be judged from the fact that, according to Holloway, Kurchatov compiled another memorandum for Pervukhin in July 1943 in which he reviewed a list of 286 reports on various topics: methods of isotope separation; the uranium-heavy water and uranium-graphite piles; transuranic elements; the chemistry of uranium. This information, Kurchatov said, was not of the same quality as the information received from Cairncross and Fuchs in Britain in 1941–2, except for the research on elements 93 and 94, in particular the work of Glenn Seaborg and Emilio Segrè at Berkeley on the fast-neutron fission of element 94.[36] We also know now that, later that year, in November 1944, the NKGB had amassed 1,167 documents on nuclear research, perhaps as many as 10,000 pages, of which eighty-eight documents came from the United States and seventy-nine from Britain and 'appeared to be of particular importance'.[37]

The Russians not only had inside knowledge, but they clearly understood that this was something the Allies wanted kept secret from them. That is, the Allies were not acting, in all respects, like Allies.

It is worth saying that neither the British nor the American scientists thought (or appear to have thought) to focus on Russian physicists, their whereabouts and their publishing activities the way they had done with their German counterparts. True, they didn't know Russia like they knew Germany, but Peierls and Weisskopf had spent time there, and were familiar with Russian physics and physicists (Peierls's wife, Genia, was a Russian physicist). Had they been more assiduous in, for example, consulting Russian journals, they would have discovered, at the least, Petrzhak and Flerov's spontaneous fission paper, which had been abstracted in *Physical Review*, Zel'dovich and Khariton's work (May 1940) on the chain reaction, the account of Kapitsa's speech to the rally of scientists in Moscow, published in a journal in 1941, where he discussed the power of atomic bombs – 'it is very probable that there are great possibilities'. And then publication stopped, as it had in the West, and British and American scientists would have been well placed to draw the correct conclusions. But it didn't happen. And this despite the fact that the *Physikalische Zeitschrift der Sowjetunion* published articles in German, French and English throughout the 1930s.[38] This imaginative lapse was both a cause and a symptom of the underestimate of Russian physics, which forms Part Four of this book.

Of course, by the time Bohr and Fuchs arrived in America, Russia, Britain and the United States were Allies, so there was not the same concern with their progress (or lack thereof) in nuclear matters as there was with German progress. Except, of course, that this was less than the whole picture. By then, Groves, Lindemann, Bush and Churchill at the least had identified Russia as the main long-term rival.

Still, Bohr saw further than the others. And, as he arrived first in Britain, and then America, and saw the progress that the Allies had made, and grasped fully the lack of progress the Germans had made, he quickly saw that Russian physics would then become a major element in the political calculations, going forward. The fact is: Bohr knew more than anyone in the West about Russian nuclear capabilities and, through Ioffe, he anticipated they had some idea of what was happening in the West. He was now ready to use his unique level of knowledge.

As for Fuchs, he knew now that Germany did not have a bomb. He had been one of those carrying out the research that revealed the true picture. Many of his colleagues in New York and then Los Alamos, especially the Americans, did not know what he knew, or hadn't taken on board the changed circumstances as he had – they were still working to beat the Germans to the punch. He realised that they were being misled and, as the purpose of the Manhattan Project began – in an unspoken way – to change, his original decision to help the Soviet Union was reinforced. He knew too that when it came to Russia, America and Britain were not acting, in all respects, like Allies.

How good were the Russian scientists and how much of a threat did they represent at the end of 1943? This brief survey shows that their nuclear physicists were of the first rank but, owing to the war, they lacked many technological capacities. We can also say that these would not have been complemented by Western intelligence, as they were, if the Allied project had itself not gone ahead, because that science wouldn't have been carried out in the first place, so that Fuchs and the other atom spies would not have been in a position to convey the secrets they did. And so ravaged by war was the Soviet Union that it could not have begun to mount a project of its own, starting from scratch, for years and years.

And by then, without the Nazi threat – once Hitler was defeated – would the appetite for a bomb have existed? Physicists everywhere were aware of the dangers of nuclear fission, but their main motivation for working on a bomb – to a man – was to prevent Hitler getting one first. So it is surely by no means true that James Chadwick's 'realpolitik' view – that once a bomb could be built, it would be built – would necessarily have prevailed. Groves himself admitted as much when he told General George Marshall, when he was secretary of state under President Truman: 'It was a project that would only be carried out in wartime.' (The Brookings Institution has calculated that the Manhattan Project cost $23 billion at 2017 values, and employed in all 130,000 people.) And we have known for years – since the summer of 1945, in fact – that the bomb was not needed, as Groves wrongly insisted, to end the war with Japan, as was discussed in the Preface.

The best we can say, then, is this:

1. There is a good chance that, had the crucial (1942) intelligence been shared, the subsequent science simply would not have been done, not then anyway. And, once the war was over, and the world divided into two blocs, as happened, but without nuclear weapons, we are entitled to ask whether the appetite for them, knowing the damage they could do, would have still existed? Most important of all, perhaps, would the scientists have agreed to take part? Groves for one clearly doubted it. This is a crucial point. As it was, at least nine prominent physicists refused to work on the bomb.* Had it been known more widely that Germany didn't have a bomb, it is quite possible that a Bohr-led movement to refrain from work on nuclear physics would have taken root.
2. If the chronology of events in 1943 reported here is accurate, then senior figures – military men, civil servants, science administrators, politicians – stand guilty of conniving in a conspiracy of silence, the true reasons for their actions concealed behind a supposed (but in fact non-existent) threat of a Nazi bomb, to secure for the Allies what they hoped would be a nuclear monopoly. Again, they could do this only by seriously misleading the scientists.

Given the nightmare scenario we are now forced to live with, this is a sobering realisation. And it adds to the drama that we are about to consider. The central point of this drama is, as the previous chapters have shown, that the coincident arrival of Bohr and Fuchs in America at the end of 1943 could not have been better timed. The balance of the world was itself in the balance. Both men knew it, and were determined to do something about it. One would fail, the other succeed.

* Peter Debye, Albert Einstein, Gerhard Herzberg, Fritz London, Lise Meitner, Erwin Schrödinger, Hertha Sponer, Max Born, Otto Stern.

PART THREE

Lives in Parallel: Niels Bohr and Klaus Fuchs

17

The Little Fox

On the afternoon of Saturday 5 February 1944, by which time he had been in America for exactly two months, and while Niels Bohr was still in Los Alamos, Klaus Fuchs went to the Henry Street Settlement on the Lower East Side of Manhattan, about three blocks from the East River. The 'settlement' was a housing and social services outlet for poor people. Fuchs was there to meet a secret contact that had been arranged in Britain the previous November.

Klaus Fuchs was a German who was born in Rüsselsheim near Frankfurt in 1911. A tall, thin, softly spoken studious type with owlish spectacles, he was the son of an outspoken Quaker minister, Emil, a pacifist who was dismissed from his position by the Nazis. Emil Fuchs identified Jesus with the poor and deprived, and conceived his Christian mission as a political struggle for social betterment.[1] This made him an out-and-out socialist and a brave and strident opponent of the Nazis. Despite this, Klaus wrote later that he had a very happy childhood and that what stood out during those years was that 'my father always did what he believed to be the right thing to do and he always told us we had to go our own way, even if he disagreed'.[2]

Klaus was raised during the social and political turbulence that followed Germany's defeat in the First World War, which brought about the downfall of the kaiser and his replacement by a republic.[3] For a nation confident of its own superiority, the turnaround was traumatic. Putsches and political assassinations ensued and then hyperinflation, which made money worthless.

While Klaus was still a boy, the family moved from Rüsselsheim to Eisenach, an industrial town of 50,000 in Thuringia to the east of the country. At the *gymnasium* there, Klaus showed himself an exceptional student, winning several prizes. 'He was unusually sure of himself and self-contained, despite the fact that the Fuchs children came in for some unpleasant remarks at school because of their father's unpopular political views. His brother was upset by this but not Klaus.' He was never troubled by the opinions of others.[4] After his *gymnasium* he began a course of mathematics and physics at the University of Leipzig.

When Fuchs was nineteen, he was hit by the first of several family tragedies. His mother, Else, had suffered regular bouts of depression and, one day in October 1931, Emil arrived home to find her slumped on the floor. She had swallowed hydrochloric acid, a distressingly painful way of committing suicide. Her last words were, 'Mother, I'm coming.' Only when he was arranging the funeral did Emil discover that Else's mother had also killed herself.

At Leipzig University, Fuchs joined the student wing of the Social Democratic Party, the SPD, and also the Reichsbanner, an SPD paramilitary outfit founded to fight the Nazis' SA, the Brownshirts. The move was also a reaction against his father's pacifist beliefs. 'Fuchs was slightly built, with thin arms and legs, and he wore glasses; he showed physical courage in joining the Reichsbanner, for they sometimes had to battle with the Brownshirts on the streets.'[5]

The Fuchs family moved a second time, to north Germany, to the 'rainy Baltic seaport' of Kiel. Emil was appointed professor of Theology at a teachers' training college, and Klaus enrolled in the local Kiel University, where he eventually became chairman of an organisation that had members from both the SPD and the Communist Party.[6] But he broke with the SPD over the 1932 presidential election. The Social Democrats supported the old president, General von Hindenburg, as opposed to Hitler. The communists preferred a united front with the socialists, against both Hindenburg and Hitler, and that made sense to Fuchs also. When the CP put up its own leader, Ernst Thaelmann, as a presidential candidate, Fuchs offered to campaign on his behalf and as a result was expelled from the SPD. Hindenburg won.[7]

Fuchs now joined the Communist Party, partly because by then he accepted the need for strict party discipline to manage the fight against the Nazis. His brother and two sisters also joined the Party at much the same time.

Fuchs was not only fearful of Nazism but also optimistic about communism. As Norman Moss, one of his biographers, says, 'It is important to say that in 1932 the most dramatic horrors of Soviet communism – the purges, the mass deportations – were still in the future. Russia was still regarded as "the great experiment"; one could have reservations about the regime and be critical of it, but there was no *prima facie* case for rejecting it out of hand.'[8]

The Depression hit Germany especially hard, and Nazis and their opponents fought pitched battles in the streets. At the elections held in 1932, more people voted for the Nazis than for anyone else, although Hitler still did not have an overall majority of seats.

Nonetheless, Hitler was appointed chancellor in January 1933, as head of a coalition cabinet that included members of other parties. For the time being, the democratic institutions remained intact. But Hitler manipulated his emergency powers to ban Communist Party meetings and called another election for March. Fighting in the streets got worse. Fuchs bravely stood up to the Brownshirts, as a result of which he was beaten and thrown into a river.[9]

'On the night of 27 February, a mentally unbalanced Dutch tramp with unknown accomplices set fire to the Reichstag, the parliamentary building in Berlin, and what was left of democracy in Germany was consumed in its flames.' Before the fire was even out, the Nazis were blaming the communists, and a reign of terror was launched against opposition parties. Four thousand communists were arrested over the next twenty-four hours.*

By chance, the following day Fuchs took an early train to Berlin. He had moved to the Friedrich-Wilhelms-Universität to read maths and physics but on that day he travelled to attend a meeting of

* Evidence was given at the Nuremberg war crimes trial that the Nazis themselves started the Reichstag fire, using the Dutch tramp as a stooge.

student communists and he read about the fire in a newspaper on the train. As he recalled later, he instantly removed the hammer and sickle from his lapel and hid it away in his pocket.

The meeting in Berlin was held in secret. In the course of it Fuchs was praised for his bravery in standing up to the Nazis and was urged to go abroad to complete his studies. One day, it was said, a post-Hitler Germany would need individuals of his quality. And so, although it had all happened in a rush, Klaus did not return to Kiel and his family. He was unable to go abroad immediately (the borders were being watched, as Leo Szilard had noted) so he hid himself in the apartment of a young female party member.[10]

From his hiding place, he could see how little opposition there was. This reinforced his conviction that liberal principles were not enough to withstand the thrust of the Nazis. From where he sat, 'only the Communist Party with its tight discipline could fight it effectively'.[11]

Klaus remained in hiding in Berlin for five months. Only in August did he escape to Paris. There he attended an anti-fascist conference chaired by the French writer, Henri Barbusse. He was twenty-one.

Outside Germany, life was very different. Many foreigners felt a revulsion towards the Nazis that went with a sympathy towards the victims. In Fuchs's case he had a cousin who was engaged to a girl working as an au pair with a wealthy English couple, Ronald and Jessie Gunn, in Clapton in Somerset. Fuchs wrote to her, laying out his circumstances. She turned the letter over to the Gunns and, being communist sympathisers, they generously invited him to stay with them.[12]

He crossed the Channel to Dover on 24 September 1933. After months on the road, he had lost weight and was pale and hungry. He told the immigration officer that he was going to live in a village near Bristol and on the spur of the moment said he was planning to study physics at Bristol University. But it was a vague hope only and he had no idea, then, where it would lead.[13]

Bristol could not match the resources of some of Britain's larger universities but it did have a sizeable and well-equipped physics department, thanks to a generous endowment from the local Wills family, heirs to the Imperial Tobacco Company. On top of that the new head of the physics department, the future Nobel Prize-winner, Professor (later Sir) Nevill Mott, at twenty-eight the youngest full professor in the country, had studied at Göttingen University under Max Born and in Copenhagen under Bohr, and spoke fluent German. He was also very firmly on the left, politically speaking.[14] Mrs Jessie Gunn was herself one of the Wills family, so she was well placed to introduce Klaus to Professor Mott.

They hit it off. Mott had a grand intellectual plan – to apply quantum mechanics to solid materials, so as to explain certain properties, such as the hardness of metals and their behaviour as semiconductors.

Fuchs, says his biographer, was now a changed personality. 'He was in a strange country, cut off from all social ties, he hardly spoke the language. He developed caution, the caution of the exile, who does not know how his behavior or opinions will be received by the strangers about him. He became reserved, withdrawn, even cold, spoke very little and kept his thoughts to himself.'[15] His sympathies were communist but in certain quarters, as he well knew now, those sympathies were dangerous. His political interests became more cerebral as he looked beyond Germany. He studied Marxism and helped refugees from the Spanish Civil War, which allied his anti-fascist sentiments with his communist sympathies.[16]

His physics was a cerebral activity too. He was not an experimental scientist but a theoretical physicist. Quantum mechanics is in some ways hardly different from pure mathematics. For a time he was a colleague of Bernard Lovell, later to become Sir Bernard Lovell, one of the founders of the new science of radio-astronomy and president of the Royal Astronomical Society. Like Ernest Rutherford, Lovell was a robust, cheery individual, and he did not take to his weedy-looking, bottled-up collaborator. 'He seems like a chap who's never breathed any fresh air,' he told others.

Mott did persuade Fuchs at one stage to attend a few meetings of the Society for Cultural Relations with the Soviet Union. Some of these gatherings explored the Moscow treason trials then catching the eye, bizarre affairs where the original leaders of the 1917 Revolution were attacked and represented as mere puppets, forced to confess pathetically to absurd charges that they were saboteurs acting secretly on behalf of Western imperialists and Nazis.

At the Bristol meetings they sometimes acted out specific episodes from the trials, and Fuchs joined in at least one of these charades, taking the role of Andrei Vyshinsky, the state prosecutor and, Mott said later, he was astonished to see a very different Fuchs. All of a sudden he was a passionate figure, fiercely eloquent in acting out Vyshinsky's denunciations of the saboteurs. 'For a few moments, and wearing another's identity, he could reveal his deeply held beliefs.'[17]

To an extent, Klaus kept up with his family, exchanging letters with his father until war made that impossible. His father was arrested in 1933 for speaking out against the Nazis and was jailed for a month. Gerhardt, Klaus's elder brother, went to Switzerland where he was admitted to a sanatorium to be treated for TB. Fuchs travelled to see him, his only trip abroad before the war. His sister Elizabeth opened a car hire business in Berlin with a friend, Kitowski, who was also a communist and whom she married. (They had a son together, also called Klaus.) The couple used their cars to smuggle communists abroad. Kitowski was eventually caught and jailed for six years in Brandenburg. But he escaped and hid in Prague, surviving the war.

In August 1939, Emil took Elizabeth with him to a Quaker conference in Bad Pyrmont, a popular spa in Lower Saxony, a journey of several hours by train. On the way home she threw herself from the carriage, leaving Emil to raise little Klaus by himself. Klaus's younger sister Kristel was awarded her teacher's diploma and in 1936 emigrated to America where she enrolled at Swarthmore College. What was left of the family did their best to keep in touch but, one way or another, the Fuchses had suffered terrible blows.

Despite the tribulations that his family had suffered, and however much he was enjoying his work with Nevill Mott, Fuchs was still a political animal. He made contact with the German Communist Party that had formed among exiles in Britain. In particular, he contacted Jürgen Kuczynski, a German communist of Polish parentage, who had arrived in Britain in 1936 and been instrumental in putting German communist refugees together.

Fuchs's biographer says that Klaus probably did not know that Kuczynski was an agent of GRU, the overseas intelligence branch of the Soviet Army. He had been recruited during a visit to Moscow, and his sister Ruth was also an agent.[18]

In 1934 Fuchs wrote to the German consulate in Bristol, asking for his passport to be renewed. This request was peremptorily refused, the local consul insisting he was entitled only to a temporary (one-way) travel document allowing him to return to Germany. The consulate also took the opportunity to tell the British police that Fuchs was a communist. Some months later, Fuchs received a demand from the consulate to report for military service, which was now compulsory in Germany. He did nothing.

His professional advancement continued, however, and after four years he earned his PhD. By then he spoke good English, albeit with a German accent that he was never to lose. A paper of Fuchs's on quantum mechanics was published in the *Proceedings of the Royal Society* in February 1936. He was proud of his achievement and wrote to his father, giving him the news. Among his colleagues was Hans Bethe, a young visiting fellow from Germany, who was to become his boss later on at Los Alamos, and would win the Nobel Prize in 1967.

By this time, however, the German-speaking Mott was employing no fewer than six German refugees and he was all too well aware that he could get tenure for just three (eventually eight Germans at Bristol were interned after war broke out). Fuchs had completed his research project, so Mott reluctantly concluded that he should be one of those who moved on. He recommended Fuchs to Max Born, one of the greatest physicists of the 1930s German diaspora, who Mott had himself studied under at Göttingen and was now an émigré professor at Edinburgh University. (Born would win the Nobel Prize in 1954.) Born found Fuchs to his taste and Klaus moved to the Scottish capital as a research assistant.

Fuchs got on as well with Born as he had with Mott. The professor recalled Fuchs later as 'a very nice, quiet fellow with sad eyes'.[19]

Born was a German Jew with no interest in religion, and he was a pacifist. He had grown up in Breslau (now Wroclaw, in Poland), the son of an embryologist. He studied or taught at several German universities, in the German manner: Berlin, Heidelberg, Göttingen, Frankfurt, and for a short time at the Cavendish Laboratory in Cambridge. He worked closely with James Franck, and taught a number of PhD students who would become famous in their own right: Werner Heisenberg, Pascual Jordan, Max Delbruck, Siegfried Flügge, Robert Oppenheimer and Victor Weisskopf, while his assistants included Wolfgang Pauli, Edward Teller and Eugene Wigner. He also got to know or worked with Fritz Haber, Peter Debye and Einstein, who became a close friend – their correspondence stretching over fifty years would later be published in its entirety.

Born was therefore right at the centre of physics in the 1920s and 1930s but in 1933, when the Nazis came to power, he was forced to resign. He was showered with job offers and chose to go to Cambridge where, shortly afterwards, Charles Darwin, the grandson of the great naturalist and someone we have already met, asked him if he would like to follow him as professor at Edinburgh. Born leapt at the chance, since it would provide a stable base from where he could organise the safe escape of friends and family from Germany.

Fuchs's gifts as a mathematical physicist complemented Born's and together they explored several different areas of theoretical physics, jointly publishing two papers in the *Proceedings of the Royal Society*.[20] At Edinburgh Fuchs earned a DSc.

His biographer says he was a little less of a loner in Edinburgh than he had been in Bristol: he helped circulate some German Communist Party leaflets, but he had no close relationships with women. 'His principal achievement was learning to live without other people rather than learning to live with them.' Later, he told his Russian controller that he avoided 'falling in love deeply'.[21]

In August 1939, the Soviet Union signed a non-aggression pact with Nazi Germany, a bewildering shock to anyone with communist sympathies. Although Fuchs went so far as to defend Russia openly in an argument with Born, stressing that the pact was a pre-emptive manoeuvre in preparation for a wider war that Russia expected, in that same month he took the momentous psychological and political step of applying for British citizenship. His application was still in the administrative works, however, when, the following month, war broke out, and overnight he became an enemy alien. Everyone had to appear before a tribunal but in Fuchs's case Born reassured them that his assistant had been in the SPD in Germany between 1930 and 1932. Fuchs himself told the tribunal he had been a communist, using this as evidence that he was a sincere anti-Nazi. He was allocated a C classification, the lowest risk.

Fuchs was sent to an internment camp on the Isle of Man in 1940, together with thousands of others (he was not even given a chance to tell Born he couldn't come in to work) but his time there didn't last long.[22] On 3 July he was taken onto the liner *Ettrick* in Liverpool, destined for Quebec. Fuchs was sent to a camp at Sherbrooke, outside Quebec City, with spectacular views of the St Lawrence River Valley. Regarded as prisoners, they were treated meanly by the guards. Fuchs later said that, despite this, most of the refugees knew that their relatives at home were faring even worse. The food they ate was better than in Britain where rationing was in force. Added to which, musical entertainment was quickly installed and a camp university established. 'One inmate was elected a Fellow of Trinity College, Cambridge while there and another received his PhD.'[23] Fuchs gave lectures on physics. These were lucid and well received and he became known as '*Fuchslein*', 'Little Fox'.

In Sherbrooke Fuchs was more fortunate than most, in fact, because he was also in touch with his sister, Kristel, now living not far away in Cambridge, Massachusetts. She had a raft of communist friends, including a few in Canada, and they sent Fuchs reading material.

All this while, Born had not been inactive, pressing the Royal Society to agitate for Fuchs's release. The pressure worked. Six months after they arrived, on Christmas Day 1940, 287 inmates of the camp were freed and sent back to Britain. Fuchs was one of them.

Not too long after his return, in the spring of 1941, Fuchs received a letter from the Professor of Mathematical Physics at Birmingham University, Rudolf Peierls.[24] They knew each other slightly, having met in both Bristol and Edinburgh.

By then, Peierls and Frisch had made quite a bit of progress in their calculations regarding the feasibility of an atomic explosive, but it involved more work than they had bargained for and Peierls decided he needed help. He recalled reading some of Fuchs's published papers, which showed the man's mathematical skill, but also a flexibility of mind that appealed to Peierls because he and Frisch were entering unexplored territory.[25]

In the first place, Peierls approached the Ministry of Aircraft Production to see whether Fuchs was eligible for employment. The ministry referred his enquiry to MI5, the domestic counter-intelligence service. MI5 had a file on him with two items of interest. One was the 1934 report from the German consul in Bristol advising that he was a communist. This was hardly objective, given the circumstances. The other was more recent. It was an account from an anonymous informant in the German refugee community, again to the effect that Fuchs was a known communist.[26]

As a result, an official wrote to Peierls, saying he could employ Fuchs, 'provided he told him only what he needed to know for the problem he was working on' (a British form of 'compartmentalisation'). Peierls replied immediately, dismissing any such notion. Any assistant he hired, he insisted, would have to be taken fully into the picture. The ministry dropped the objection.

Peierls offered Fuchs the job. He didn't say what the nature of the work was – only that it was related to the war effort – but Born's biographer says that both Born and Fuchs knew it involved atomic work.[27]

Fuchs accepted and took the train south 'through blacked-out Britain'.

Because of the Blitz, housing was not easy to come by in Birmingham, and Rudolf and Genia Peierls tactfully invited Fuchs to share their large house in Culthorpe Road in Edgbaston. He took the room Otto Frisch had vacated only recently.

Rudolf and Genia Peierls were complete opposites. Rudolf was quiet, self-effacing, academic. Genia was outgoing, warmly exuberant and voluble, with a heavy Russian accent. Fuchs, says Norman Moss, was easy to live with. 'He gave his ration book to the Peierls and ate his meals with them'.[28] He helped with the dishes, was tidy in his room and quiet.

Genia's gregariousness, and despite the fact that her husband and she had to double up their daytime jobs as fireman and nurse, meant they often gave dinners or parties. Fuchs was always present at these events but rarely took a prominent part, so much so that Genia began to describe him as a 'penny-in-the-slot person'. As she explained: 'Put a question in and you get an answer out. But if you don't put anything in, you don't get anything out.'

Fuchs was in fact largely self-contained. This was true both emotionally and intellectually. And when scientific papers came in from America, Rudolf found that the Little Fox could display a certain arrogance: 'Let's see if the Americans are working along the right lines.'[29]

Now came a further irony. Fuchs was appointed assistant to Frisch and Peierls in the spring of 1941, when Britain had set up its highly secret committee of scientists and politicians to consider the Frisch-Peierls report (the MAUD Committee). Fuchs was part of this team.

It will be recalled that Hitler had signed his notorious non-aggression pact with Stalin in August 1939. Yet on 22 June 1941, not long after Fuchs joined the Birmingham team, the Fuhrer suddenly invaded Russia without warning, with millions of Nazi forces storming across the border. Overnight Britain and Russia became allies.

On the day that Germany invaded Russia, Winston Churchill, Britain's prime minister, went on the radio to proclaim that 'the Russian danger is therefore our danger'. Despite Churchill's own 'persistent opposition to communism', he said, he nonetheless offered Russia 'any technical or economic assistance that is in our power'.

By the autumn the British were sending the Soviets fighter planes, radar equipment, bombers, anti-aircraft guns, destroyers, ammunition and three million pairs of boots. It is doubtful if Vyacheslav Molotov, the Russian foreign minister, or Stafford Cripps, Britain's ambassador in the Soviet Union, thought that this exchange would include details about Britain's work on an atomic bomb (for one thing Cripps knew nothing about the atomic bomb: Churchill, as we have seen, kept the news secret even from his wartime cabinet). But that is exactly what Klaus Fuchs decided to do. He had barely signed the Official Secrets Act when events conspired to force his momentous decision.

Momentous historically but, as Norman Moss, one of his biographers, argues: 'It was probably less momentous for him than it would be for most people.' What Moss meant was that, although Fuchs had taken out British citizenship, he was still, at root, a German and the country he had grown up in was now changed out of all recognition. As a result, his ties were to 'the abstraction of communism; and because of this to a country that was, for him, hardly less abstract, the Soviet Union'.[30]

On a visit to London later in 1941, he looked up his old acquaintance, Jürgen Kuczynski, saying he had some information that he felt could be of value to the Soviet Union. Kuczynski quickly arranged for Fuchs to meet someone he only ever knew as Alexander. We now know that this was Simon Davidovitch Kremer, who was part of the staff attached to the Soviet embassy's military attaché.

At the time he made his decision Fuchs was working on nuclear fission and on uranium diffusion and he took carbon copies of his typed reports, involving complex mathematical calculations, written out by hand, to his first meeting with Kremer. This was held at a Soviet 'safe house' near Hyde Park.

Fuchs met Kremer three times in the next six months, and each time handed over copies of his reports, some typed, some handwritten. Peierls and Fuchs were working on two things.[31] One was the theoretical calculations on nuclear fission reactions, which were designed to show, crucially, how much uranium-235 would be needed for a bomb. The other was tackling how to isolate uranium-235 from uranium-238: the size of this problem was already becoming apparent.

Peierls was well satisfied with his new assistant.[32] Fuchs showed exactly the flexibility of mind Peierls had wanted. As we have seen, a 'Tube Alloys Directorate' was set up with a small office in Old Queen Street, London, and all the scientists working on the different aspects of the project were required to send in monthly reports. Fuchs was always prompt and his reports were always lucid – he wrote English easily by now.

The growing appreciation of what Fuchs was telling the Russians became apparent in the autumn of 1942, when Kremer told Fuchs that he was passing him on to another contact, a woman he would know only as Sonja. She lived in the village of Kidlington, near Oxford, which meant they would be able to meet in Banbury, about 35 miles from Birmingham. The advantages were plain. Travelling to London was not easy in wartime and in fact Fuchs had to fake a number of illnesses and pretend he was seeing a specialist doctor so Genia Peierls would not be suspicious about his movements. Sonja was in fact Ursula Kuczynski, Jürgen's sister, though Fuchs never knew that. She was a seasoned Soviet agent, having previously worked in China and Switzerland.

For the next eighteen months (through the rest of 1942 and all through 1943) they met in and around Banbury, and each time Fuchs would hand over papers. He also told her about the general theoretical background to the work he was involved in – that it was all part of a project to build an atomic fission bomb and that the theoretical work was being reinforced by the construction of a model diffusion plant. He added that similar work was going on in the United States and that the UK and the USA were collaborating.

Kurchatov found the material to be 'of immense value . . . its value cannot be overestimated . . . informing us of new scientific and technical approaches and enabling us to skip labour-intensive phases of development.' Moreover, the Russian files confirm that Fuchs also, for example, provided a complete description of Francis Simon's proposed gaseous diffusion unit. This, Kurchatov judged, was perhaps 'the most valuable' piece of information. The Anglo-American choice for gaseous diffusion as a means of separating U-235 from U-238 was unexpected; the Soviets until that point had opted for centrifuges.[33]

In 1942 Fuchs applied again for British citizenship. He needed two sponsors: his were Nevill Mott and the Tube Alloys Directorate, which said that, although an enemy alien, he was doing work that was valuable for the war effort. Fuchs became a British citizen on 7 August 1942 and took an oath of allegiance to the Crown.

When, out of the blue, Fuchs was asked by his Russian contacts whether he knew anything about the use of electromagnetism to separate uranium-235, it surprised him. It was something he knew nothing about. But he did know that the Tube Alloys group in Oxford had made a preliminary examination of the matter and that Ernest Lawrence was working on an electromagnetic separation project at his Berkeley Laboratory in California.[34] So that told Fuchs that he wasn't the only physicist passing information to Moscow and that there must be some leaks from America.

Towards the end of 1942 a handful of Tube Alloys scientists visited the US to see the work being done there. Peierls was among them and Fuchs knew that. We can be certain of this because crossing the Atlantic in wartime was by no means straightforward, and Peierls left instructions on how his work should be continued if he should be torpedoed. So this gave Fuchs an insight into his boss's thinking about the future direction of the project. On that visit the British team found that although they were still ahead of the Americans on the theoretical side, the Americans had made excellent progress across a wide range of experimental work. The balance of the effort was beginning to change, and Fuchs learned this through Peierls.[35]

Back in Birmingham, Rudolf and Genia Peierls – in their genial manner – gave a New Year's Eve party to usher in 1943. Alcohol was never rationed in Britain during the Second World War and that evening it was consumed in generous proportions. In particular, Genia Peierls had a lot to drink and, as she sometimes did on such occasions, she began to sing Russian ballads. As she sang she suddenly noticed that Fuchs was gazing at her 'with a look of such extraordinary intensity, such as she had never seen on him before, with an expression of what could have been adoration'. It occurred to her that Fuchs was falling in love with her and she made a mental note to discourage any such attachment. Later, when Fuchs's true colours were revealed, she realised that the look of adoration on his face that New Year's Eve was directed not at her but at the reminder of the Russia that lay in the songs that she was singing, 'a land that was the repository of his hopes for the world'.[36]

Later that year, in August, Churchill and Roosevelt signed a secret agreement on Anglo-American (and Canadian) collaboration on building an atomic bomb during their meeting in Quebec. Churchill, by all accounts, was on his most dynamic form, performing far into the small hours and leaving Roosevelt 'nearly dead'.[37] Among the give-and-take at the meeting, Churchill finally agreed that an Allied landing in France could take place the following year under US command (he had previously sought to bomb Germany into submission, fearing an invasion would be a strategic disaster). And at a private meeting Roosevelt confirmed that the two nations would once again collaborate on building the atomic bomb, after the unpleasant interval when they had fallen out, though Churchill had had to concede – to Cherwell's irritation – that the US would have control over Britain's post-war access to nuclear technology.

The biggest single task in the development of the bomb would be separating enough uranium-235, a subject where British scientists – Fuchs included – had done a great deal of work and had a lot to contribute. This is how the decision came about to send twenty or so scientists to New York to help a team working on uranium diffusion.

Before he left for America, the information Fuchs passed to Moscow had done enough to 'confirm', in the words of the KGB officer who was Fuchs's control in London after the war, 'that, first, the corresponding research in Hitler's Germany had reached a dead-end; second, that the USA and Britain were already building industrial facilities to make atomic bombs'.[38]

There are three things to say about this. First, how did Fuchs know that the German research had reached a dead-end? We don't know whether Fuchs had access to the Paul Rosbaud intelligence – about the meeting between Heisenberg and Speer – but it seems extremely unlikely. Had Fuchs known, then information about the meeting would surely have spread to the other scientists in the Manhattan Project. Instead, what happened with Fuchs was that, having been one of those who was first charged with keeping an eye out for what the German physicists were engaged in, he kept at it, knowing how valuable it would be not just for his British masters and colleagues, but also for the Russians as well. And of course he was also on excellent terms with Max Born, who was in touch with Lise Meitner in Stockholm. That was almost certainly one route for atomic information leaking out of Germany.

What we do know is that Fuchs and Peierls had examined the German literature in 1941 and concluded that all the most important physicists were in their normal places of work and so had probably not been put together on a major project. Surveys of German literature were carried out yet again in January 1942 and March 1943, and these too had shown not only that the physicists were still where they ought to be, but that more of them had begun publishing again after a gap, as R. V. Jones noted independently.[39]

Second, thanks to Fuchs, by the end of 1943 Stalin knew enough about the progress of research in other countries that he didn't have to worry that a bomb would be crucial to the outcome of the war. This was an item of strategic information of incalculable value to the Russians but it also changed fundamentally the diplomatic situation that Bohr would try to influence in the weeks and months ahead (see below, pp. 297–300).

The third thing relates to Fuchs's communism. When General Groves received the list of names of the British scientists who were to cross the Atlantic and take part in the Manhattan Project, he observed that no mention was made of their reliability, and enquired of his opposite number in the British set-up as to whether the men had been properly vetted. He was told that they had all been investigated and cleared. Groves accepted this assurance at the time but later came to feel that the British had not been very thorough and had misled him. Margaret Gowing, however, in her account of the British role, argues that the imputation is unfair. Washington had been told by London that especially strict security clearance had been used and it was 'the most comprehensive' used in the UK up to that point.[40]

At this distance it is difficult to know if Groves had been deliberately misled, but it is probably true that the British were less thorough than they might have been. For example, in Fuchs's case, they never seem to have uncovered the fact that, as he later told one colleague in an unguarded moment, while he had been in Paris he had worked with Henri Barbusse. Barbusse was a prominent French communist, a biographer of Stalin and an unwavering supporter of the Russian leader, who refused to write anything critical of the Soviet Union and even argued that 'given the lies peddled in the capitalist press about the USSR, it was legitimate for Communists to tell lies in defence of the Soviet regime.' And Barbusse notoriously refused to rule out violence 'in the revolutionary cause'.[41] This shows that Fuchs was a much more committed communist than MI5 – or anyone in Britain – realised.

But of course this cuts both ways. Had Groves made known to the British what had gone on in his own house, sharing his intelligence that communist sympathisers in the US were doing all they could to infiltrate the various sites of the Project, both sides would immediately have seen that Fuchs fitted *exactly* the classic profile of those who were seeking to penetrate the various locations where research and development were taking place. Fuchs had told an aliens board meeting as recently as 1941 that he was a communist in order to convince them of his anti-Nazi credentials so that he would be allowed to take up war-related work.[42] MI5 had received an anonymous tip-off that Fuchs was a communist and they knew from the German consul in Britain that he had been one since at least 1934. The warning signs were there, had the Allies compared notes.

In fact, had the security services been truly thorough, they would have seen the red danger signs glowing brightly. We have already seen that Fuchs, when he decided to work secretly for the Russians, had contacted Jürgen Kuczynski, a German-speaking Polish exile and well-known communist, who had put him in touch with his contact at the Russian embassy in London. The security services couldn't have known that but they were aware that Fuchs had befriended Kuczynski since his days in Bristol, and that Kuczynski was a militant communist. What can be revealed here is that Kuczynski had been on the FBI's radar as well, and since 1941.

In April of that year, one Harald Gumbel, travelling from Marseilles, was held by the United States Immigration and Naturalisation Service at Charlotte Amalie, on St Thomas in the Virgin Islands, pending clearance, because he was suspected of being a German secret agent with a 'front' as a communist. In Gumbel's possession was found a long list of names, including that of Jürgen Kuczynski. And once in the system, Kuczynski's name kept recurring. Another time was when the American censor intercepted a letter by Dr Hans Gaffron, a long-standing and well-known member of the Communist Party and a chemist at the University of Chicago. He had left Germany in 1937 after being interviewed by the Gestapo. They had not followed up the interview, however, which was hardly characteristic and the FBI again felt that, with his profile, Gaffron might be a Nazi agent in communist clothing. In fact, when the Bureau looked into the case in more detail, they found that when Gaffron left Germany he had given power of attorney to Marguerite Kuczynski, Jürgen's wife, so she could look after his affairs. He and Jürgen were close friends, who wrote to each other regularly, and in April 1941 they both became joint suspects in an FBI investigation, called 'Enemy Alien Control – Foreign Funds'. Gaffron had been born in Peru, and the FBI now suspected the two men of arranging for funds – for use in subversive activities – to reach the United States via South America.[43] From his censored letters, the FBI knew that Gaffron had a log cabin in Tennessee, to which he invited Kuczynski. The geography of Gaffron's life – Chicago and Tennessee – closely paralleled that of two major sites of the Manhattan Project. Kuczynski did not travel to the United States during the war, but he was again the subject of an FBI investigation, again into internal security, this time the chief suspect being Robert Dunn, a researcher at a labour organisation with offices on Broadway in New York

City and again suspected of being a communist front. This investigation was opened in July 1944, and Kuczynski was once more a correspondent.[44]

Had Kuczynski's name been flagged up by MI5, and made known to the FBI, the Bureau would surely have asked more searching questions about Fuchs having the access he had in the United States.

Fuchs's departure for America marked a major change in affairs and it was accompanied by his realisation that the Germans did not have a bomb, from which it followed that Russia was now in the sights of the Manhattan Project.

He had told Sonja he was going to America at their meeting in October. She acted quickly and at their next meeting, in November, only weeks before he sailed, she told him how to make contact with her opposite number in Manhattan. The meeting would take place on either the first or the third Saturday of every month. The contact was called 'Raymond'.

And so, on that cold Saturday, on the Lower East Side of Manhattan, near the Henry Street Settlement, Fuchs was to be seen carrying a tennis ball and a green-covered book, as Sonja had instructed him.

Raymond, he was told, would be wearing gloves but would carry a spare set. He would approach Fuchs at four o'clock in the afternoon and ask the way to Chinatown. Fuchs was to reply, 'I think Chinatown closes at five o'clock.' As someone said, they had all been reading bad thrillers.

Fuchs's contact was one Harry Gold, of Russian-Jewish extraction (original name Golodnitsky), who worked as a chemist at a sugar factory in Philadelphia. Turned down more than once by the military, scarred by the Great Depression, shocked by the rise of the Nazis and dispirited by the anti-Semitism he saw all around him, Gold was first persuaded in April 1934 to provide to a communist friend copies of some industrial secrets at the sugar company where he worked. After that, he was gradually drawn in, in classic fashion.

He was a small, round-faced loner, who worked all the hours he could at the sugar factory, every so often collapsing in 'exhaustion'.

In reality these 'collapses', like Fuchs's fake illnesses, were surreptitious devices designed to earn him a break from the factory routine, when he could carry out his spying activities. He was a bit of a fox himself, mainly a low-level courier for a number of people (such as the Rosenbergs) who provided crucial intelligence, Gold acting as a cut-out link-man to his Soviet control. At his meetings with Fuchs, Fuchs used his real name but Gold only ever identified himself as 'Raymond' (though with other contacts he was known variously as 'Arno', 'Charl'z' and 'Kessler').

With contact safely made that February day, the two men took a cab up Lexington Avenue, in the forties, and walked to the East River. Then they walked up 3rd Avenue to 49th Street, where Gold suggested they have dinner at Manny Wolf's Steak House.

They didn't speak much in the restaurant, but afterwards went for a walk where various elements of their routine and tradecraft were sorted out. (There must always be a back-up location for a meeting; no one was to wait more than four minutes.) Fridays were best, said Fuchs, Fridays being reserved at the lab for staff meetings when he was left pretty much to himself. At other times in the week, he was invited to dinner by colleagues and it would draw attention if he didn't participate.

And Fuchs said there would be two kinds of meeting. One kind, where he brought files, would be very short, a minute or two at most, where he would hand over handwritten documents of his own work. At the other meetings there would be no documents but Fuchs said he had an excellent memory (in fact he had a phenomenal memory) and at these rendezvouses he would be describing the work of others in the Manhattan Project, work that was too risky for him to have on him in written form.

At this first meeting Fuchs said something about atomic energy, and realised that the other man must know something about the subject because he did not ask any questions of interpretation or explanation. Fuchs also discussed the comparative strength of an atom bomb in relation to conventional explosives.[45]

Gold confessed to being 'dazzled' by Fuchs, and although the other man may have regarded him as an 'inferior', Raymond doesn't seem to have taken offence: 'I like this tall, thin, somewhat austere man ... with huge horn-rimmed glasses ... from the very first, and in his stuffy, repressed British manner he reciprocated.' In fact, Gold described Fuchs as a 'genius (a word I always use with caution)'.

After they parted following the first meeting, Gold rendezvoused later that same evening with his control, Sam, and relayed what had taken place. In the report he wrote later he said that Fuchs had described the 'factory', a code they devised to describe the Manhattan Project and that a diffusion method was being used as a preliminary step in separation, with the final work being done by 'the electronic method', developed at Berkeley, California. He also said he thought 'Camp Y' was in New Mexico and he outlined the watertight compartments that Groves had devised for security. He also told Gold that the two allies had fallen out for several months and that even then there was much that was being withheld from the British. 'Even Niels Bohr, who is now in the country incognito as Nicholas Baker, has not been told everything.'[46]

The Russians were relieved that contact with their golden source had been re-established and his information was quickly channelled to Kurchatov, who responded by studying the information and then sending out more questions to *rezidents* in Berlin, London, New York and San Francisco.

What can also be revealed here is that the American intelligence services missed a link between Fuchs and the investigations they were carrying out in San Francisco (see chapter 11). It was a long shot, but it was there and it was exactly the sort of trail intelligence services are set up to follow.

Louise Bransten, the wealthy socialite who was very close to (and may have been the mistress of) Grigori Kheifets (officially Soviet vice consul), was the woman who had introduced Oppenheimer to Steve Nelson at a party in her home. The FBI closely monitored Bransten throughout the Second World War, in one report describing her as 'the hub of a wheel, the spokes thereof representing the many aspects of her pro-Soviet activities running from mere membership in the Communist Party ... to military and industrial espionage and political and propaganda activities'. But they missed one vital spoke in the wheel – Elizabeth Bentley, a Soviet spy with eighty individuals in her network not unveiled until 1945, when she defected. Bentley had been in the same class at Vassar as Bransten, and they had been together at several meetings of the Communist Party in the run-up to war.

Had Bransten led the FBI to Bentley, she would in turn have led them to Abraham Brothman, an industrial chemist and a long-time Soviet spy, whose contact with the KGB Bentley was for a time. But not always. One of Brothman's other liaisons with the Soviet intelligence services, and to whom he gave blueprints for various pieces of technology, was Harry Gold.[47]

18

Lunch in the Supreme Court

When Bohr first arrived in America, he almost straight away bumped into Felix Frankfurter at the home of the Danish ambassador. Frankfurter and Bohr were old friends. They had first met as long ago as 1933, at Oxford University, and again in 1939 on several occasions both in London and the United States. Bohr had been impressed by Frankfurter's breadth of knowledge and interests, and noted his connection with President Roosevelt, who had appointed him a Supreme Court judge on the eve of the war in 1939.

And so, on Tuesday 15 February 1944 – ten days after Fuchs's first meeting in Manhattan with Harry Gold – when Bohr returned from Los Alamos to Washington, one of the first things he did was to have lunch with Frankfurter at the Supreme Court building, on 1st Street.

Frankfurter was a dapper, round-faced, silver-haired man who was born in Vienna and emigrated to the US when he was twelve. His family had been rabbis for generations but he was a non-practising Jew. As a young man he worked for the Tenement House Department of New York City (not so far from the Henry Street Settlement) to raise money for college.

He succeeded to the extent that he eventually graduated from Harvard Law School and later became a professor there. He was assistant to Henry Stimson, US attorney for the Southern District of New York and later secretary of state for war, and worked for Franklin Roosevelt in his first attempt to become president. Frankfurter was sympathetic to labour issues and Zionism, serving in 1919 as a Zionist delegate to the Paris Peace Conference. An

Anglophile, he studied at Oxford, and in the New Deal years he became a trusted friend of President Roosevelt.

Bohr was anxious to meet Frankfurter, and the feeling was reciprocated. Their first encounter took place over lunch in the justice's judicial chambers on the third floor of the four-storey neoclassical Supreme Court building overlooking the Capitol. The justice confided that he wanted to discuss sensitive matters where he had put two and two together to make a very interesting four.

Over Christmas, he said, he had been approached by a young friend of the family, a physicist, who was very troubled and was seeking advice. The young man couldn't describe what work he was doing – it was classified – and would refer to it only as Project X. But he was so troubled that he had tried to reach the president in a variety of ways. Not getting anywhere, he had approached the judge. He had told Frankfurter that he thought Project X was making scientific mistakes, so that Hitler might get to a 'conclusion' first.

Frankfurter told his lunch guest that of course he found it difficult to give advice on something so vague but it had then occurred to him that his young friend was a physicist, and so was Bohr. It was therefore natural to ask himself if that was why Bohr was in America – was he too involved in Project X? He said that, without going into specifics, he could more or less work out what was going on. And he wanted to know whether what troubled his young friend was troubling others.

No actual details were ever mentioned that day but even so Bohr was very keen to have the discussion they did have. As Frankfurter put it later: 'It soon became clear to both of us that two such persons ... could talk about the implications of X without either making any disclosure to the other.'[1] Bohr confirmed that many in Project X *were* troubled by what they were doing. They weren't troubled by the pace of development, and they didn't think that Hitler would get to a 'conclusion' first (he knew by then there was no German bomb). But he added that they *were* troubled in another way, and they had more or less authorised him to speak on their behalf in the corridors of power. He therefore welcomed the lunch and hoped Frankfurter could get word to Roosevelt about what was troubling so many involved in Project X.

He went on to say that, perhaps understandably, the US had rushed into building the bomb without considering the long-term implications of nuclear power.[2] And the more progress they made and the nearer they came to making it work, the more they were worried by those long-term consequences.[3]

Bohr impressed on his host that the open traditions of science, the unfettered nature of inquiry, and the collective checking of all advances, in a spirit of honest co-operative rivalry, was an ideal model for the way society should work. The scientific method was itself an ideal, preserving the fundamental values of Western civilisation: individual freedom, rationality and good fellowship. He argued that the discovery of fission was a good example of this model: 'Knowledge is itself the basis of civilization,' he said, '[but] any widening of the borders of our knowledge imposes an increased responsibility on individuals and on nations through the possibilities it gives for shaping the conditions of human life.' Every person, as he put it to his host, was obliged 'to confront the historical process, not merely as an observer, but as an active participant'.[4]

Bohr's record stood for itself, as Frankfurter was well aware.[5] It was now time, Bohr felt, for another intervention. The world would soon find itself in a totally unprecedented situation, when all traditional understanding was being overtaken. In his view, all economic, ideological, territorial and military questions needed to be reconfigured within a new framework.

In short, a new international order was needed. Atomic weapons were so powerful that their control was possible only in an 'open world'. Only with such openness could each nation be confident that its potential enemies were not engaged in amassing nuclear weapons. Scientists were always free to check each other's results – as with fission, that is how convincing progress occurred. This openness was more necessary than ever, if the recent achievements of physicists were not to destroy the world. For Bohr it was not a problem that there was no precedent. The threat of atomic annihilation was also unprecedented.

This was the essence of the context that Bohr gave to Frankfurter. But Bohr wasn't naïve, as some critics later accused him of being. On specifics, for example, he never suggested that the actual use of the bomb in anger might influence post-war relations with the Soviet Union. On the contrary, 'What role it [the atomic bomb] may play in the present war,' was a question 'quite apart' from any post-war considerations. He recognised it as a military rather than a political matter. In respect to the post-war world, however, Bohr argued that international agreement could be achieved only if the Soviets were brought in *before* the bomb became a certainty and *before* hostilities were over.[6]

Bohr took two assumptions virtually for granted. First, the bomb was out of all proportion to anything else in human experience. Second, it could not be monopolised. Rival nations would get there sooner or later. And Bohr saw clearly that the bomb would drastically limit the way politics operated: there would only be co-operation or disaster – there was no middle ground.

The heart of Bohr's analysis followed directly from this. Stalin must be informed about the Manhattan Project *before* its development had proceeded so far as to make any approach seem more coercive than friendly. It must to be made clear to the Russian top brass that an Anglo-American atomic monopoly was not being formed against them. Timing was crucial.

He well understood that an approach to Stalin did not guarantee co-operation. At the same time, he thought that co-operation would be impossible without some such proposal. He knew there were risks but thought the stakes were great enough to justify such risk.

His proposal was that the Soviets be told simply that the Manhattan Project existed, but no more. No details of the bomb's construction should be revealed. If that were done then the Soviet response to this limited disclosure could be evaluated. If it was positive, then further co-operation could be pursued. But to begin with, certainly, no information about technical details should be exchanged. Rather, the opportunity should be taken to tell the Soviets that all such information would be withheld until wider safety issues had been agreed.[7]

Finally, Bohr advanced the view that the international community of scientists might well have an important role to play beyond a purely technical one. As he put it: 'Helpful support may perhaps be afforded by the worldwide scientific collaboration that for years had embodied such bright promises for common human striving. On this background personal connections between scientists of different nations might even offer a means of establishing preliminary and non-committal contact.' This was a most interesting aspect of his proposal, for reasons we shall come to, and is often overlooked.

In other words, Bohr envisaged a repeat – on a much grander scale, no doubt – of the collaboration between scientists of all nations (including Russians, Germans, Italians and Japanese) that had existed at his own institute in Copenhagen before the war. He was certain, he said, that among the leading scientists in Russia, 'one can reckon to find ardent supporters of universal cooperation'.

The whole context of Bohr's discussion with Frankfurter was not about Germany and the bomb, the problem that had so concerned the justice's young friend of the family, and the reason Project X had got under way in the first place. This confirms that most Project physicists were still in the dark in late 1943/early 1944 about the way the purpose of the bomb had changed. For Bohr, Russia was the chief focus of attention. To him, the switch appeared complete, but not to others.

After lunch, Bohr left the building by the west entrance. As Frankfurter and he shook hands, standing on the grand steps of the west façade, overlooking the Capitol and James Earl Fraser's 1935 monumental sculpture, 'The Contemplation of Justice and Authority of Law', the Supreme Court judge hinted that he would pass Bohr's message to Roosevelt. What he actually said was, 'Let us hope that this will be a memorable day.'

Aage Bohr, who had been to Los Alamos and back with his father, recalled Niels's pleasure after he returned from his meeting with Frankfurter: 'It was an exciting day. The result of the meeting corresponded to his best expectations.'[8]

19

'The Mission He Had Been Waiting For'

One of the best-known landmarks on the fashionable Upper East Side of Manhattan in New York City is Bloomingdale's department store, on Lexington Avenue between 59th and 58th Streets. There is a subway station on the uptown corner and the proximity of the 59th Street Bridge means the traffic is always heavy. This is where the second rendezvous took place between Fuchs and 'Raymond' on a wintry Friday early in March.

Before this second meeting, Gold was told by his control, Sam, that he was being passed on to another man, 'John'. John was in fact Anatoli Antonovich Yatzkov, trained as an engineer (so he should be able to follow what Gold was told by Fuchs) and had been in the US since 1941, officially working as a clerk in the Soviet Consulate in New York.[1]

Gold met Fuchs on the northwest corner of 59th Street and Lexington Avenue, by a bank and the subway entrance. Their intention was to walk across the bridge into Queens but that night it was closed to pedestrian traffic so they walked along 1st Avenue instead.

At this meeting the first thing they discussed was their alibi should they ever be tackled on how they were friends. They decided that they had met at a concert of classical music at Carnegie Hall when they had by chance sat next to each other and struck up a friendship based on music and chess. Gold was tasked with finding out the details of a recent concert.

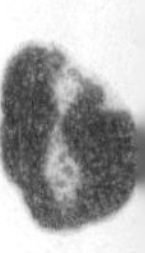

The son of Samson Golodnitsky, a woodworker, and Celia Ominsky, who had various jobs, including rolling cigars, Harry Gold's parents had both found it prudent to leave their native Ukraine on economic grounds and, being Jewish, on racial grounds too. Samson and Celia met in Bern, Switzerland, where Heinrich Golodnitsky was born on 12 December 1910, with 'long and very blonde curls' and an early liking for chocolate.

Life in Switzerland was better than in Ukraine but still not to Samson and Celia's liking and, in July 1914, they emigrated to the United States. On Ellis Island the immigration officer suggested to Samson – who barely spoke English – that he change his name to something simpler and so the Golodnitskys became the Golds. They travelled first to Chicago, then to Philadelphia where, from 1917 until well into the 1930s, the Gold family lived in a series of rented homes, in a mainly Russian, Jewish and Irish area of 12ft-wide row houses.[2]

Sam Gold found work as a cabinetmaker. He was said to be a master craftsman but suffered aggressive anti-Semitism, especially from Italian immigrants, who put glue on his chisels and clothes and stole his tools. They complained that he worked too fast, a sentiment the shop foreman agreed with, so that 'Harry watched his father "suffer uncomplainingly for years".'[3] Fortunately, Celia was a good teacher who spoke several languages, including Hebrew, and was known – and loved – locally as *die Rebbetzin*, the rabbi's wife (not strictly true, but a mark of respect).

Harry grew up a shy child. He didn't have many friends, but those he did have recognised that he was smart. He was good at chess and read a lot – Dickens, Milton, Shakespeare, Browning, O. Henry, Jack London, Conan Doyle, Zane Grey – and fell in love with the movies, then just hitting their stride. An interest in music and opera would follow.

He loved sports too but his stature – 5ft 3 in high school, and weighing around 100lbs – was against him. His schoolmates laughed at his efforts. 'They wouldn't even let me put on a uniform.'[4] He became a target for bullies, a dangerous element when there were Jewish and Irish neighbourhoods in close contiguity. His father found it necessary to accompany his son to the library and would wait outside while Harry returned his books and then borrowed more. But he recovered some respect by tutoring less bright schoolmates and helping them pass their exams.[5]

He graduated from school as third in a class of 160, then got a job, through a friend of his father, in a woodworking plant. It wasn't what he would have preferred but it was 1928, it was a job and the Great Depression was just around the corner. He secured a position as a laboratory assistant with the Pennsylvania Sugar Company, 'where some significant science was undertaken'.[6] By September 1930 he had saved enough money to enter the University of Pennsylvania.

Then the Great Depression hit and the family had to dip into its savings. He voluntarily withdrew from university so he could reclaim that semester's fee, which he gave to his mother. By December 1932, just before Christmas, father and son were out of work, and they were forced to return some new furniture they were still paying for.[7]

Some families in Philadelphia, including Jews, considered going to Russia, where they had heard things were better. Some of the Golds' friends actually went. The Golds were never tempted but they were facing economic ruin.

Then, out of the blue one day, a neighbour stopped by the house with exciting news. A mutual friend, Tom Black, someone Gold had known at university, was leaving his current job and transferring to another, and he was pretty sure Harry could take his old position. Which is how it turned out.

Black was leaving the Holbrook Manufacturing Company, a soap firm in Jersey City, and moving to the National Oil Products Company in Harrison, New Jersey. Both firms carried out some chemistry, and the pay was better than Harry had been receiving. The move took Harry back into the company of Black, a politically activist communist who thought that 'capitalism was doomed' and that working men owed their allegiance to the Soviet Union. Harry was not at all keen on Black's world view but, under relentless pressure and in gratitude for having kept him and his family from the relief rolls, he succumbed to Black's entreaties and attended party meetings.

He didn't enjoy the early sessions, feeling that the communists were a mix of 'lazy bums and windbags'.[8] What Gold didn't know, but would find out soon enough, was that Black was by then in contact with a certain Gaik Ovakimian, an Armenian and a Soviet spy interested in scientific industrial processes being pioneered in the United States. The FBI later claimed they knew about Ovakimian from the moment he arrived in America in the 1930s, but if so they never followed him to Gold, or discovered the identity of one of his informants, codenamed 'Kvant'.

In September 1933, Gold heard that Pennsylvania Sugar was hiring again, and he was delighted when he got his old job back. In turn that enabled him to re-enrol at university and it seemed as if his career was now back on track. Black was no longer pestering him.

That was true for about six months, but then Black reappeared. This time, though, he no longer tried to recruit Gold into the Party: Black said there were more practical things Gold could do to help. Specifically, the sugar company had developed processes – industrial finishes and lacquers, for example – the details of which could be of the greatest help to the people of the Soviet Union. Gold agreed to steal them.

He still felt he owed a debt of gratitude to Tom Black. He was aware that the Soviet Union had outlawed anti-Semitism – a definite strong suit in his mind – and communism was the most reliable bulwark just then against Fascism. For Gold, Nazism, Fascism and anti-Semitism were identical. As he was to write later, he also had acquired a 'definite lack of faith in the Democratic process'. The 'shenanigans' in the stock market, he felt, increased homelessness and mass unemployment.

And so, by the spring of 1935, Gold was lifting documents from the sugar company he was working for and handing them to Black. He seems to have taken to it like a duck to water. Later he swiped a dry-ice process that prevented ice cream from melting and even materials used in biological warfare. In fact, he was filching documents at such a rate that Black found it difficult to keep up in copying them, so much so that, at the end of the year, Gold was told that Amtorg, the Russian trading organisation in America, would put its copying facilities at his disposal, provided he could get the documents to New York.

Once in New York he was introduced to a 'handler', a man he

only ever knew as Paul Smith, someone who knew how to strike up a proper relationship, much more so than did Tom Black. Smith asked Gold to write out his family history, thereby showering him with sympathetic attention. Then he got Gold to write out a list of what processes he had access to and which *he* thought were important. In no time, Gold was meeting Smith every three to five weeks, and made to feel he was doing important work.

Of course, there are only so many processes – and documents describing those processes – that can be developed by any one company, and by the end of 1935 the supply of material from the sugar company was beginning to run down. This coincided with the appointment of a new handler. The new man, known as 'Fred', was very different from Smith. He was much less personable, a small moustachioed man in his early thirties 'with black eyes and a swarthy complexion', someone moreover who appeared to have cultivated the Bolshevik virtue of *tverdost* (hardness).[9] He set Harry a new task, coming up with the names of people like him whom the Soviets could recruit to secure fresh information from other companies.

This was not to Gold's liking at all and for a time he made up names to keep Fred satisfied. There then followed a number of abortive missions. One was to join a team tasked with assassinating Trotsky, a proposal Gold found absurd and loathsome. Then, in August 1940, he met yet another new handler, someone whom he only ever knew as 'Sam', but who was in fact Semyon Markovich Semenov, an MIT-trained mechanical engineer and mathematician, who worked for Amtorg, 'ostensibly purchasing oil refinery equipment'.[10] Sam would be Gold's longest-lasting and most important handler.

While all this was taking place, the world kept turning. Gold had been distressed and puzzled by the Molotov–Ribbentrop Non-Aggression Pact in August 1939. Fred had laughed at Harry's naivety, saying that Russia 'needed time'. When Germany had invaded Russia in July 1941, and America and Britain and Russia had become Allies, Gold was much happier, and had consented to a series of projects to act as a liaison with certain scientists at companies such as Eastman Kodak and Du Pont to obtain secret scientific processes, which he passed on to Sam.

And then one day, in early January 1944, he received a phone call ordering him to New York for an important meeting.[11] 'Something has come up,' said Sam, more excited than Harry had ever heard him before. The Soviet Union, Sam whispered, was assigning Harry a mission 'of the utmost significance and importance'. He must drop everything else and apply himself 'single-mindedly' to this one thing.

Gold agreed immediately. Maybe, he thought, this was the mission he had been waiting for, something that would really make a difference.

Sam said he would be meeting a man recently arrived from England, who would be working with a group of fellow scientists in the New York area. The man would be providing information on a new kind of weapon. In fact, 'devastating' was the word he used. It would be Gold's job to receive the information and pass it on to Sam.

The first meeting went exactly as planned. In the second meeting, during their walk up 1st Avenue, Fuchs explained to Gold why he was in America. It was the first time Gold had heard the words 'The Manhattan Project' and Fuchs named all the members of the British mission, and some of their American colleagues, too. He explained that the Project was divided into tight compartments, and that he did not have an overview – hardly anyone did – but he said that his own work had to do with the separation of the fissionable isotope of uranium-235, through either gaseous diffusion or electromagnetic separation. Fuchs added that he did not think the advances they were making were enough to build a bomb quickly enough to be ready before the war ended. But it would matter later.

He informed Gold on this occasion about the role of General Groves. Fuchs had attended at least two meetings, on 22 December 1943 and 5 January 1944, at which Groves presided. Just giving the Russians Groves's name would have told them a lot. Here was the man who had built the Pentagon – no secret there. The very mention of his name conveyed the sheer size of the Project.

By that time, too, Fuchs and Peierls had attended at least five meetings, between 10 and 21 December, on plant control systems, time-dependent problems, by-pass systems, all technical problems that, as the frequent meetings showed, were being urgently pushed forward.[12] Fuchs also told Gold that the contracted entities for this part of the Project were the Kellex Corporation, Columbia University and the Carbide and Carbon Chemical Company.

No less important, Fuchs had attended 'a stormy meeting' with General Groves, also in January 1944, when the British scientists had supported his proposal for a new and superior barrier material at Kellex. The importance of this lay in the fact that the preparation of the new barrier material would delay the start up of the Oak Ridge gaseous diffusion plant then under construction but also carried the implication – as Harold Urey independently appreciated – that the US was 'pursuing a postwar nuclear-weapons capability, not simply trying to beat the Germans to the bomb'.[13] (Note that, in January 1944, Harold Urey, Nobel Prize-winner for the discovery of heavy water and a senior Manhattan Project scientist, thought that Germany was the main bomb threat.)

At this second meeting, Fuchs described to Gold some of the problems he – and the others – faced, in particular the fact that Project scientists 'worked in extremely tight compartments, and that one group did not know what the other group was doing'. As Gold told the story after he was caught and interrogated, 'This I can verify by the fact that he told me that he thought there was [the] possibility of a large-scale installation for isotope separation projected for future development somewhere, he thought, down in Georgia or Alabama.' This, of course, turned out to be Oak Ridge, Tennessee, the very plant Fuchs was helping to design.[14]

Gold knew enough chemistry to understand what isotopes were and Fuchs came away from the second meeting believing his American contact 'had some scientific knowledge'. He promised that at the next meeting he would deliver a package that would contain 'a complete written account of who was working on the project ... a general physical make up of it, and how far it had progressed'.

Gold diligently wrote down all that Fuchs had told him as soon as they parted. He met his new control 'John' (Yatzkov) later that evening in mid-town Manhattan and handed over his notes, keeping physical contact to a minimum. Then he took a train home to Philadelphia.

While he was doing that, John was sending the information back to Moscow Centre in coded cables, using Fuchs's codename 'Rest' (later 'Charl'z'). These cables were sent out over commercial telegraph lines, which meant they could be intercepted – and they were.

Following the Japanese attack on Pearl Harbor on 7 December 1941, the State Department had developed a system, called a 'drop copy' programme, under which the cable companies held up transmission of messages long enough to copy them and pass them on to the Office of Wartime Censorship. These messages were available to both the army and the FBI. However, the Russian messages were coded using one-time pads, matching similar pads kept in Moscow. This process meant that they were in theory indecipherable. Fuchs's messages were kept, along with thousands of others, but no one could read them. Not then.

20

A President 'Eager for Help'

In the last week of March 1944 most of the war news was grim. On the 24th, hundreds of Italians were massacred by Germans at the Ardeatine caves in Rome in reprisal for a huge bomb that had killed many Germans. On the 30th the RAF took huge losses in an air raid over Nuremberg. The only respite was that the Japanese were being forced to retreat in Burma, though that would in time reveal the atrocities that had gone on there.

Bohr had perhaps overestimated Frankfurter's influence with the president, because he didn't manage to get in to see him at the White House until that grisly week. But then Roosevelt allowed his old friend an hour and a half so the justice was able to give the president a detailed account of Bohr's analysis, concentrating on his central worry: 'that it might be disastrous if Russia should learn on her own about X rather than that the existence of X should be utilised by this country and Great Britain as a means of exploring the possibility of an effective international arrangement with Russia for dealing with the problems raised by X'.[1] Frankfurter also made it plain to the president that Bohr was convinced that Soviet physicists would – probably sooner rather than later – acquire the necessary information to build their own atomic weapons. He relayed the gist of Ioffe's 1940 letter to Bohr, hinting that Russian physicists had noticed that Allied scientists had stopped publishing fission work in the professional journals, and that he had drawn appropriate conclusions. The Russians were laying out feelers, letting slip hints about what they knew.

Of course, Roosevelt was not surprised by the Soviet interest in atomic science. As we have seen, he had been alerted the year before, by General Groves, who had briefed him on Russian attempts to infiltrate the Berkeley Rad Lab, Oak Ridge and even Los Alamos. The president knew about the Russian requests for uranium and heavy water.[2]

In early 1944, however, when Frankfurter went to see the president, events had moved on. The president was well aware that Germany didn't have a bomb. He couldn't know that Stalin knew it too, at the time. But he did know, or should have known, that the Russians were aware that an Allied bomb project was well under way. And he surely grasped Bohr's point, carried to him by his friend Frankfurter, that the existence of such a powerful weapon would have a decisive impact upon Soviet–American relations after the war was over.

Bohr had put it clearly. The central issue was coercion. There were undoubted military and diplomatic advantages for the Allies to keep the bomb to themselves. But how long would those advantages last and would they offset the advantages of telling the Russians in good time, with the good will that that would (theoretically) bring? And if the bomb was kept secret, and the Allies secured an advantage, exactly how would that advantage work? Would the Soviets succumb to such coercion?[3] An approach to Stalin promised benefits at least in theory. A proposal for the international control of atomic energy could form the basis of a great-power collaboration after the war. And, as Bohr had also pointed out to Frankfurter, if the Russians did *not* agree to such collaboration, it would at least confirm the limitations of Soviet intentions.

In London, Anderson was aware of Bohr's meeting with Frankfurter through Halifax, the British ambassador in Washington. Knowing that Bohr was due back in London, he decided to set the stage and put before the prime minister some thoughts on nuclear strategy but without mentioning Bohr's name.

He was summarily rebuffed. When he suggested to Churchill that there should be 'collaboration' with the Russians, the prime minister was crudely dismissive, circling that word in the memo and scribbling in the margin 'on no account'.[4] Anderson never told any of this to Bohr and so, still kicking his heels in Washington, awaiting Frankfurter's meeting with the president, Bohr began to put his thoughts together more formally, preparing a long report for Anderson with the title 'Confidential comments on the project exploiting the latest discoveries of atomic physics for industry and warfare'. The document was marred by Bohr's lockjaw English syntax but, as Graham Farmelo has said, 'his circumlocutions contained a powerfully original argument for thinking about the Manhattan Project as a global opportunity rather than as a threat'. And it concluded: 'Such an initiative, aimed at forestalling a fateful competition about the formidable weapon, need in no way impede the importance of the project for the immediate military objectives, but should serve to uproot any cause for distrust between the powers on whose harmonious collaboration the fate of coming generations will depend.'

Some two weeks after Bohr sent this report to Anderson in London, he heard from Frankfurter. The justice had spent an hour and a half with Roosevelt and when he left the White House he remarked to others in the know that the president was 'plainly impressed by my account of the matter'. He even said that when he had suggested to the president 'that the solution to the problem of the atomic bomb might be more important than all the plans for a world organization [the United Nations, then being mooted]', Roosevelt had agreed.[5] Frankfurter also said that Roosevelt had confessed that he was 'worried to death' about the bomb. 'He was very eager for all the help he could have in dealing with this problem.'

Most interestingly, and most controversially in retrospect, Frankfurter also said that the president had authorised him to tell Bohr that when he returned to Britain, as was planned, he might 'inform our friends in London that the President was most eager to explore the proper safeguards in relation to X'. In another version, Roosevelt had told the justice he was receptive to Bohr's ideas and 'would welcome any suggestions to this purpose from the Prime Minister'.[6]

Vastly encouraged, Bohr prepared to leave for London. He was found a seat by Lord Halifax on a military aircraft.

The third meeting between Fuchs and Raymond took place at almost the same time as Bohr's second meeting with Frankfurter, except that it was late on Friday 31 March – at nine o'clock in the evening. In Europe it was already April. General de Gaulle was preparing to take command of all French forces, and the Allies were beginning their bombing of Hungary and Bulgaria ahead of the advancing Red Army.

It was a cold dark night as the two men met along Park Avenue in the Seventies or Eighties, and Gold later remembered that they both wore overcoats. Fuchs was walking slowly north, on the west side, when Gold spotted him. He strolled up behind the slim outline and, when he drew close, the two men instinctively veered on to an 'extremely deserted' side street and continued walking in the direction of 5th Avenue and Central Park.

There, in a matter of seconds, Fuchs handed over 'a very bulky package' of fifty pages or so, which Gold immediately hid in the inner pocket of his overcoat. Fuchs asked how his earlier information had been received and Gold said it had been satisfactory with one drawback, that a general guide to the overall plant was needed. Fuchs was annoyed by this, saying he had already provided that and it was Moscow's fault that they hadn't connected the different parts that he had supplied. He also complained that his work was being 'deliberately curbed' by the Americans, who were still not co-operating wholeheartedly. The two men then parted, with hardly anything else being said.

That time Gold committed a serious lapse of procedure. He took out the package and thumbed through its pages. He judged there were twenty or twenty-five closely written sheets with vital information about the design, progress and participants involved in the construction of the world's first atomic bomb.

We now know that among the information that changed hands that night were nineteen papers, at least thirteen written by Peierls and Fuchs, or by Peierls, Fuchs and Skyrme, on pressure problems in gaseous diffusion, formulae for the membranes used in the process, and the titles of nine other papers by different authors.[7]

The procedure in the Kellex offices was for Fuchs to write out by hand the papers where he had done some or all of the work, and have them typed up by Mrs Gertrude Crosby Rowen, the secretary he shared with Peierls. He then showed the typed version to Peierls for approval before it was circulated more widely. He would then take back his handwritten originals, and it was these that he passed to Gold. He didn't dare take away papers produced by others, but he could take down their titles and the names of their authors.[8]

The information disclosed that night also discussed various corrosion problems that were being encountered, and identified 'S-50' as the code for thermal diffusion, in which, it was said, an attempt was being made to enrich uranium from 0.71 per cent U-235 to 0.85 per cent.[9] These codes were crucial, of course, because other spies, elsewhere in the Manhattan Project, could use the codes to link work in different locations.[10]

By then, Fuchs had attended three special meetings devoted to gaseous diffusion, the latest on that same day, 31 March. That is how up to date the information was.

Gold replaced the package in his overcoat and walked quickly towards midtown where, after twenty minutes or so, he met his Soviet contact, John, in another quiet cross-town street.

21

The Letter from Moscow

Emboldened by what he now felt was his presidential mission, Niels Bohr arrived in London a few days later, in early April. Once there, however, he was kept waiting by the prime minister *for weeks.*

To be fair to Churchill, he was then in the throes of preparing the D-Day invasions that he knew (but Bohr didn't) were set for 5 June, later put back by one day. Not only that, in March US bombers had begun their daylight attacks on Berlin, the Battle of Monte Cassino was raging, the Labour Party in Britain was agitating for a relaxation of the wartime regulations regarding strikes (which might affect wartime production), and the Germans had discovered the Allies' secret talks with the Hungarians, as a result of which they occupied Hungary. The prime minister had a lot on his plate and he didn't understand why, as he was told, Bohr had a message from Roosevelt. Churchill and Roosevelt got on famously, were in touch virtually daily by telegram or telephone, and the president had never sent emissaries before. Churchill felt very strongly that prime ministers should deal with presidents and vice versa: an emissary like Bohr was 'not fitting'.

Anderson was of course well aware that Churchill's views and Bohr's hardly coincided, and he was privy to many other, non-nuclear-related secrets as well, so he knew there were all sorts of reasons why Bohr was not given easy access to Downing Street. But he still tried to help. About two weeks after Bohr had been in London, the chancellor wrote again to Churchill, having been informed by Halifax in Washington, who had been talking to Frankfurter, that

Roosevelt was 'giving serious thought' to international arms control and 'would not be averse' to discussing it with him.[1]

Anderson went so far as to draft a memo for the prime minister to send to the president, in which he concluded that 'the matter seems to me to require deep thought'.

Churchill was as unimpressed as he had been before and replied immediately, saying 'I do not think that any such telegram is necessary.'

Anderson still didn't tell Bohr about this further exchange, not wanting to take the wind out of his sails. He still thought the Dane's views had considerable merit.

For his part, Bohr was more confident of his case than ever. This was because, while he was in London, he had become more certain that the Russians were working on a bomb of their own, so that it would be fruitless to try to keep it secret from them. He based his growing confidence on the fact that while he was in London, he received a curious note from the Russian embassy to say that it was holding a letter for him, from Russia, and would he please collect it. Who even *knew* he was in London?

The letter turned out to be from Pyotr Kapitsa, his old Russian atomic physicist friend, whom Bohr had known from Cambridge, from the time when Rutherford was alive, and the man on whose behalf he had campaigned when he had been forced to remain in Russia. Kapitsa had in fact written the letter the previous October, and sent it to the Russian embassy in Sweden who had forwarded it to London.

'I want to let you know,' Kapitsa wrote, 'that you will be welcome in the Soviet Union where everything will be done to give you and your family a shelter and where we now have all the necessary conditions for carrying on scientific work. Even the vague hope that you might possibly come to live with us is most heartily applauded by all our physicists: Ioffe, Mendelshtam, Landau, Vavilov, Tamm, Alihanov, Semenov, and many others.' And it also contained these words: 'At our institute we hold a scientific gathering every week where you will find a number of your friends. If you come to Moscow you will find yourself joined with us in our scientific work ... We have very little information about English physicists. They, like us, are hard at work fighting for our common cause against Nazism . . . We scientists have done everything in our power to put our knowledge at the disposal of the war cause.' Abraham Pais, a Dutch-born American physicist, who worked with both Bohr and Einstein, says it is not easy to decode the letter but that it is now known (from a 1989 report in *Moscow News*) that by that time Stalin had called in a number of physicists, Kurchatov included, to enquire about developing atomic bombs.[2]

David Holloway adds that Kapitsa had obtained permission from Molotov to send this letter and had explained to the minister of foreign affairs that Bohr was 'well-disposed' to the Soviet Union, having visited the country three times. 'I think it would be very good and correct,' he wrote, 'if we were to offer hospitality here in the Union to him and his family for the period of the war.'[3]

At one level this letter may be no more than it appears to be, a friendly gesture to someone forced to leave his home. There are, however, several aspects to it that suggest it is more than it seems, which is perhaps why Molotov had to be consulted.

The first is its timing. If we take it at face value, Kapitsa found out fairly promptly that Bohr had escaped from Denmark, though in fact the letter is dated 28 October, by which time Bohr had long left Sweden (he arrived in London on the sixth of that month). However, the Russians must have known, in autumn 1943, that Bohr was out of the loop. Given that Denmark was occupied, he could never have done any practical work on a bomb. He could have thought about it, yes, but nothing more.

However, by the spring of 1944 the picture was very different. Bohr had not only left Denmark, he had gone on to Britain and the United States and we know that the newspapers had reported at least his basic movements. And so, when he arrived back in London, it was as someone who had now caught up with what the Allies were doing. He was no longer out of the loop but very much part of it.

And so an invitation to the Soviet Union would now be much more likely to be useful for Russia's own atomic effort. However, a too-frontal approach might not work, risking making him suspicious. Instead, a dissimulation was needed. We only have the Soviet embassy's word that the letter was originally sent to Stockholm in

October 1943. But perhaps the Russians only *pretended* that the letter was first sent to Stockholm, as a way to reassure Bohr that the invitation had been made *before* he was brought up to date on the Allied bomb effort. This is not as cynical as it at first appears because of the other elements in the letter.

The second was its authorship. It did not come from Ioffe, the man who had sent the previous letter, though he was mentioned, in passing. It came from Kapitsa and Kapitsa, it will be remembered, was the man who had featured prominently in the *Daily Telegraph* report in 1941, when Soviet scientists had mounted 'a veiled appeal for Soviet inclusion in the atomic bomb programme as a full member of the Grand Alliance'. Kapitsa had been quoted personally, discussing the future development of an atomic bomb by the Allies.[4]

The third element was the omission from the list of names that Kapitsa gave of Igor Kurchatov. By October 1943, when Kapitsa actually wrote the letter, or spring 1944, if the interpretation being given here is correct, Kurchatov had had months of access to the espionage files – since February 1943 in fact – so he would have known that Bohr's name hardly featured in the earlier documents. However, now that Bohr was back in London, he was fully in the picture so far as the bomb project was concerned.

The timing is certainly suggestive. Is that why Kapitsa's letter reached Bohr in April (although the letter was dated October 1943)? Was that the real reason he was invited to Russia?

This is at least possible because of the fourth element in the letter, that it contained the phrase, 'we now have all the necessary conditions for carrying on scientific work'. On the face of it this is a bland phrase. Bohr had been to Russia, and had a high opinion of Russian physics and knew that it was up to date on theoretical matters. What could it mean, then, to say 'we *now* have all the necessary conditions for carrying on scientific work'? Russian physicists had been carrying on scientific work for years – decades – as Bohr knew very well. So what was different *now*? Could the use of this normally innocuous word in such circumstances mean that Russian physicists were *now* fully informed about fission, that proper budgets were *now* in place, and the necessary equipment *now* available? This meaning seems supported by Kapitsa's other sentence: 'We scientists have done everything in our power to put our knowledge at the disposal of the war cause.'[5]

The reference to English physicists was also carefully worded. In one breath Kapitsa says they had very little information; in the next he says they are working for the war effort. In the circumstances, this was more than a statement of the obvious.

A final element in the mix is that, at precisely this time, Roger Makins, a career diplomat, who was in on the Manhattan Project from his office in the British embassy in Washington, wrote to General Groves with the news that a Danish friend of Bohr's had been approached, 'a few days later', by a Russian scientist visiting Denmark who said he had a letter from Kapitsa to Bohr, but the letter was never delivered.

The wording of the Makins letter is itself ambiguous but Russian scientists could not have visited Denmark in October 1943, when the letter was dated, because, following Stalingrad, Russia and Germany were vicious enemies. The chronology of this episode is far from clear, in the archive, but to an extent it does seem to support the idea that the Russians were interested in inviting Bohr only *after* he had been to Los Alamos.

Whatever the exact truth of all that, the Kapitsa letter convinced Bohr that the Russians had an atomic weapons programme of their own, and that therefore an approach by Churchill and Roosevelt to Stalin was all the more urgent and timely. More even than that, Bohr himself, with his status and independent nationality, would have been an ideal go-between in setting the ball rolling.

Then there was the fact that, as David Holloway says, Bohr's presence in the Soviet Union would be a feather in the cap of Russian physics 'and would lay the basis for international ties after the war'.[6]

From this side of the Cold War we must never forget that, in 1944, Russia was an ally. Kapitsa was one of many who wanted to re-establish contact with foreign scientists and also argued that scientists had a role to play in securing international peace – an exactly parallel thought to what Bohr had argued with Frankfurter. In a speech to the third Anti-Fascist Meeting of Soviet Scientists in June 1944, Kapitsa argued that, with the end of the war now in sight, 'our task, the task of all scientists, must not be limited to gaining a knowledge of nature in order to harness it for the benefit of mankind in peaceful construction. It is my opinion that scientists must also take an active part in the establishment of a sound and lasting peace.'[7]

Here again the wording and the timing are what attract our attention. Kapitsa's argument was published openly in Moscow *and in English.* Publishing Russian material in English was not unknown, of course, not by any means, but in this case is it too much to ask whether, taken alongside his approach to Bohr, Kapitsa was once again putting out feelers? Kapitsa was in good standing, having developed during the war a new technique for producing liquid oxygen for the metallurgical and chemical industries, and was in touch – and got on well – with Molotov.[8] His phrase about the 'harnessing' of nature suggests that he had nuclear energy in mind and was suggesting, as was Bohr, that nuclear scientists, who all knew each other, should play a central role in international relationships.

Kapitsa did not have the ear of Beria or Stalin that Kurchatov did but he would become a Hero of Socialist Labor, the highest civilian honour, so it is certainly possible his June 1944 speech was sanctioned as a diplomatic overture, and was designed to show that Stalin had an inkling of what the Anglo-Americans were up to and was using Kapitsa's friendship with Bohr as bait.

Bohr politely refused the invitation to Russia, after first showing the text of his reply to MI6. But his proposal – about sharing information – was more urgent than ever. To have any force, an approach must be made while the Allies were still ahead.

The fourth meeting between Fuchs and Raymond took place in late April, around the time Bohr was replying to Kapitsa's letter and a month before D-Day.

They met in the Bronx, at a cinema near Grand Concourse and Fordham Road, near the library, and began walking. Gold later recalled, 'we went for a walk partly along the Grand Concourse ... during which time we discussed the next meeting ... at which a second transfer of information was to take place ... it was a wet and somewhat chilled night for April, and as I recall, he had a bad cough, and I did not wish to expose him to the elements any more than was necessary.' Suspending the rules of engagement they had originally agreed upon, they went to Rosenheim's Restaurant, near 180th Street. Over a light snack they discussed topics ranging from contemporary world politics to music and chess.[9]

Fuchs explained how he had 'fallen into' espionage when, in 1941, he felt that Britain and America were letting Nazi Germany and communist Russia fight it out, in the hope that both would be mortally weakened, after which the Allies could, in effect, pick up the pieces cheaply. This, he felt, was unfair and he wanted to even up Russia's chances. Balance was on his mind.

It was over this snack that Fuchs began to open up personally. Gold had come to know Fuchs, he said later, as a 'very shy, reserved individual', who never disclosed emotion. At that meeting, however, Fuchs revealed something of his earlier life, in which he had displayed strident partisanship directed against the Nazis and Fascism. He told Gold how he had been lucky to escape the Nazis after the Reichstag fire, being on a train for Berlin when the Gestapo had come for him. He said that the fire was a 'set up', something not generally known, that the communists had been branded as fire-setters when in fact the conflagration had been started by the Nazis themselves and used as a pretext to round up their political opponents.[10]

He told Gold about the bravery of his own father and the suicide tragedies that had befallen his grandmother, mother and his sister. We can't be certain that he told Gold about his escape to Paris and his attendance there at the International Youth Congress where Henri Barbusse had been the focus of so much attention. But it is more than likely. In a sense Barbusse – with his unflinching belief in the Soviet Union and his conjoining of anti-Fascism and communism – was doing exactly what Gold and Fuchs would have longed to do, if they had had the chance.

For his part, Gold shared with Fuchs the hatred he had for the faults of the West, the crude and vicious anti-Semitism his father and he had faced, the fact that they had briefly considered leaving America and emigrating to the Soviet Union where, they had been told, life was better for people like them. He explained how he had been drawn into the role he now occupied with Fuchs. He never told the other man his real name, or gave away any details that could be used to identify him – nothing like the names of the schools and universities he had attended, or even the name of the sugar company where he worked.[11]

They were doing something they shouldn't have done, but the two men bonded that day. They were not natural friends – neither was especially sociable – but they did have music and chess in common, as well as the lonely, secretive and dangerous world of espionage. The snack at Rosenheim's must have been one of the warmest moments they shared in their cold isolated lives.

That day, Gold found himself admiring this man 'who had played tag with the Gestapo with his life as the game – as the forfeit – and who did not need very much in the way of teaching regarding precautions'.[12]

And Gold now knew he was dealing with the most important spy he had ever met.

22

The Prime Minister's Mistake

Over in London, Churchill still wouldn't see Bohr and the 'Great Dane' continued to kick his heels. Growing worried, Anderson, Halifax and even Cherwell petitioned the prime minister, in minutes and notes and casual conversations, to see Bohr. They did so – and this is the central point – because *they too all agreed with Bohr.* All thought that, between them, Churchill and Roosevelt had a unique opportunity to shape the future of the world in an unprecedented way.

As if all these figures were not enough, Sir Henry Dale, the president of the Royal Society, was prevailed upon to add his word to those of the others. He wrote directly to the prime minister, drawing attention to Bohr's eminence and adding, 'It is my serious belief that it may be in your power, even in the next six months, to take decisions that will determine the future course of human history.'[1] Dale thus joined a select group who knew about the bomb and were familiar with Bohr's arguments. It is important to say that none of these men were idealistic young scientists or naïve: Dale, for example, was sixty-nine and a Nobel Prize-winner in Medicine, one of the team who had discovered/invented penicillin.

Finally, the indefatigable Anderson brought in Churchill's friend, Jan Smuts, the 74-year-old prime minister of South Africa, who was hugely impressed by Bohr, likening him to 'a Shakespeare or Napoleon – someone who is changing the history of the world'.[2]

R. V. Jones was, perhaps, not in quite the same class as these other luminaries but he was an accomplished physicist who was playing an important role in the war, converting scientific intelligence into military policy. He was no naïve ideologue either and it fell to him to look after Bohr while the wait went on.

Niels and Aage were accommodated in a flat in St James's Court in Buckingham Gate and they were allowed contact with only a strictly controlled group of individuals. (The secrecy couldn't have been that thorough though, given that Kapitsa's letter to Bohr had arrived in London.)[3]

Jones says he and his deputy, Charles Frank, took the opportunity to have several 'tutorials' from Bohr about nuclear and quantum physics.[4]

Jones also says that Bohr asked him to try to set up a meeting with Churchill though, as we have seen, Anderson, Cherwell and Dale were already pressing this on the prime minister. According to Jones, he was very ready to try to arrange such a meeting,

> because I could see in it a chance to make Churchill realise just how vital the issue was. I thought that he had been inadequately briefed by Cherwell, because the latter did not really believe that nuclear energy would ever be released. In fact, he had said as much to me, as also had Tizard. It seemed that each of them had become so alarmed at the destruction that atomic bombs could cause that they clung to the hope that God had not so constructed the Universe that he had put such power in the hands of men.[5]

Jones therefore calculated that, if Cherwell wasn't certain, a greater physicist like Bohr would finally convince the prime minister to take the whole matter more seriously. In the event, Jones wrote that he felt it was *his* intervention that was crucial, more so than those of Cherwell himself, Anderson or Dale. 'At any rate, the message that Churchill would see Bohr came through Cherwell to me, with the request that I should tell Bohr.'[6]

No sooner had Jones done so than, he says, Bohr asked for his help. He knew what he wanted to say but was aware that his English was hardly good enough. He proposed he would write it all down and have Jones tidy it up. Jones says he was not entirely happy with this procedure but agreed and, after two or three drafts, Bohr said he was content with what had been achieved.[7]

What we don't know is what Jones told Bohr, by way of preparation, about the situation with regard to atomic intelligence, or if Bohr showed Jones the drawing that had been the object of so much discussion in Los Alamos. Nor do we know directly if Jones and Bohr discussed the fact that they both knew by then, and for some time before, that Germany did not have a bomb.

But, given Bohr's reason for being in London, and the number of times they met, and the fact, mentioned above, that Bohr had met with more or less all Britain's wartime intelligence chiefs, it is inconceivable that they did not. In the book he published long after the war was over, in 1978, Jones made it quite clear that he had always got on very well with Churchill, that he had advised the prime minister on many matters of scientific intelligence and, as just mentioned, felt that it was *his* intervention that finally persuaded him to see Bohr. He also made it clear in his book – via two extensive footnotes, one about neutron release and capture in U-235, and the other about the bomb's dimensions – that he fully understood the physics determining explosions. Having Bohr to himself, so often and for so long, in the six weeks between Bohr's return and the Downing Street meeting, knowing what he knew about the German programme, and knowing that Churchill knew it too, Jones briefed the other man fully, not just on basic details but on how best to approach the prime minister.[8]

The half-hour meeting between Churchill and Bohr was finally scheduled for three o'clock on the afternoon of 16 May 1944, a Tuesday. It has to be said that, among those who knew of the meeting, hardly anyone – save Bohr himself – expected too much from it.

Anderson had tried to interest the prime minister in the subject of sharing information with the Russians twice, and been summarily rebuffed both times. Cherwell had argued with Churchill over the Quebec Agreement, something that still rankled (the professor thought the prime minister had given too much away). Dale's letter to the prime minister was, in Anderson's view, 'not a happy effort' – two pages 'of verbose pleading that made the serious error of drawing attention to Bohr's attempts to win the ear of Roosevelt in Washington', which Churchill, the chancellor thought, would see as 'presumptuous dabbling'. Henry Dale himself was braced for the way the prime minister might react to Bohr's 'inarticulate whisper' and 'mild, philosophical vagueness'.[9]

The state of the war obviously didn't help, either. It was three weeks before D-Day and the prime minister was nervous that the whole thing could be a disastrous bloodbath. Churchill could be forgiven for being on edge. Everyone was.

Finally, there was Russia. By the time Bohr saw Churchill, victory over Germany was increasingly likely and the prime minister was anxious that Roosevelt did not seem to quite share his apocalyptic view about Soviet intentions. Churchill had been shaken and chilled at the Big Three Conference in Teheran when Roosevelt had paid Stalin more heed than Churchill himself and even, on one or two occasions, seemed to take a mischievous delight in embarrassing the prime minister in front of the Russian leader, who had lapped it up. Following Teheran, it was now all too clear that Britain, having promised and provided Russia all the help it could in 1941 following the German invasion and before the US was even in the war, was now very much the junior partner. But one thing the prime minister still had up his sleeve, so to speak, was the great secret he shared with the president about the atomic bomb.[10]

What is not clear is whether Churchill grasped that, just then, it was still unknown how much fissionable material was needed for a bomb, and scientists did not yet have the means for constructing a test device, let alone a deliverable weapon. Any role an atomic bomb might play in the war was still uncertain.[11]

The choreography of the encounter was that, on the afternoon of 16 May, Jones took Bohr to Cherwell's office and Cherwell then took him on to Downing Street.

The two physicists arrived a little early for the three o'clock start, a time when the prime minister would normally have had his afternoon nap. He walked in and found them sitting together.

He had apparently just read – or re-read – Sir Henry Dale's letter, which Graham Farmelo describes as a 'windy appeal on behalf of the scientific community'. If there was anything guaranteed to get up Churchill's nose it was scientists interfering in politics, despite the fact that he himself was arguably the most scientifically literate prime minister the country had ever had. What happened next is that he ignored Bohr and launched an attack directly against Cherwell, whom he accused of arranging the meeting 'to reproach me for the Quebec Agreement'. This was a reference to a fight that had been festering since the previous summer, when Cherwell – and R. V. Jones come to that – thought the prime minister had given away more than was justified, and allowed the president more or less carte blanche to decide just how much access to nuclear technology Britain would have after the war.

When Bohr did at last begin to speak, Churchill took immediate exception to his detailed systematic approach, ending in his conclusion. Churchill was running a war, and wanted people to start with the conclusion and then justify that conclusion if an explanation were felt to be needed. He cut Bohr short before the Nobel Laureate could even get to his main point. 'I cannot see what you are worrying about,' he grumbled. The atom bomb was only a bigger version of existing bombs, he insisted, and 'made no difference to the principles of war'. And, in his first put-down, he said that there were no long-term problems with the bomb that could not be 'amicably settled between myself and my friend President Roosevelt'.[12]

There was no discussion of the German bomb, or non-bomb, no discussion of what the Russians knew, or might know, and when they might have known it, no discussion of how the target had changed, or was in the process of change, no discussion of how scientists themselves might help initiate the process of co-operation, and no discussion of the need to avoid the appearance of coercion in relation to Stalin.

At the end of the meeting, which arrived all too soon, when Bohr asked if he might write to the prime minister, outlining the ideas he hadn't yet got to, Churchill, in a notorious second put-down, said he would be honoured to hear from the Nobel Laureate, 'but not about politics'.

Bohr was devastated and afterwards he wandered around London 'like a wounded lion'. At about five o'clock that afternoon, R. V. Jones bumped into him in Old Queen Street. Jones said Bohr had his eyes 'cast heavenwards' and seemed to be in a daze, so much so that he walked right past him. 'Fearing that something was wrong, I went back and stopped him, to ask how he got on. All he could say was, "It was terrible! He scolded us like two schoolboys!"'[13]

Churchill had been shaken by the meeting, too. A little over a week later he sent Cherwell a note in a sealed envelope in which he returned yet again to the Quebec Agreement. He wrote that, 'I am absolutely sure we cannot get any better terms by ourselves than are set forth in my secret Agreement with the President. It may be that in [later] years this may be judged to have been too confiding on our part. Only those who know the circumstances and moods prevailing beneath the Presidential level will be able to understand why I have made the Agreement.'

This was of course a backward reference to the fall out that had existed between the Allies prior to Quebec, a predicament that Churchill, to his credit, had partly rescued. 'There is nothing more to do now but to carry on with it . . . Our association with the United States must be permanent and I have no fear that they will maltreat us or cheat us . . . The great thing is to get on with the job and keep it absolutely as secret as we can.'[14]

In fact, even as Bohr was meeting Churchill, British and American officials were surveying the supplies of uranium ore around the world, calculating what they were likely to need after the first bomb(s) had been dropped, and whether such ore could be denied to other nations. The British were in a strong position here, because America had little in the way of raw materials but Britain had ready access to large deposits in the Commonwealth. Vannevar Bush, James Conant and in particular General Groves were most aware of the strategic importance of these deposits and put the British under pressure to make available as much as possible. Churchill was kept abreast of these negotiations which, he felt, went some way to securing for Britain a future place at the top table where nuclear power was concerned.

Bohr remained in Britain while the D-Day landings were completed, and throughout June. The success of the invasion was clearly important because it then became clear to most people that the war in Europe was entering its final stages, and that the Allies would win.

But that only made the need for a nuclear agreement of the kind Bohr was proposing more urgent than ever. The window of time when disclosure would not look like coercion to the Soviets was narrowing all the time.

The next meeting between Fuchs and Gold occurred in mid-June, while Bohr was still in London, and took place in the Sunnyside section of Queens, the New York suburb which, among other claims to fame (Ethel Merman, Lewis Mumford and Bix Beiderbecke were all residents), was home to William Patrick Hitler, the Führer's nephew. The D-Day invasion had just got under way in northern France. Hitler had hit London with his V-1 rockets, or flying bombs, the long-rumoured 'secret weapon', an exotic idea that proved highly inaccurate in practice.

There had been fears that the V-1 weapons (V for *Vergeltungswaffe*, or 'Vengeance Weapon') would carry radioactive materials, but those worries proved groundless. So too did the parallel concerns that the German army would try to block the Allied invasion route with fission products. In both cases, British intelligence had shown its worth. French and Belgian resistance had identified seventy-four 'ski jumps', from where the V weapons would be launched, and sixty-six had been bombed.[15] Enigma traffic had shown no mention of radioactive materials, or troops being trained to use such products.

The June meeting in Sunnyside lasted for only three or four minutes because once again Fuchs's main purpose was to hand over a package of information, 'some 25 to 40 pages'.[16] And once again Gold's curiosity got the better of him and, soon after they parted, he snatched a glance at the material. 'I still had about five minutes to wait and I recall stopping near a drug store and taking a glimpse ...

The material was in a very small but distinctive writing; it was in ink, and consisted mainly of mathematical derivations. There was also further along in the report a good deal of descriptive detail.' It was another series of formal reports that Fuchs had had a role in preparing, with still more on gaseous diffusion.

By this time, too, Christopher Kearton, his colleague who had arrived in America on the *Andes* with Fuchs, had over the previous month visited a number of other facilities across America. These included the Allis Chalmers Manufacturing Company in West Allis, Wisconsin and the Houdaille Hershey Company. This outfit had moved from New York to Oak Ridge, so this may have been how Fuchs learned that the huge plant in the south was not in Georgia or Alabama but in Tennessee.

Fuchs also identified his American opposite number in the gaseous diffusion (K-25) work. It was Dr Karl Cohan, a theoretical physicist, who left the New York area in May 1944. This was evidence, Fuchs concluded, that Cohan was moving on to Site Y. He still didn't know were Site Y was but it was clear the Project was moving up a gear.[17]

23

The Bomb in Trouble: Mistakes at Los Alamos

In London, Bohr had some satisfaction at last in that he met with Sir John Anderson and the South African leader, General Jan Smuts. Although the meeting between Churchill and Bohr had gone badly, Anderson had subsequently prevailed on the prime minister to allow the chancellor himself to inform Smuts about the Manhattan Project and asked the general to begin a study of post-war control. Bohr was the first person Smuts consulted.

'This time Bohr talked to a man willing to listen. If ever there was a matter for international control,' Smuts agreed, 'this is it.'[1] However, Smuts felt that this control should be installed only *after* the war was over. He was not convinced that any immediate approach should be made to other nations and in any case the approach should be made, when the time came, from the US president.

Bohr had achieved, perhaps, a half step forward, but no more. He now left London for Washington, half hoping that Anderson and Smuts would prevail on Churchill to change his mind but also intending to refresh his friendship with Frankfurter.

At about this time, Fuchs and Gold had their sixth meeting, on this occasion near Borough Hall in Brooklyn. This rendezvous was quite unlike any other. Fuchs talked about his sister Kristel as they walked around. Kristel, who was also a communist, like Klaus, had married Robert Heinemann, a member of the Communist Party but by all accounts not as avid politically as his wife. Her marriage was rocky, Fuchs told Gold, and she might have to leave Boston. He was fond of his sister and was exploring whether she could go back and live in Britain. Another alternative was that she should move to New York with her two children and live with him. Fuchs was aware also of the tendencies to suicide in the family, especially among the women. More, given the sensitivity of his position, he wanted to know from Gold how the Soviets would react to his taking his relatives in. Gold, who hadn't come across this kind of problem before, said it would take a while for him to have an answer.

He then said that, prior to this meeting, he had been given 'several typewritten pieces of paper' by John, 'about 3 by 9 inches, or irregular size, which had contained a number of questions relating to atomic energy. The phraseology of these questions,' Gold later said, 'was extremely poor, and I had great difficulty in making any sense out of them.'[2] Gold got the impression that the translation had suffered in decoding from the Russian, and tried to explain this to Fuchs. But 'Fuchs seemed to take offense at being instructed and said very briefly that he had already covered all of such matters very thoroughly, and would continue to do so.'

As soon as possible after he returned to Washington, Bohr saw Frankfurter and told him of the disastrous meeting in London. Frankfurter lost no time in informing the president who was highly amused by Churchill's reaction, which he thought typical of the prime minister in one of his belligerent moods. But he said he would now like to meet Bohr himself.

In order to prevent the kind of meltdown that had occurred in Downing Street, Frankfurter told Bohr that he should prepare a carefully worded memorandum that the president could read and digest before they met.

Through a very hot July, the memorandum went through many drafts, dictated by Bohr and typed up by his son, Aage, since the matter was too secret for a secretary. So sequestered were they that Bohr even darned their socks while Aage typed.[3]

The seventh meeting between Fuchs and Gold began on 5th Avenue, near the Metropolitan Museum of Art. During an hour-and-a-half-long stroll 'in the many winding roads and small paths' around Central Park on a very warm July day, at exactly the time Bohr was at work on his memorandum for the president, Fuchs again talked with Gold about his personal problems. He had a brother, Gerhardt, convalescing in Switzerland after escaping from Torgau prison in east Germany. His daring and imaginative escape had been plotted with the aid of his wife, who had hidden a cache of civilian clothes near the prison so that Gerhardt could disguise himself as an ordinary workman. Fuchs was proud of what his brother and sister-in-law had achieved, but naturally distressed at his tuberculosis.[4]

At that meeting Gold took it upon himself to tell Fuchs that it would be perfectly in order for him to bring his sister to New York and live with him, as he had asked at the previous rendezvous. In fact, Gold was acting off his own bat, because he hadn't discussed it with John. But in any case, the problem with Fuchs's sister appeared to have gone away, for Fuchs had some momentous news. He had been to Washington to meet James Chadwick, the most senior British scientist on the Manhattan Project. Chadwick had told Fuchs that it appeared more than likely he was being transferred. He, Fuchs, was being relocated, 'somewhere to the South West, possibly New Mexico'.[5]

In fact, Los Alamos was – to an extent – in trouble. Until then the detonation mechanism for a bomb had been planned as a so-called gun mechanism. One subcritical mass of nuclear material would be fired up the barrel of a cannon to amalgamate it with another subcritical ring fitted around the muzzle. The problem with this was that both uranium and plutonium fissioned spontaneously, as Flerov and Petrzhak had been the first to show. This meant that secondary neutrons might start a chain reaction prematurely, and cause predetonation, what someone called 'the fizzle of the century'.*

* Moscow knew this already thanks to intelligence provided by other atomic spies in America, in particular Theodore Hall and David Greenglass.

This discovery of what one expert called 'the nasty properties of plutonium' had emerged in the summer and provoked a crisis at Los Alamos. No one had foreseen that when plutonium was produced (by exorbitantly expensive nuclear reactors), it would contain an unstable isotope, PU-240, in which one gram of this impurity produced no fewer than 1.6 *million* spontaneous fissions every hour.

This was such a devastating development that Oppenheimer, who was so worried that he had lost thirty pounds, to the point where his clothes were hanging off him, even considered resigning. But then saner counsels prevailed and it was decided 'to throw the book' at the problem. By July 1944, when Fuchs saw Chadwick in Washington, Los Alamos had decided the gun mechanism had to be scrapped and another method – implosion, the rapid compression of a plutonium sphere (to the size of an almond) by a symmetrical detonation of conventional explosives – developed to replace it. But this was a complex process that needed the science of explosions to be developed to a more detailed level than ever before, and required a great amount of preliminary theoretical work – calculations about fluid motions, the correct placement of detonators, the geometry of the shell. Hans Bethe, who was head of the theoretical division and had worked with Rudolf Peierls at Manchester in the 1930s, asked Peierls to help with the theory. Peierls agreed, provided he could bring Tony Skyrme and Klaus Fuchs.[6]

Bohr finally had his meeting with Roosevelt on Saturday 26 August, following which his spirits quite revived. The two men spent an hour and a quarter together, during which time the president laughed out loud at Bohr's description of Churchill's treatment of him, saying it was nothing other than the prime minister's 'black dog' depressions, which were well known.

Bohr found the president 'very agreeable' and learned that he shared his view that contact with the Soviet Union 'had to be tried along the lines he suggested'. The president went so far as to suggest that he was optimistic such an approach would have a 'good result' and open 'a new era'.[7] He said that in his opinion Stalin was enough of a realist to grasp the 'revolutionary importance' of the bomb and its consequences. No less important, Roosevelt told Bohr he was confident he would win the prime minister around to this view. 'They had disagreed before, he said, but in the end had always succeeded in resolving their differences.' He suggested that he and Bohr might meet again after the upcoming Quebec Conference (the second such meeting there and codenamed 'Octagon') but, in the meantime, if Bohr had any further views to communicate, 'he would welcome a letter'.[8]

In fact, the meeting between Roosevelt and Bohr was so successful that the president later told Frankfurter that the Dane was one of the most interesting people he had ever met. Reinvigorated, Bohr was therefore hopeful that Roosevelt would talk Churchill round, and convince the prime minister of the unique opportunity that lay within their grasp, when they had their second Quebec meeting – now not far off – in September.

How realistic was Bohr being in all this? On 13 June 1944, weeks before Bohr met the president, Roosevelt and Churchill had signed an Agreement and Declaration of Trust, which specified that the United States and Great Britain would work together in seeking to control available supplies of uranium and thorium ore during and after the war. The blunt truth is that the Agreement was designed to ensure that, if an arms race *did* develop after the war, the United States would always have a great enough supply of raw material 'to ensure superiority'.[9]

Fuchs's eighth meeting with Gold was scheduled to be near the Brooklyn Museum of Art on Eastern Parkway in exactly the same week that Bohr met the president. Gold stood at the Belle Cinema and Brooklyn Library for quite some time, far longer than the four- or five-minute waiting period that had been agreed upon. But Fuchs failed to show. And he failed to show at the back-up location also, near the junction of 96th Street and Central Park West.

Gold was concerned. 'On this second occasion I became very worried, particularly since the area is very close to a section of New

York where "muggings" often occur, and also the fact that Klaus was of slight build and might seem an inviting prey.'[10] Gold had no way of contacting Fuchs other than at these agreed meeting places. He didn't know how his Russian masters would take to the news that contact had been broken.

He met with John, more than once, to discuss Fuchs's disappearance. John instructed Gold to visit Fuchs's apartment. On the way he bought a second-hand book, Thomas Mann's *Joseph the Provider*, and wrote Fuch's name and address in it. He pretended he was returning it to its owner but at the building the superintendent told him that Fuchs had left town.[11]

It would be seven months before Fuchs and Gold were in touch again. But there was a good reason for Fuchs's silence. He had indeed been relocated, to 'Site Y', the top secret central location of the Manhattan Project at Los Alamos in New Mexico. The Russians had a spy who was a first-rate mathematical physicist at the very heart of the Allies' atomic bomb facility.

24

The President's Mistake

Some historians have suggested that the atom bomb played a crucial role in Churchill's psychology because he was all too well aware, as the war's end came in sight, in 1944, that Britain's power and role in world affairs would soon be eclipsed by the United States and Russia, who would be the only two post-war 'superpowers'. Possession of the bomb, and its denial to Russia, would at least enable Britain to hang on to some semblance of influence.

Having made his mark as a belligerent, the prime minister had little taste for the task of post-war reconstruction that lay just over the horizon. The atom bomb, on the other hand, held out the promise of continued superiority, and continued post-war co-operation between Great Britain and the United States, and as such it was a forceful item on the agenda for the second conference held in Quebec, Canada, in September 1944 between Roosevelt and Churchill.

Held in the 'afterglow' of the liberation of France and Belgium, this was the tenth time the leaders had met in wartime. In fact, with both, their age was beginning to tell. Roosevelt, in particular, had grown frail. 'You could put your fist between his neck and collar,' noted Charles Moran, Churchill's doctor, who accompanied him to Canada.[1]

Before he had embarked for Quebec, Churchill had asked for a briefing note from Cherwell. Cherwell was not encouraging. 'The extraordinary American ideas on security', he said (he meant Groves's mania for compartmentalisation), made it difficult to arrive at a fully considered verdict about progress on the bomb. But he doubted at that stage that the weapon would be built in time to be of use in the war. In another telegram to the prime minister, on 12 September, the very day the conference started, Cherwell urged the prime minister to get from the president some specific idea about his country's post-war energy policy. 'Do they wish,' Cherwell had written, 'as we should like, to go on collaborating after Japan is defeated – and could the two countries continue to cooperate in developing such a vital weapon unless they were united by a close military alliance?'[2]

As things turned out, the bomb was not considered in the second Quebec Conference proper, but was saved for a subsequent meeting, at the president's country retreat, Hyde Park, in the Hudson River Valley, held over a few days immediately after the Quebec gathering ended.

And no quarter was given. As a result of those more private deliberations, an aide-memoire was drawn up, which the two leaders signed on 19 September. Churchill completely won over the president, convincing him of the need for secrecy, arguing that Bohr's attitude, though shared by many other scientists and many officials surrounding himself and Roosevelt, was little short of treason. The second paragraph of the agreement read that: 'Full collaboration between the United States and the British government in developing tube alloys for military and commercial purposes should continue after the defeat of Japan unless and until terminated by joint agreement.'

The aide-memoire also contained wording that indicated what was in Roosevelt's mind at that time. 'When a bomb is finally available, it might perhaps, after mature consideration, be used against the Japanese, who should be warned that this bombardment will be repeated until they surrender.'

But Churchill's hostility to Bohr warped his judgement so much that this hostility crept into the agreement. 'Enquiries should be made regarding the activities of Professor Bohr and steps taken to ensure that he is responsible for no leakage of information particularly to the Russians ... The suggestion that the world should be informed regarding Tube Alloys, with a view to an international agreement regarding its control and use, is not accepted. The matter should continue to be regarded as of the utmost secrecy.' This was, in effect, a (deliberate?) misrepresentation of what Bohr was saying. He never at any time argued that 'the world' be informed about the bomb. He focused on the Soviet Union only and on the bomb's existence only, so as to avoid its possession and use being seen as coercive.

As Martin Sherwin phrases it, 'The agreement bears all the marks both of Churchill's attitude toward the bomb and of his distrust of Bohr.' By general agreement among historians, Churchill himself wrote the document, or at least edited it. The word usage is British English. In a draft of the aide-memoire, the phrase 'might perhaps, after mature consideration' was written in Churchill's handwriting in place of the single word 'should'.[3]

Churchill left Hyde Park the following day and by all accounts was still raging about the threat to secrecy he felt Bohr represented. In a letter to Cherwell, which he wrote the next day, he dismissed the world's greatest physicist as a publicity seeker, arguing that he had deliberately leaked intelligence matters to Frankfurter and was in 'close correspondence' with a senior colleague in Moscow. 'What is this all about?' growled the prime minister, adding that both he and the president were 'much worried'. 'It seems to me that Bohr ought to be confined or at any rate made to see that he is very near the edge of mortal crimes.'[4]

This was too much of a misrepresentation for those in Whitehall (and Washington, come to that) who knew Bohr to be a man of the utmost integrity and probity, and their views were forcefully made known to Churchill. Cherwell took the initiative here in putting the record straight, informing the prime minister that although Bohr had 'some rather woolly ideas' about how nuclear weapons could persuade countries that built them to live in mutual trust (ideas he himself had appeared to share until Quebec), Churchill need not worry about the professor's loyalty.[5]

But the damage had been done. According to Frankfurter's memoirs, the president was so overwhelmed by Churchill at Hyde Park that he even denied his earlier enthusiasm for Bohr, though there were several witnesses who could vouch for it. And the follow-up meeting that Roosevelt had promised Bohr never took place.[6]

A few days after the Hyde Park conference, Roosevelt did have a follow-up meeting in Washington with Admiral William Leahy, his chief of staff, with Vannevar Bush and with Cherwell. Here the earlier aide-memoire was more formally underlined, that the two governments should share all their discoveries in nuclear matters and that collaboration should continue after the war, though it was for later administrations to decide to what extent such co-operation should be maintained. Roosevelt added that he thought the only thing that could get in the way 'would be if he and the prime minister, Bush and Cherwell were all killed in a railway accident since we all saw eye to eye'.[7]

Churchill would have been grimly gratified by this remark. The partnership in nuclear energy, despite its unpleasant setbacks in 1942 and 1943, had come to represent for him a reliable cornerstone of the collaboration that had been so successful in other areas of wartime endeavour. Nothing must be allowed to interfere with this basic lineament of national policy.[8]

After Hyde Park, Lord Cherwell had an aircraft placed at his disposal by the Americans, so he could tour the various Manhattan Project sites. On his return to Washington, at a meeting with the president's other aide, Harry Hopkins, Hopkins put American policy clearly: 'It was vital for the United States to have a strong Britain because we must be realistic enough to understand that in any future war England would be on America's side and America on England's. It was no use having a weak ally.' Even more to the point, in a document in the American Atomic Energy files, dated 25 September 1944, less than a week after the Hyde Park meeting, Vannevar Bush told James Conant that, 'The President evidently thought he could join with Churchill in bringing about a US-UK postwar agreement on this subject [the atomic bomb] by which it would be held closely and presumably to control the peace of the world.'[9]

As Sherwin phrases it: 'The underlying idea, the concept of guaranteeing world peace by the amassing of overwhelming military power, remained a prominent feature of his [Roosevelt's] postwar plans.'[10]

This view fitted neatly with Churchill's own and went directly against what Bohr was advocating. We don't know what, exactly, Roosevelt told Churchill at Hyde Park but we do know that he was fully aware by then that the Russians knew about at least the existence of the Manhattan Project, its several sites and their respective sizes, together with the names of many of the physicists involved and the new processes they were working out. So Roosevelt knew that the US-UK would have 'overwhelming' military power for a limited amount of time only.

And what did Churchill know? He knew that the original reason for building a bomb no longer existed: he could be certain Germany didn't have a bomb, courtesy of Jones, Cherwell and Anderson at the least, and he had known it for some time. But did he know what Roosevelt knew, that the Russians had penetrated the Manhattan Project, that, as John Lansdale had told Oppenheimer as long ago as September 1943, the Russians knew about Los Alamos, Oak Ridge and Chicago? Had Kapitsa's various interventions in the matter been fully explained to him, and had their implications registered? He made no mention of any of this in his memoirs. If the president had told the prime minister what he knew, would it have changed Churchill's mind about what secrecy could achieve? To be blunt again, did Churchill put the interests of Britain before the interests of the world? If so, could he be blamed for that?

As the British historian A. J. P. Taylor said of another phase of history: it was a turning point at which the world failed to turn.

PART FOUR

The Underestimate of the Russians

25

Bohr and Stalin

Niels Bohr's ideas about sharing the secrets of the atom bomb with the Russians were notoriously labelled 'naïve' by Donald Cameron Watt, professor of history at the London School of Economics, and the author of several books on the Second World War. Frederick Lindemann, Lord Cherwell, called Bohr's views 'woolly' though for months on end he himself appeared to agree with them. R. V. Jones thought that many of the details needed to be thought through more than they had been.[1]

These remarks and observations may be true so far as they go but they would hold more force if it were not for the fact that many others – Sir John Anderson, Lord Halifax, Felix Frankfurter and Sir Henry Dale, president of the Royal Society – agreed with Bohr wholeheartedly and still more soon came round to his view, albeit rather late in the day. Late in the day because, even without knowing what we know now, it soon became clear to most people who were in on the great secret that some form of international control was needed to protect the world from ruinous catastrophe and that, as both Bohr and Fuchs were among the first to realise, the central ingredient in that protection was *balance*. Only if the two great post-war powers – the USA and the USSR – were evenly matched in military capability, could an arms race be avoided.

This may sound obvious to us but it was anything but obvious to many people at the time. To some – Groves and Churchill come immediately to mind with Secretary of State James Byrnes and Truman not far behind, and Stimson vacillating – the idea of an American, even an Allied, monopoly on atomic weapons was mouth-wateringly appealing. It would, they thought, give America – and by implication the Western Allies – a means by which the main post-war enemy, Soviet Russia, could in effect be managed. She could be compelled to abandon territories she had acquired during the onslaught of war and the menace of world communism could be severely constrained if not removed altogether. (Stimson at one stage early on envisaged threatening Russia itself if she did not democratise as he wished her to.)

The first moves in following up the Bohr approach came from Bush and Conant. Neither had been in a hurry to 'precipitate' post-war issues but, in the summer of 1944, just ahead of the second Quebec Conference, they formed the view 'that the time had come for earnest thoughts about arrangements for both domestic and international control'.[2] Even as late as November 1944, says Martin Sherwin, scientists – some scientists – on the Manhattan Project believed that the 'German atomic energy project was probably at about the same stage as the Anglo-American project and "it would be surprising if the Russians are not also diligently engaged in such work"'.[3] They were still (being kept) in the dark about the lack of progress in Germany.

In fact, it was the scientists at the Metallurgical Laboratory in Chicago who were the immediate stimulus to action. Since most had initially joined the project out of fear that Hitler might build an atomic bomb, and as it became increasingly evident that there was no Nazi bomb, they began to worry about what they were now doing.

In June 1945, at the Chicago laboratory, a committee on the 'Social and Political Implications' of the bomb was formed, headed by Nobel Prize-winner (and German-Jewish exile and friend of Lise Meitner) James Franck, and tasked with putting the scientists' views in writing.[4] His report recommended a demonstration of the bomb in an uninhabited area, and that there be no surprise attack.

> If the United States were to be the first to release this new means of indiscriminate destruction upon mankind, she would sacrifice public support throughout the world ... It will be very difficult to persuade the world that a nation that was capable of secretly

> preparing and suddenly releasing a new weapon, as indiscriminate as the rocket bomb and a thousand times more destructive, is to be trusted in its proclaimed desire of having such weapons abolished by international agreement.[5]

Franck's report was completed on 11 June and passed to Arthur Compton, who made sure it was considered by the Interim Committee, specifically set up by the administration to consider the use of the bomb. But there was little chance the report would have an effect on the committee – Compton admitted that for its members dropping the bomb was 'a foregone conclusion'.[6]

In response to the Franck report, Groves asked Compton to survey the views of the Chicago scientists in more detail, and present them with a series of options ranging from military use with no warning to keeping the weapons secret and not using them at all. One hundred and fifty scientists were canvassed, with 46 per cent – the largest group – opting for a military demonstration, followed by the opportunity for Japan to surrender.[7]

There is no evidence that the petition was ever seen by the president and so Groves achieved his aim of appearing to collaborate but, in effect, keeping the scientists' views from reaching Truman. The general himself was dismissive of the scientists. None of them had relatives at risk, he said, which made them 'soft'. He was especially dismissive of Franck and Szilard who were, he said, 'strongly affected by their Jewish background. Once Hitler was defeated, they were not interested any more.' As the organisation of the petition shows, this was patently untrue.[8]

On 19 September 1944, three days after the Quebec 'Octagon' Conference finished and when Churchill was worrying about Bohr being a Russian spy, Bush and Conant drew up a letter for Stimson in which they addressed the domestic control of atomic energy within the United States and urged the importance of a treaty with Britain and Canada to ensure equivalent domestic controls within those countries too. They were also aware that the three allies would need

to control the harvesting of raw materials and they foresaw difficulties – which would recur – in that America's close relationship with Britain should not compromise, down the line, her relations with Russia. They recognised that Russia would at some point acquire a weapon of her own.

Roosevelt, for his part, felt that in the wider scheme of things Britain needed America's continued support after the war. Thanks to six years of fighting, some of the time spent struggling alone, the country was close to bankruptcy. The president felt that there were two ways America could support her principal ally – with economic aid and by having her become a nuclear power. This is why he proposed the full exchange of information on atomic matters even after the war was over. Though Bush sympathised, he still felt this might at some stage interfere with America's best interests vis-à-vis Russia, simply because an Anglo-American monopoly on the bomb would be impossible to maintain 'and therefore disastrous to pursue'.[9]

A few days after their letter, on 30 September 1944, Bush and Conant sent Stimson two papers on the future international handling of atomic weaponry.[10] They advised the secretary for war that there was every reason to expect the first bomb to be ready by 1 August 1945, with a blast equivalent to 10,000 tons of high explosive. Though terrifying in itself, they emphasised that not far into the future lay the hydrogen bomb, which might be as much as 1,000 times more powerful. 'This promised to place every nation in the world at the mercy of whoever struck first.' And was undoubtedly a new and threatening development in warfare.

They then went on to say that the lead then enjoyed by the US and UK was 'temporary'. Any nation with 'good technical and scientific resources' could catch up and even overtake the Allies 'in three or four years'. 'It would be folly for the two English-speaking countries to assume they would always be ahead. Accidents of research might give another nation as great an advantage as they currently enjoyed.'[11]

The Bush and Conant papers went on to argue that it would be 'foolhardy' to try to maintain security by preserving secrecy. 'Physicists knew all the basic facts before development began.' Therefore they advocated 'full disclosure' of all but the manufacturing and military details of the bombs *as soon as the first one was demonstrated.* They also thought that it was unwise to rely on the control of raw materials (as Groves was arguing).[12]

The best chance of forestalling a 'fatal competition', they said, lay in the free interchange of all scientific information via an international office deriving its authority from whatever international association of nations was arrived at after the war. It should comprise a technical staff with unimpeded access to laboratories, industrial plants and military establishments across the world, exactly as Bohr had argued. Since no bomb had been exploded yet, they thought it might be worthwhile to try to create in the first place an international organisation in regard to biological weapons, as a sort of practice run.

Although this provided a basis for subsequent discussions, Henry Stimson in particular was not in a hurry to make up his mind. While not implacably against giving information to the Russians (he thought they would get the bomb sooner or later anyway), and though he even considered a 'freeze' on the production of more bombs while a basis for co-operation was worked out, Stimson felt that such information should not be divulged until the United States could work out what it wanted in return, such as an agreement by the Soviet leadership to allow the liberalisation of Soviet society. Still others in the administration thought that continued total secrecy was the best policy, combined with a continuing research effort.[13]

Leo Szilard's views had developed too. In January 1945 he wrote to Bush arguing that the bombs should be brought into being as quickly as possible because unless and until they were used 'the public would not comprehend their destructive power'* and only such understanding would lead to a stable peace.[14] He too thought that the time to approach the Russians was immediately after a successful military use, because only if America's lead could be shown to be 'overwhelming and unapproachable' could it have a hope of success.[15]

* This was echoed, in a way, by Norris Bradbury, the director of Los Alamos who followed Oppenheimer. He thought that 'The occasional demonstration of an atomic bomb – not weapon – may have a salutary effect on the world.'

The Yalta Conference, codenamed 'Argonaut' and held in the Crimea from 4 to 11 February 1945, was the last time the original 'Big Three' met. Everyone was aware that the end of the European war was in sight, bringing nearer the invasion of Japan and, with it, the risk of massive loss of life on both sides. The atomic bomb was not discussed in this context, not then, though Churchill proposed that a general ultimatum be delivered by the US, UK, China and, if possible, Russia – which was at the time neutral in the Far East – calling for 'unconditional surrender' by Japan.[16]

Also at Yalta Churchill called on Stalin to honour an agreement he had made at Teheran to abandon neutrality and declare war on Japan 'within two-to-three months of the end of the fighting in Europe'. By then, many felt that the sheer shock of Russia abandoning its neutrality might be enough to bring Japan to its knees.[17]

Yalta was interesting for rather more than this, however, important as all that was. It was at this conference, held in the Livadia Palace, the summer retreat of the last tsar, Nicholas II, that Stalin was at his most amenable. Ahead of Argonaut, both Churchill and Roosevelt had felt that the Balkans, save for Greece, were lost to the Bolsheviks, as was Poland. In fact, they found Stalin agreeing to free elections in Poland 'as soon as practicable', and signing a 'Declaration of Liberated Europe', which included the right of national self-determination. Later, Yalta would come to be regarded as a major betrayal and a 'sell-out' by Roosevelt and Churchill (and soon after it major differences began to appear in regard to Soviet ambitions in Europe), but it interests us for what it reveals about the personal relationships between the Big Three as the first nuclear explosion approached.[18]

For example, Churchill went on record at that time as asking why the Soviet leader would go to such lengths to accommodate the Anglo-Americans 'if he were not anxious to work with the two English-speaking democracies' after the war was over. Kevin Ruane, professor of history at Canterbury Christ Church in the UK, says that both Churchill and Roosevelt 'over-sold' what was achieved at Yalta, one reason being that each seems to have been charmed by the

Russian leader. The prime minister told Parliament after he returned to London: 'The impression I brought back from the Crimea ... is that Marshal Stalin and the Soviet leaders wish to live in honourable friendship and equality with the Western democracies. I feel also that their word is their bond.'[19]

He further told the Cabinet that 'Stalin meant well to Poland and the world', that the promised elections would be 'free and fair', and that although Prime Minister Neville Chamberlain had been wrong to trust Hitler, 'I don't think I'm wrong about Stalin.'[20] He also added that Roosevelt felt much the same way as he did. 'If only I could dine with Stalin once a week, there would be no trouble at all ... We get on like a house on fire.' Churchill even expressed the view that, in working alongside the Western democracies, Russia 'might be in the process of purging itself of its revolutionary outlook' and was giving grounds for believing 'that the Soviet Union would settle down and become a peace-loving member of the international community'.[21]

At the same time, however, he didn't like the way the president's thoughts about the atomic bomb appeared to be changing. Roosevelt, it seemed to Churchill, was 'wavering' in his view about keeping the bomb secret. One reason was because Russian diplomats at Yalta were asking pointed questions about weaponry. Another was that keeping Stalin in the dark risked damaging long-term relations concerning international control. A third was because he knew the secret was hardly a secret any more (he had been kept apprised of the FBI reports). And finally there was always the possibility that atomic knowledge could be traded for other political quid pro quos.

Churchill didn't see any of this, or want to see it. From his point of view, international control threatened, above all, the Anglo-American monopoly and the special relationship that went with it, which was his main strategic goal. Roosevelt, like Bush however, was not entirely convinced that Anglo-American co-operation was in America's best interests. He was sympathetic to Britain, yes, but he appears to have been 'troubled' by the 'incompatibility' of withholding information from Stalin while anticipating so much further collaboration with the Russians in the future.

About a month after Argonaut, in March, the Canadian prime minister William Mackenzie King visited Washington. On some things in their discussions, Roosevelt echoed Churchill. 'He did not think there was anything to fear particularly from Stalin in the future,' King wrote in his diary. Specifically on atomic matters, however, Roosevelt told Mackenzie King that he 'thought the Russians had been experimenting and knew something about what was being done'. As Mackenzie King observed, if this was true then continued secrecy could be counter-productive. '[T]he time had come to tell them how far the developments had gone.' But, the president added, Churchill 'was opposed to doing this'.[22]

Following Roosevelt's fatal stroke on 12 April that year, Stimson briefed the new president, Harry Truman, eleven days later, on the 23rd. He presented the new commander-in-chief not just with intelligence about the bomb but with the arguments for and against international control. These had been prepared by himself and his aide, Harvey Bundy, on the advice of Bush and Conant. Groves was also present at the meeting. Truman was not entirely overawed by what he was told, remarking that 'the bomb might well put us in a position to dictate our own terms by the end of the war'.[23]

On the same day that the new president was first briefed, Vannevar Bush raised a new question for Stimson: 'What was the best time to tell Soviet leaders about the bomb and the hopes for its control?'[24] By now Stalin was being more and more difficult, especially in regard to Poland, consistently sabotaging Allied plans and contributing to a rising tide of mistrust.[25] So much so that by the end of April Stimson feared a 'head-on collision'.[26] Still others urged that, in regard to Poland, Russia needed a cordon sanitaire in Eastern Europe, and could not be bullied into acquiescence, and that only friendship could 'beget' co-operation.[27]

By this time Bohr had refined his views in another memo, which Bush warmly endorsed, and in which he said it was all the more urgent to tell the Russians because, with the progress they were making in Eastern Europe, they might soon get their hands on the German physicists working on whatever nuclear project they had going. 'It was important to raise the question of international control while it could be done in a spirit of friendly advice. If the United States delayed to await further developments, its overtures might seem an attempt at coercion no great nation could accept. Indeed, it was important to start consultations before the weapon made its debut in actual warfare. This would permit negotiation before public discussion aroused passions and introduced complications.'[28]

It bears repeating that Bohr had thought through the problems of nuclear collaboration more than anyone else. He was far from being naïve.

The Interim Committee to advise the president came into being in May 1945, eight long months since Bush and Conant had drawn up their initial memorandum. They were crucial months because at the end of that time the European war was drawing to a close, and there were still substantive questions to address. Should the United States seek its security in Anglo-American atomic solidarity, 'or should it take daring measures to allay Soviet distrust and win Russian cooperation in a system of postwar safeguards?'[29] Given Russia's increasingly obdurate behaviour and policy in Eastern Europe and the Balkans, there was no easy solution. At its meeting on 21 June 1945, the committee revoked Clause Two of the 1944 Quebec Agreement, which required the USA to seek UK approval for use of the bomb, meaning America could now go it alone if circumstances warranted such a course of action. The committee also recommended that, if the opportunity arose, the president should 'advise the Russians that we were working on this weapon with every prospect of success and that we expected to use it against Japan. The president might say further that he hoped this matter might be discussed some time in the future in terms of insuring that the weapon would become an aid to peace.'[30] The committee also stipulated that should the Russians ask for more details, they should be told no additional information would be provided – not then, anyway.

The selection of targets in Japan was a fraught one. There was some disagreement among specialists as to whether the atomic bomb was a new type of weapon or whether it was much the same as other explosives, only more powerful. The Target Committee had originally selected seventeen potential cities for study, later narrowed to five, labelled AA (most desirable), A and B, the reserve list. Part of the point in the selection of targets was to prove that the weapon was a 'shocking' innovation, and for this purpose Groves and his scientific counterparts wanted a 'clean' background against which to judge the effects of the bomb – 'an undamaged target was more likely to fit into the "shock" strategy by showing the significance of this single weapon'.[31] The fact is that Groves was worried that the extensive firebombing raids, carried out over Tokyo in March that year, which had killed tens of thousands of civilians, would mean that the atom bomb might not stand out as a shocking weapon and not bring about surrender, as was hoped.

For this reason he was keen to bomb Kyoto, Japan's ancient cultural centre, with many beautiful buildings and gardens, and where, it was thought, the 'highly intelligent and cultured citizens' might push the government towards surrender. He was, however, overruled – more than once – by Stimson, who thought such a plan barbaric.

The prelude to the attacks on Hiroshima and Nagasaki left quite a lot to be desired, according to Michael Gordin. Only at the last minute were written orders required – until that point the bomb might have been dropped on the basis of verbal orders alone, little more than a conversation, with all the confusion that might have engendered. This, he says, was mainly Groves's doing. 'Groves stacked the deck so that everyone involved in the decision to use the atomic bomb had a vested interest in seeing those bombs used as quickly and as often as necessary.' Gordin also shows that a concern with radioactivity – perhaps *the* defining aspect of atomic bombs – only became a concern *after* surrender. 'During the war radioactivity was consistently underestimated or ignored.'[32] There was even a plan to follow the atom bomb with a firebombing raid which, on account of radiation, would have been suicidal for the pilots. Groves, for instance, so little understood the dangers and spread of radiation (despite the fact that he said nuclear physics was not difficult) that he told General Marshall American troops could enter the area within 'about thirty minutes' of the detonation![33] The firebombing raid was abandoned but not because of the risk of radiation, rather because Groves insisted on a single atomic bombing so that the effects of the blast could be evaluated 'cleanly'.[34] Groves was unprepared for the news of radiation sickness 'and sincerely puzzled by it'.[35] Stimson's undersecretary of war, John J. McCloy, admitted later 'that when the bomb was used, before it was used and at the time it was used, we had no basic concept of the damage it would do'.[36]

At this time, ignorance was compounded by obfuscation – to put it lightly. The day after Hiroshima, Truman released a statement that said in part, 'By 1942, we knew the Germans were working feverishly to find a way to add atomic energy to other engines of war, but they failed.' If not an actual untruth, this statement was seriously misleading. Since the middle of 1942, the Allies – some of the Allies – had known that Germany was not doing any such thing. Churchill released an equivalent statement: 'By God's mercy British and American science outpaced all German efforts. These were on a considerable scale, but far behind. The possession of these powers by the Germans at any time might have altered the result of the war and profound anxiety was felt by those who were informed.' This was a deeper distortion, both statements presumably being made to justify the great cost and appalling damage inflicted by the bomb and to forestall any awkward questions.

Nine days later, Churchill addressed the House of Commons. He told Parliament that he was in 'entire agreement' with the president 'that the secrets of the atomic bomb should so far as is possible not be imparted at the present time to any other country in the world'. He added that there would be 'at least three and perhaps four' years before the US could be overtaken and in those years, 'a sublime moment in the history of the world ... we must remould the relationships of all men, wherever they dwell, in all the nations'.[37]

This was a public stance. But privately, three weeks before, at Potsdam, the last meeting of the three leaders of the USA, Great Britain and Russia, he had argued for a diplomatic 'show-down' (a favourite word of his), in which he thought that the bomb could be used as a key to breaking the Soviet hold on Eastern Europe.

Bohr, as we have seen, was criticised for being naïve. Was Churchill any less so? Belligerence can be naïve, too. What seems clear is that the prime minister was by no means as nuanced as Roosevelt was. It is not certain, for example, that FDR would have embraced the view that many did who were in on the secret – that, once the bomb was built it would inevitably be employed. The Quebec Agreement had referred to 'mature' consideration of the problem and Roosevelt's general approach to the Soviet Union was both more subtle and more oriented to co-operation than either Churchill's or Truman's.[38] Most important of all, as Gar Alperovitz has pointed out, if FDR had lived, James Byrnes would never have become secretary of state. Alongside Truman himself, and Churchill and General Groves, Byrnes was by far the most belligerent of the Allied leaders at the end of the war and the beginning of the Cold War. These men were all for using the bomb as quickly as possible. Even James Chadwick, perhaps the only scientist of either nationality to get on with Groves, eventually came to the view that the general had 'connived' to keep the British out of the project and, with Byrnes, was a 'high-level obstructionist'.[39]

But the military and diplomatic aspects of the weapon were just as clear to Stalin as they were to Churchill and Truman. Stalin had actually comported himself cleverly at Potsdam, when Truman had mentioned – almost in passing – that the US had a new, especially powerful explosive, and the marshal had feigned indifference.

He had not forgotten, as Churchill appeared conveniently to forget, that Britain had an agreement with Russia to share all developments on new war technology, and she was not playing fair.* And when Stalin was finally told, at the conference, Churchill again misinterpreted his reaction.

On the final day, the prime minister rounded on Stalin over Russian behaviour towards the Balkan nations and in Eastern Europe. 'An iron fence has come down around them,' he hissed. 'Fairy tales!' Stalin spat back. Then, as the awkward meeting began to break up, around 7.30, Truman approached the Soviet leader. Churchill was watching, knowing what was coming. By now they had both digested Groves's report on the Trinity test of the plutonium bomb in the Alamogordo desert, received the day before and showing that the bomb not only worked, but was more powerful than had been thought (see p. 3).†

> I knew what the President was going to do. What was vital to measure was its effect on Stalin. I can see it all as if it were yesterday. He seemed to be delighted. A new bomb! Of extraordinary power! Probably decisive in the whole Japanese war! What a bit of luck! This was my impression at the moment, and I was sure that he had no idea of the significance of what he was being told. Evidently in his immense toils and stresses the atomic bomb had played no part. If he had the slightest idea of the revolution in world affairs that was in progress, his reactions would have been obvious ... But his face remained gay and genial ... As we were waiting for our cars I found myself near Truman. 'How did it go?' I asked. 'He never asked a question,' he replied. I was certain therefore that at that date Stalin had no special knowledge of the vast process of research upon which the United States and Britain had been engaged for so long.[40]

* The Russians did what they could to capitalise on this. In an instruction to Kvasnikov in New York in August 1945 Moscow urged him to use the statements by Truman and Churchill – to the effect that the bomb's technicalities would remain secret – as an argument to recruit 'agents among scientists'.

† Not included were details of Groves's instructions to the governor of New Mexico that he might have to impose a state of emergency if the explosion was bigger than calculated, or that he gave instructions for three press releases for three different outcomes: a story of an explosion without any casualties or damage, a second about severe damage – and a third consisting of obituaries of all those present at the test. This too shows the uncertainty about the bomb's destructive power. Trinity was banner headlines in El Paso, six lines in the *Washington Post*.

Truman's own version is equally misguided and revealing.[41] 'He didn't know what I was talking about. I told him we had been experimenting with this tremendously high explosive in New Mexico and that it had been successful ... [H]e had a pleasant expression on his face, and I don't think he knew what I was talking about, and I didn't care whether he did or not.'[42] Like Churchill, Truman was convinced that Stalin 'knew no more about it than the man in the moon'.

Their confidence suggests that neither Churchill nor Truman knew what Roosevelt had known: that the Russians had penetrated the Manhattan Project.

But of course they were both mistaken, and grievously so. Moreover, after he got the news about the Alamogordo test, Truman's behaviour changed. He became more assertive at Potsdam which, of course, Stalin noticed.

In fact, Stalin – we now know – had been expecting some sort of approach from the Americans at the conference and had agreed with Beria that he should 'pretend not to understand'. Although Stalin was expecting something at Potsdam, he seems nevertheless to have been shaken. Michael Gordin makes the interesting observation that, on 6 August, the day of the Hiroshima bombing, the Soviet leader refused to see anyone, 'which paralleled his reaction to the Nazi invasion of the Soviet Union, on 22 June, 1941, indicating severe shock and depression'.[43]

Stalin had been unnerved but perhaps not as much as he might have been. Marshal Zhukov later reported,

> Stalin ... pretended he saw nothing special in what Truman had imparted to him. Both Churchill and many other Anglo-American authors subsequently assumed that Stalin had really failed to fathom the significance of what he had heard. In actual fact, on returning to his quarters after this meeting, Stalin in my presence told Molotov about his conversation with Truman. 'They're raising the price,' said Molotov. Stalin gave a laugh. 'Let them. We'll have to have a talk with Kurchatov and get him to speed things up.'[44]

Molotov's own recollection was no less interesting. 'Truman didn't say "atomic bomb", but we got the point at once. We realized they couldn't yet unleash a war, that they had only one or two atomic bombs ... but even if they had some bombs left [so few weapons] could not have played a significant role.' As Richard Rhodes points out, 'the most important point in this recollection' is that the Soviet leadership grasped how many bombs the US had.[45] Fuchs, we now know, was partly responsible for this.

Despite everything, as David Holloway contends, it seems that Stalin didn't really grasp the full significance of the bomb until Hiroshima. Yakov Terletsky was a physicist on the staff of the NKVD and he recorded that after the explosion at Hiroshima, 'Stalin had a tremendous blow-up for the first time since the war began, losing his temper, banging his fists on the table and stamping his feet.' Terletsky conceded that his leader had something to be angry about. 'After all, the dream of extending the socialist revolution throughout Europe had collapsed, the dream that had seemed so close to realization after Germany's capitulation.'[46] In mid-August, barely a week after Hiroshima, Stalin summoned the People's Commissar of Munitions, Boris Vannikov and some of his senior staff, Kurchatov among them. 'A single demand of you, comrades,' Stalin growled. 'Provide us with atomic weapons in the shortest possible time. You know that Hiroshima has shaken the whole world. The equilibrium has been destroyed.'[47]

'War is barbaric,' Stalin also remarked at this time. 'But using the A-bomb is a super-barbarity. And there was no need to use it. Japan was already doomed!'[48] With Japan already defeated, Molotov for one had no doubt why the bomb was used. The two bombs 'were not aimed at Japan but rather at the Soviet Union. They [the Americans] said, bear in mind you don't have an atomic bomb and we do, and this is what the consequences will be like if you make the wrong move!' Stalin thought along similar lines. From Potsdam, he called Marshal Aleksandr Vasilevskii and instructed him to bring forward the planned invasion of Japanese territory.[49] And he told Beria, 'Hiroshima has shaken the whole world. The balance has been destroyed. That cannot be. A-bomb blackmail is American policy.'[50]

Stimson's view had begun to change. At Potsdam he had advised Truman that the US might use its discovery to help force the Soviet Union to abandon its communist system and come to resemble a Western-style democracy (this is why Churchill thought it was a 'sublime' moment in history). In a memo written in September 1945, a month after the bombs were detonated, and as a result of discussions with Averell Harriman, the US ambassador to Moscow, Stimson now conceded that no such change was going to happen, that any attempt to enforce such a tactic was likely to backfire and that, following Hiroshima and Nagasaki,

> the temptation will be for the Soviet political and military leaders to acquire this weapon in the shortest possible time ... Britain in effect already has the status of a partner with us in the development of this weapon. Accordingly, unless the Soviets are voluntarily invited into the partnership upon a basis of cooperation and trust, we are going to maintain the Anglo-Saxon bloc over against the Soviet in the possession of this weapon ... Such a condition will almost certainly stimulate feverish activity on the part of the Soviet toward the development of the bomb in what will in effect be a secret armament race of a rather desperate character ... the question then is how long we can afford to enjoy our momentary superiority in the hope of achieving our immediate peace council objectives.

Stimson also foresaw that the advent of the atomic bomb risked the destruction of 'the Grand Alliance', which had performed so well against the Axis countries.[51]

Bohr could not have put it better, nor what followed.

> Those relations may be perhaps irretrievably embittered by the way in which we approach the solution of the bomb with Russia. For if we fail to approach them now and merely continue to negotiate with them, having this weapon rather ostentatiously on our hip, their suspicions and their distrust of our purposes and motives will increase ... It is my judgment that the Soviet would be more apt to respond sincerely to a direct and forthright

> approach made by the United States on this subject than ... if the approach were made after a succession of express or implied threats or near threats in our peace negotiations.[52]

Truman did not agree. Early in the following month, on a visit to Tiptonville in Tennessee, he made it clear to a posse of reporters accompanying him that the US would keep control of all atomic technological information. 'I don't think it would do any good to let them in on the know-how,' he said, 'because I don't think they could do it anyway.'* He even acknowledged the beginnings of an arms race, in which, he insisted, 'we would stay ahead'.[53]

An interesting sidelight was provided by Eisenhower, who visited Moscow shortly after Hiroshima and, according to Edgar Snow, in the course of a private conversation there, spoke as follows:

> Before the bomb was used, I would have said yes, I was sure we could keep the peace with Russia. Now, I don't know. I had hoped the bomb wouldn't figure in this war. Until now I would have said that we three, Britain with her mighty fleet, America with the strongest air force, and Russia with the strongest land force on the continent, we three could have guaranteed the peace of the world for a long, long time to come. But now I don't know. People are frightened and disturbed all over. Everyone feels insecure again.[54]

A very different sidelight takes us back yet again to General Groves. Gar Alperovitz, in *The Decision to Use the Atomic Bomb*, says that Groves's incessant worrying about Congressional criticism of the huge cost of the Manhattan Project was an added factor in the rush to bomb Hiroshima. 'It seems clear today that the rush to produce the active materials and to drop the bombs on Japan as soon as possible was driven largely by a fear that the war might end before both types of fission bombs could be used.'[55] Stanley Goldberg, in

* This sentiment appears to have lingered, despite the evidence. In late January 1953, just as he was leaving office, Truman told a reporter: 'I am not convinced Russia has the bomb. I am not convinced the Russians have achieved the know-how to put the complicated mechanism together to make an A-bomb work.'

a paper given before the American Historical Association in 1995, went further. The three most important factors in dropping the bomb, he says, were: 'Momentum, protection of the reputations of the civilian and military leadership of the project, and the personal ambitions of some, especially General Leslie R. Groves.'[56]

Goldberg was at the time planning a biography of Groves but died before it could be completed.[57] The project was taken over by Robert Norris and, according to Barton Bernstein, in a review of Norris's book, 'To Goldberg, Groves was feverishly eager to use both bombs at virtually all costs.'[58] Goldberg, says Bernstein, was unreliable in many ways, but this does place into context a memorandum sent to the adjutant general by General Kenneth D. Nichols, Groves's deputy, written as Nichols was preparing to retire on 27 October 1953. The subject of the memo was 'Custody of Manhattan District Classified Files' and read: 'The files contain highly classified information concerning the Manhattan District. Some of the information is of considerable historical importance and should be preserved for that purpose. Other parts of the files are of particular value for protecting the interests of the War Department, General Groves and General Nichols.'[59]

Stimson, meanwhile, had changed his mind again. By now he thought there was a 'mountain' of evidence showing that Stalin and his fellow leaders were 'committed' to a policy of expansion and 'dictatorial repression'.[60] In these circumstances, 'we should stand by with all the bombs we have got and can make ... until Russia learns to be decent'. Sean Malloy, in his book on Stimson and his 'ordeal' over the atom bomb, makes it clear that the secretary of war, more than anyone in the Roosevelt and Truman administrations, wrestled with his conscience about when and what and how to tell the Russians. In the end, the tragedy was that Stimson failed to evolve a coherent *and durable* viewpoint. And that failure was writ large in American policy, right through until 1949.

The extent to which not everyone was thinking straight is shown by the publication of the so-called Smyth Report, *Atomic Energy for Military Purposes*, only days after the bombs exploded. Henry De Wolf Smyth was a Princeton physicist whose 264-page report was designed to make the basic principles of the bomb's operations 'available to all-comers'.[61] In practice, it was Groves's way to show what could be talked about and what couldn't. When Chadwick read the draft, he was perturbed but Groves argued that the information it contained could help an adversary by at the most three months.[62] The report was a runaway publishing success, going into nine editions and being translated into forty languages. But there seems no doubt that it did help the Russians to an extent. As we shall see, it was a small own goal.

Disappointed that they had had so little influence on government policy – despite their intimate connection with the bombs – and that their (Franck) report had been ignored by high officials, a group of scientists, mainly from Chicago, formed the ASC, the Atomic Scientists of Chicago. In their most noteworthy initiative, they met at Stineway's Drugstore on 57th Street in December 1945 and conceived the idea of the *Bulletin of Atomic Scientists*, a journal to warn the general public on the dangers of atomic warfare, and to advocate international control of nuclear weapons. In a still more imaginative exercise, in 1947 they initiated the so-called 'Doomsday Clock', a clock face in which the hands were set nearer to or further away from midnight (signifying the apocalypse) to indicate how close the world was, in the judgement of the atomic scientists, to nuclear catastrophe.

As imaginative exercises the *Bulletin* and the clock were a success. But scientists were never allowed to play a central role in arms control, and this is certainly one area where Bohr's ideas were stillborn. He had always thought that the international agreement of physicists would be easier to achieve than that of politicians or military men. Mindful of the small number of nuclear physicists that then existed, and only too well aware of the close collaboration of those individuals in the 1920s and 1930s, before the war, and that they all knew each other, had shared laboratories and prizes, had attended the same conferences, and in many cases – despite their different nationalities – were friends, he thought this was an asset to the world that could be fruitfully exploited.

And he was not alone. Heisenberg once said, admittedly after the war, 'In the summer of 1939, twelve people might still have been able, by coming to mutual agreement, to prevent the construction of atomic bombs.' He added that 'he had just such an agreement in mind when he went to see Bohr, and that he hoped to propose the possibility that Bohr might serve as an intermediary in arranging a secret agreement among German and American physicists to use their influence at this delicate moment to stress the difficulties of making a bomb and thereby avoid its use in the war – *by either side*.'[63]

One of the earliest accounts of Heisenberg's post-war claims is contained in a letter from Francis Simon to Michael Perrin, written in March 1948. Heisenberg had been in Cambridge but visited Oxford for a couple of days and he had a number of meetings with Simon. He was, apparently, much disturbed by the book that Samuel Goudsmit had written about Alsos – in fact, says Simon, he was 'infuriated'.

> Heisenberg claims that German scientists had no other wish than to prevent Hitler from getting the bomb. They knew about everything, including the fast neutron reaction and the possibility of using plutonium, but all their actions were determined by their aim to mislead Hitler and the 'high ups' about the possibility of a bomb. He said that if he had gone to Hitler at the beginning of the war and told him what he knew, then he was quite sure that Germany could have developed the atomic bomb just like the Allies!
>
> I replied that judging from the reports that came into our hands, this did not seem very plausible and we had more or less the same impression about the German effort that was expressed in Goudsmit's book. In particular, no serious mention of the plutonium possibilities was made in the German reports, nor about the possibility of a fast neutron bomb. The German scientists thought of a bomb as a whole heavy water pile that one had to drop – this incidentally was confirmed by Bonhoeffer when I saw him in Oxford last year. Heisenberg tried to ridicule this. In particular he emphasized that he always knew that the bomb would be the size of a pineapple.

> Naturally, I could not tell him about some of our sources of information, from which we knew for certain that his story does not correspond with the facts ... At the end of our lengthy discussion ... he admitted that ... the German scientists had not behaved very well except for a few, for instance Hahn and Laue, and more or less himself.
>
> I am quite sure that Heisenberg, like many other Germans, is a strictly honest person in his private life, but as soon as the great glory of the 'fatherland' is involved – and perhaps also his glory as a scientist – it is quite a different matter. Whether he now deliberately tells these falsehoods I cannot say. It is quite possible that he has constructed post festum a picture of the state of affairs as he would have liked to see it and that he has so persuaded himself that this picture is correct that he now seriously believes in it. One should also compare what Heisenberg says in his 'Naturwissenschaften' article. There he also emphasised that the German scientists did not bother much about the bomb, but he gave quite a different reason, namely, that they knew that the industrial effort was too much for Germany ... Two things are clear: (a) That what he says does not correspond with the facts, and (b) that he tells these stories with an air of complete sincerity and conviction ... I am sending a copy of this [to] Cockcroft, who I know will be interested.[64]

Weizsäcker said much the same in a letter to Thomas Powers in March 1988. 'Bohr was the great moral authority for all of us and Heisenberg wanted first to find out whether perhaps under the guidance or help of Bohr the physicists of the world could come to a mutual agreement on the way in which the horrible responsibility that was posed upon them by the possibility of nuclear weapons would be carried by the community of physicists.'[65] He put it more strongly in a letter to the editor of *Die Zeit* in 1991: 'The true goal of the visit by Heisenberg with Bohr was ... to discuss with Bohr whether physicists all over the world might be able to join together in order that the bomb not be built.'[66]

Powers was not impressed by Weizsäcker's argument, feeling that it was an exceedingly dangerous course for any physicist to follow, and of course it is at dramatic odds with what Bohr remembered. But Heisenberg and Weizsäcker's idea of the *community* of physicists playing a uniquely informed role in the control of developments in nuclear research, or something very like it, was certainly at the back of Bohr's mind when he went to see Churchill and Roosevelt.

Something very similar was also at the back of Pyotr Kapitsa's mind, too. His very approach to Bohr, in the letter Bohr received in London in April 1944 while waiting to see the prime minister, reflected his belief that scientists had a role to play in securing international peace.[67] Bohr had held himself ready, while he had been in London, to travel to the Soviet Union as a Western emissary. Though well aware that decision-making must rest with elected politicians, he believed that personal contact between scientists of different countries 'might be a means of establishing preliminary contact and working out a common approach to security'.[68] In October 1945 Kapitsa wrote to Stalin calling for closer relations between scientists and politicians, and greater trust.[69]

By all accounts, Stalin was as opposed to scientists 'interfering' in politics as Churchill. As David Holloway puts it: 'All attempts to imagine alternative courses of postwar international relations run up against Stalin himself . . . His malevolent and suspicious personality pervades the history of those years.'[70]

While undoubtedly true, we can never forget that Churchill, Roosevelt and the governments around them – Allies in the war – had deceived Stalin throughout where the atom bomb was concerned. Or they *thought* that they had. Stalin had in fact known for years that he was being deceived, and had plenty of opportunity to think through his response.

Had the Russians been approached early enough, would it have made the difference Bohr expected? We shall never know but, despite Holloway's reservations, Stalin is, as Simon Sebag Montefiore says, no longer an enigma. Since the archives have been opened, we have a much better idea of his character. We know he talked constantly (often enough about himself), that he was super-intelligent, a nervy intellectual who manically read history and literature, a fidgety hypochondriac who suffered from psoriasis and rheumatic aches from his deformed arm. He was garrulous, sociable and a fine singer, but this lonely and unhappy man ruined every love relationship and friendship in his life by sacrificing happiness to political necessity and cannibalistic paranoia ... abnormally cold in temperament, he tried to be a loving father and husband yet poisoned every emotional well, this nostalgic lover of roses and mimosas who believed the solution to every human problem was death, and was obsessed with executions. This atheist owed everything to priests and saw the world in terms of sin and repentance, yet he was a convinced Marxist fanatic from his youth.[71]

The foundation of his power in the Party was not, however, fear – it was charm. 'While incapable of true empathy on one hand, he was a master of friendship on the other. He constantly lost his temper, but when he set his mind to charm a man, he was irresistible.'[72]

Churchill, at the end of his visit to Moscow in 1942, during which he felt he had been insulted and even mocked by the Russian leader, nevertheless concluded, 'Stalin had been splendid ... What a pleasure it was to work with "that great man".'[73] The Soviet leader had a genuine fondness for Roosevelt and the American president 'always expressed a high opinion of Stalin', according to Beria's son, Sergo.[74] And he told others how much he liked Roosevelt.[75] Although he preferred Roosevelt to Churchill, Stalin admired the British leader the most. 'In the war years he behaved as a gentleman and achieved a lot. He was the strongest personality in the capitalist world.'[76]

Was such a man *so* impossible to deal with? Stalin would certainly have wanted a bomb of his own but he could also see the advantages of limiting possession to an atomic club, in which nuclear physicists would play a central role. Does this confirm Bohr (and Kapitsa and Heisenberg) as naïve? All three won the Nobel Prize – they were not nobodies. There was a decent chance that where they led, other scientists would follow. Or does the failure to involve Stalin, via the world community of physicists, count as a most imaginative opportunity lost?

26

Fuchs: Shining in the Shadows

Over the next years, at the very beginning of the Cold War, both American officials and Churchill – now out of power, following his surprise General Election defeat in late July 1945 – considered the idea of a pre-emptive attack. More than once – far more than once – Churchill aired his view that Russia should be 'atom-bombed into submission', that the country could be cowed by, say, one bomb on Moscow, which would destroy the leadership and leave the country rudderless (reflecting a view that a highly concentrated dictatorship was especially susceptible to atomic bombing). Or that Russia should be told to withdraw from the Balkans or Poland, say, or face the consequences.

Much the same attitudes also informed the plans to set up a United Nations Atomic Energy Authority of some kind. The Allies, with the bomb, might see the authority as a force for good. But the Russians, then without the bomb, could see it only from a coercive point of view. Despite all the arrangements for inspection and the control of raw materials, the basic reality was aimed at maintaining the status quo and stopping the Soviet Union from acquiring the bomb, and therefore being forever at the Allies' mercy. Bohr understood this more than Churchill or Truman seemed willing to.

Alongside this, Gregg Herken and others have chronicled how difficult it turned out to be for the Anglo-Americans to make the most of their monopoly of the bomb so as to achieve practical political advantage. Several military plans were formulated in the late 1940s – codenamed Pincher, Broiler, Grabber, Intermezzo, Fleetwood, Dropshot, Offtackle – some of which, in retrospect, sound more or less desperate. One conceded that America did not have bombers capable of delivering nuclear weapons deep into Russia so that the objective could only be achieved by one-way 'suicide' flights into the Soviet heartland.[1]

The real problem, Herken writes, was that, at least to begin with, the US had far fewer bombs than most people thought (no more than a dozen by spring 1947), that the different services fought among themselves for 'custody' of the weapons, while at the same time, and as mentioned, they did not have an effective means of delivering them. Added to that was a fundamental conceptual muddle as to whether American policy should centre around a 'pre-emptive war' or use of the bomb as a deterrent. What was also true – and maybe not enough is made of this – was that while some of the plans were aggressive and bloodthirsty in the extreme (100 targets were selected at one point, some of them well beyond the range of American bombers!), several politicians and military leaders thought it was simply immoral to be the first to use such weapons, with their indiscriminate targeting of civilians. There was also the fact that, as American diplomats found at successive international conferences, the Russians didn't scare easily.[2]

The one success that possession of the bomb did bring about was that the United States' nuclear monopoly in all probability forestalled a Soviet invasion of Western Europe (as mentioned in the previous chapter, the Soviet leadership grasped this immediately). Following the rapid demobilisation of American troops, in both the Far East and in Western Europe, from a strength of ten million down to three million, the Red Army was maintained at colossal strength in Eastern Europe. Such an invasion was expected for several years after the Second World War ended, with plans for the Allied troops to fall back, either to the Rhine or the Pyrenees, and for the Middle Eastern oilfields to be lost for a year or more until the Western powers could regroup. But it didn't happen.

An added complicating factor was that, in Herken's words, although the bomb was a 'wasting asset', as time went by a form of complacency set in, because many in Washington could not bring themselves to believe that the Russians would have the bomb any time soon.

As we shall now see, the underestimation of the Russians matched the earlier overestimation of the Germans. And, just as he had played a major role in overestimating the Germans, so did Groves play a central role in underestimating the Russians.

Groves claimed at various times that 'there was no uranium in Russia' and that it would be at least twenty years, 'if ever', before they could produce a bomb. On another occasion he said it would take them 'ten, twenty, or even sixty years'.[3] He refused to discuss the basis for his view, and why it differed so much from that of the scientists, such as James Conant, but managed to convince his political masters that he knew more about the subject than anyone else.[4] He pointed out that the government had concluded agreements with the Belgian Congo to buy up their uranium for the next thirty-three years. Even if Russia did get the bomb after twenty years, Groves invariably added that their raw materials were so poor that the US would always be ahead.[5] Groves envisaged an 'American-administered Pax Atomica' in which, if need be, pre-emptive nuclear strikes should be carried out against any nation that looked likely to be working on its own nuclear weapon. David Lilienthal, chairman of the Atomic Energy Commission, thought that a 'miasma of misunderstanding' about Russia as a backward country was engendered by Groves, whom he found 'impossible' to work with.[6] But the general clearly had Truman's ear.[7]

At the same time and in contravention of his oath as a military officer, Groves openly campaigned for the May–Johnson Bill, which proposed a committee – joint military and civilian – which would have control over the use of atomic weapons, rather than the president, who insisted on civilian control.

Throughout the monopoly period Groves remained a key figure in sustaining the belief that that monopoly would last a long time. He retained his power of veto over who was told what, and who controlled what, where the bomb was concerned.[8] Even the Joint Chiefs complained that he was excluding them from intelligence on Russian progress with their bomb while Vannevar Bush regarded

Groves's ideas about the worldwide control over raw materials as impossible to police.[9]

But the general retained his influence with many senior members of the Truman administration, even going so far as to argue that the Russians were 'technologically and even psychologically unequipped' to produce a bomb and that they wouldn't be able to do it, 'even with possession of the Manhattan Project's blueprints'.[10] When he eventually retired from the Project in 1947, Groves continued to insist to a group of reporters in October that year that 'it would still be from fifteen to twenty years before Russia got the bomb'.[11]

The sublime moment in world history, as Churchill put it, totally changed on 3 September 1949. Ten years to the day since the Second World War broke out, a specially equipped US Air Force B-29, on a mission over the north Pacific, recorded atmospheric signs that suggested that an atomic explosion had recently occurred 'somewhere' in Soviet Asia. It was later confirmed that the Russians had indeed detonated their first nuclear weapon – a plutonium device similar to that dropped on Nagasaki and codenamed *Pervaia Molniia*, 'First Lightning' – at a test site on the Kazakhstan steppes, with a blast yield of 22 kilotons (Hiroshima was 10 kilotons). When the Russians, prompted by a statement about the bomb from the White House, finally acknowledged what had happened, Stalin added, 'Atomic weapons can hardly be used without spelling the end of the world.'[12]

Most people were surprised by the speed with which the Russians had achieved their own bomb. The initial reaction in Washington, says Gregg Herken, was 'shocked disbelief'. Churchill had made different predictions at different times: he told Parliament that he thought the 'sublime period' in history, when the West could force its views on Soviet Russia, would be about three years. In private he told his doctor, Lord Moran, that it would be eight years. Bush and Conant, in their advice to the president, had given it as their opinion that Russia 'might' catch up in three to four years. In London, Harold Nicolson was distraught. 'We are all feeling depressed,' he wrote. 'We were told they would not have one for five years, and they have got it in four.'[13] James Reston, then working in the *New York Times* bureau in the American capital, said that the nuclear monopoly was ended 'about two years earlier than Washington expected'.[14] Other estimates ranged from two to twenty years to never.[15] Henry Stimson himself thought it would take the Soviets 'from four to twenty years' to catch up.[16]

To begin with, the various post-war intelligence agencies were as varied in their views as anyone else, though later a consensus emerged. Intercepted letters in 1947 showed where several of the German atom scientists were working in Russia, and four German atomic scientists defected also in that year, having travelled to the Soviet Union for job interviews but returned home. One of these reported that isotope separation was being carried out at Sukhumi, in Georgia, on the Black Sea, while Manfred von Ardenne's institute was located nearby, producing heavy water and uranium metal. Von Ardenne had worked on fission during the war but as a maverick – he was not part of the Uranverein and found funding from, of all places, the German post office. Fuchs saw the report on Sukhumi, so the Russians knew what the Allies knew.*

This same defector also revealed that the former director of the Joachimsthal mine was now heading a group near Tashkent in Uzbekistan, Central Asia, exploring for uranium. British intelligence had also found out, in 1946, that one 10-ton freight car of uranium was being shipped every ten days from Joachimsthal to Elektrostal, about 40 miles east of Moscow. On top of which it was known that the Russians required the former Bitterfeld plant of I. G. Farben to produce 30 tons of highly pure metallic calcium each month, 'enough for the manufacture of sixty tons of uranium metal'.[17] A CIA contact produced a bill of lading for three freight-car loads of calcium from Bitterfeld to Post Box 3, Elektrostal, confirming the earlier information.

This told the scientists that there was a uranium factory at Elektrostal, producing the metal in substantial quantities and pointed to the

* Fuchs also told his Russian masters that, in 1947, the British had had little success in developing an atom bomb of their own.

fact that the Soviet programme included a functioning or planned reactor to make plutonium, since the metal being produced was not needed to enrich uranium. A year or so later, the British interrogated a former German prisoner of war who disclosed the existence of just such a plutonium facility – Chelyabinsk-40, as it was called – near Kyshtym, a day's drive east of Moscow, on the far slopes of the Ural mountains. Several other former prisoners of war confirmed the existence of Kyshtym. A separate report in 1948/9 provided new information on uranium mining in Czechoslovakia and showed that the Russians were extracting as much as six times what the British had initially thought.

Interception of communications was productive to an extent, partly because, across the vast distances of the Soviet Union, communication was quite primitive, and this enabled American intelligence services, mainly, to identify many of the locations where nuclear work was going on. In fact the bomb was designed and built at Arzamas-16, near the small monastery town of Sarov, 230 miles south-east of Moscow. Arzamas-16 later became known affectionately (and ironically) by Soviet physicists as Los Arzamas, as a tribute of sorts to Los Alamos.[18]

Knowing *where* the work was taking place, and what sort of work was being done there, was one thing, but it did not help answer the all-important – the burning – question: when would the Russians have a bomb?

Michael D. Gordin tells us that the Americans did not have a *single* agent working inside the USSR providing human intelligence about Soviet plans, and so had to rely entirely on technical surveillance which, almost by definition, meant they could only detect an explosion *after* the event – if then.[19]

Although the various intelligence agencies had their 'inside' information, it didn't stop them giving varied answers to the crucial question. In 1946, the Central Intelligence Group, forerunner of the CIA, while conceding that its 'real information' was in fact 'relatively meagre', advised that the Soviets would first test a bomb 'sometime between 1949 and 1953'.[20] In the same year, the air intelligence element of the Air Staff suggested that the Russians might detonate a bomb 'toward the end of 1949'. In the following year a consensus of sorts emerged between three of the major intelligence services. Their analysis concluded that 'some details' of the Hanford plutonium production plant may have been leaked to the Russians by spies, and then went on to argue that the Soviet Union did not yet have a working reactor, that it lacked the necessary engineering skills and other technical personnel, and that uranium ores available from within the Russian heartland were of low uranium content. The intelligence chiefs of the army and navy went on to give it as their considered view that the Russians 'could not have atomic weapons now', and would 'most probably' have them during 1952. The assistant chiefs of the Air Staff had by now moderated their view, but still differed from the other agencies, saying that the Soviet Union 'might already have' an atomic weapon and would 'most probably' have one between 1949 and 1952.

In 1948 the Joint Chiefs of Staff Joint Intelligence Committee estimated mid-1950 as 'the earliest date by which the Soviets [might explode] their first bomb', and mid-1953 as 'the probable date'. In July 1948, the *Estimate of the Status of the Russian Atomic Energy Project*, provided to President Truman by Rear Admiral Roscoe Hillenkoetter, the first director of the CIA, echoed this analysis. 'On the basis of the information in our possession, it is estimated that the earliest date by which it is remotely possible that the USSR may have completed its first atomic bomb is mid-1950, but the most probable date is believed to be mid-1953.' By early 1949 it was more or less agreed by most insiders that 'mid-1950 is the earliest possible date and mid-1953 the most probable date'.[21]

But General Groves was never part of this consensus. He 'answered' the question about how long it would be before Russia had the bomb several times and his answers always had in common the fact that his estimate was on the high side. Within the span of three months in late 1945, Groves gave three different government audiences three different numbers. Addressing the War Department in September he opted for ten years; before the House of Representatives in October he chose 'from five to twenty years'; by November, before the Senate, he was more certain it would be twenty years. After that he settled on fifteen to twenty years.

He and others slighted Russian engineers, who compared badly with the 'genius' of the Americans, in some cases concluding the country was as much as twenty-two years behind the United States. Groves even went so far on occasions to say that, given that the US now controlled 97 per cent of the world uranium supplies (thanks partly to the British Empire), 'the Soviets could not catch up'.[22] As if that were not enough, in July 1949, fewer than two months before the Russians did in fact detonate their first device, Groves went on record as saying 'that scientists are not in the best position to make sound estimates. The real problems faced in such a development are those of management and engineering, rather than purely scientific ones.'[23] He told Truman that the Soviets would not have a bomb 'on his watch' (that is before 20 January 1953) and that it would be a problem 'for their grandchildren', not themselves.[24] After the first Soviet bomb had been dropped, he admitted that he had 'completely underestimated' the value of German uranium to the Russian effort.[25]

The scientists were closer than Groves but still some way off. Chadwick thought it would be five years before the Russians had a bomb, Szilard testified to Congress in October 1945 that it might be six years before Russia got one, Bethe wrote in 1946 that it would be another five years, the Germans being held at Farm Hall thought it would be ten years, while Arthur Compton predicted 1952, though he said he would not be surprised if they didn't get it until 1970.[26]

All this being the case, the Americans and, to a lesser extent the British, had worked on several techniques to spot the presence of radioactivity at great distances from a potential explosion, on the grounds that, when it came, the detonation would be in central Russia, in a deserted, remote location far from her borders. Various techniques were tried, including specially equipped meteorological balloons, flying at specified altitudes, seismological sensors, both at ground level and underwater and, most colourful of all, looking for bursts of light reflected from the moon, which it was conjectured would follow nuclear explosions. The most successful turned out to be the most prosaic: high-flying aircraft fitted with a specially adapted paper filters that could capture minuscule amounts of debris in the high atmosphere. Specially equipped US Air Force planes covered the western hemisphere from the equator to the North Pole and RAF planes flew between Scotland and Singapore, essentially covering Russia's southern border. There were sometimes three reconnaissance flights a day.[27]

One hundred and eleven recordings of atmospheric anomalies were made, all of which proved negative – they were volcanic eruptions, earthquake debris, chemical waste and dust of all kinds. But then Alert No. 112, made by aircraft WB-29, piloted by 1st Lt Robert C. Johnson, flying from Japan to Alaska at 18,000ft, on 3 September 1949, detected 'the real thing'.[28] More flights, from Hawaii and California, and RAF flights from Scotland, confirmed the high radioactive content of the debris. Louis Johnson, US secretary for defence, at first refused to believe it.

But it was true and there can be little doubt therefore that the Soviet bomb arrived before – well before – most people, and that included most insiders, expected it.

Michael Gordin has examined why this 'egregious' error was made, resulting in 'egg on American faces'.[29] One reason was the publication of the Smyth report, which told – or confirmed – for the Russians a lot that they weren't sure of: for example, the importance of gaseous diffusion and the other methods of isotope separation, which had been cleared for release by 'scientifically ignorant censors'. More subtly, or perniciously, one other reason for the error was that a lengthy estimate discouraged the need for a pre-emptive strike and what that might unleash. In other words, it diminished fear.[30]

By the same token, this was undoubtedly one major reason why the Russians didn't announce the detonation of their bomb. They were frightened of an American pre-emptive attack while they had more bombs than Stalin did. 'If the Americans knew in late August 1949 that the Soviet Union had detonated its entire supply of plutonium, would they not certainly mount an attack before the Soviet Union could properly arm itself?'[31]

Before the Truman administration could respond properly, however, and to make matters worse, certainly from an Allied point of view, the People's Republic of China came into being on 1 October, only a few weeks after the first Russian atomic detonation, adding 500 million to the roster of souls in the world governed by communism.

In response to this dual darkening, instead of making a preemptive strike, the Truman administration hurriedly undertook a thorough review of its national security strategy. By the end of January 1950, the president announced that the US was pressing ahead with 'work on all forms of atomic weapons, including the hydrogen or so-called super-bomb'. All arguments against this development were rejected. Edward Teller, who had been in favour of constructing a hydrogen bomb from the very first, now said he was prepared to bet that he 'would be a Russian prisoner of war in the United States within five years' unless the H-bomb was built.

To some non-Americans, the Russian bomb was actually good news. Otto Hahn, who had been awarded the 1944 Nobel Prize for his discovery of fission, told reporters that he had not expected quite such a quick end to the American nuclear monopoly but that he was relieved because he thought no one would now ever use such weapons. Not everyone was so sanguine: after all, the Russian success showed that it was possible for the West's adversaries to build an atomic bomb in secret.[32]

At which point, Churchill changed his tune again. Instead of arguing for the use of atom bombs to force the Russians to make concessions, he now argued, in a widely reported speech at the Usher Hall in Edinburgh in February, as part of a general election campaign, for a final effort, in the short interval between Russia's first atomic explosion and the delivery of its first usable weapons, to seek an agreement 'to bridge the gulf between the two worlds, so that each can live their life, if not in friendship at least without the hatreds of the cold war'.[33]

But this all has to be put alongside the National Security document that landed on President Truman's desk in April that year, which he approved before the spring was out. This celebrated report, NSC 68, called for a huge increase in military spending, not just by the US but by its allies as well, in order to build a level of superiority in both conventional and nuclear weapons, specifically designed to combat 'the Kremlin design for world domination'.[34] At that point the US had 1,591,232 men under arms. The army could field ten divisions but only one was combat-ready. The navy had 331 combat ships and the air force had forty-eight groups. The Russians had also reduced their forces but the Soviet strength of its land army in Europe was variously estimated in the late 1940s as between 2.8 and 5.3 million men.

And by now it was becoming clear that the nuclear alliance that had once seemed so hopeful was emerging as a threat – certainly from a British point of view. Once Russia had deliverable weapons, in the first years it would not have aircraft capable of dropping them on American soil. But its nuclear-armed aircraft *would* be able to reach Britain. General Groves saw this as proof that the British should never have been part of the Manhattan Project. 'England could be brought to her knees by the delivery of five atomic bombs in the right places.' In these circumstances, he thought, Britain might be forced to give Russia such nuclear information as she didn't have.[35]

Nor was this all. With awful timing, in early 1950 it also became clear why the Russians were able to achieve their nuclear goal more quickly than most people appreciated. In January, Klaus Fuchs confessed to being a Soviet spy.

The British had been tipped off in the second half of 1949, following the American decryption of the so-called Venona intercepts, Soviet Russia's secret Second World War cables. In Fuchs's case, the revealing telegram had been transmitted as long ago as 1946 (see above, p. 281) but had been fully transcribed only in 1949.* After first denying the charge, Fuchs eventually confessed, the news reaching Stalin the same day.

There was some evidence in his confession that Fuchs had grown disillusioned with his Soviet masters and fond of his British colleagues, especially those at Harwell, Britain's Atomic Energy Research Establishment, set up in the wake of war. But despite this he found no mercy in court, was quickly tried and sentenced to fourteen years in prison, the maximum term for breaches of the Official Secrets Act. Because Russia was an ally at the time, he couldn't be tried for treason but even so, this hardly helped relations between Britain and the United States.

The most recent evidence, using Soviet-era files since declassified, shows that Fuchs's work saved the Russians more than a year's work, and very possibly two, in constructing their bomb.[36] Richard Rhodes confirms that all the Los Alamos materials that reached the Russians matched what Fuchs made available via Harry Gold.[37] In Los Alamos, after he had left New York, Fuchs had worked in the theory group and therefore was at the very heart of discussions on how to make a plutonium implosion bomb work. He was in fact the theoretician who analysed the photographs of the imploding shells.[38] By all accounts Kurchatov had not even thought about implosion until he saw the documents provided by Fuchs and in the minds of some it was such a counter-intuitive idea that it may have taken the Russians as much as a decade to arrive at the answer unaided. Fuchs reported on the high spontaneous fission rate of plutonium, that the critical mass for plutonium was smaller than that for U-235 – about 5 to 15 kilograms – and that there would at that time be a hollow core for the plutonium bomb, with sixteen detonation points. He also reported on the outer dimensions of the bomb, the timing

* The Americans had failed to fully decipher the telegram but the British managed to do so. Fuchs's name was also found in the diary of one of the suspects in the Nunn May case. Groves later claimed that had he been shown this, he could have got on to Fuchs much earlier.

sequence for implosion and the rate at which bombs were expected to be produced.[39] Fuchs himself thought that this was the 'worst' of his crimes, though he also told the Russians how much explosive the Allies had, so they knew that America only had enough for two or three bombs.[40]

Kurchatov described the Fuchs material as of 'exceptional importance', in particular the conditions by which 'to achieve a symmetrical explosion' and suggested that several pages of his plutonium report ('pages 6 to the end save for page 22') be shown to others in the Russian set-up.[41] Fuchs also delivered to Gold a sketch of the plutonium bomb, with details about the explosives to be used for the implosion, details of the so-called 'Urchin' initiator ('containing fifty curies of polonium') and the tamper (which reflects escaping neutrons back into the pile), together with the method of calculating efficiency. The sketch and other documents, importantly, gave the exact thickness of each of the shells, plus the fact that an aluminium shell had been inserted between the explosive layers and the uranium tamper 'to dampen the hydrodynamic instability that would otherwise have developed when the light explosive mixed with the heavy metal, a phenomenon known for its English delineator, Geoffrey Taylor, as Taylor instability'. (Taylor was also part of the British contingent at Los Alamos.)[42]

The Soviet bomb, as it turned out, looked 'an *awful lot*' like the one the Allies dropped on Nagasaki, while the tower in which the trial bomb was exploded was 'almost an exact analogue of the 100ft tower of Trinity'.[43] This was not surprising given that Beria had told his scientists to build an 'exact copy' of the American bomb, according to the specifications provided by Fuchs and others, while a Russian design was rejected for the moment.*

Fuchs had been very popular at Los Alamos, a reticent bachelor who was welcome at parties because of his 'nice manners'.[44] He was a hard worker, close to Teller and his wife among others, and was well placed to learn many different aspects of weapons work. In February 1945, after a seven-month silence, he had met Harry

* The Russian design was tested in 1951 when it proved to be twice as powerful, while only half as heavy.

Gold in Boston, at the home of his sister, Kristel, and passed across a report several pages long summarising the problems of making an atomic bomb. (Gold had also visited him in Santa Fe, feigning throat trouble as cover for visiting a nearby clinic.) Most importantly, Fuchs told the Russians that a plutonium bomb would not work by the gun-assembly method – an implosion assembly was required. In a second report, delivered in June 1945, he described the implosion method in more detail (this was also when he disclosed there was an upcoming test of the procedure at Alamogordo). We now know that Kurchatov evaluated this as possibly the most important single item of information made available by espionage and agreed it would save 'quite a bit of time'. Fuchs himself was surprised by how quickly the Soviets produced a bomb.

After the war, Fuchs had returned to Britain and worked in the theory division at Harwell, near Oxford, helping to build Britain's first atomic device. There he renewed contact with Aleksandr Feklisov, his Russian contact in London.[45]

Fuchs was first detected through a simple error of duplication, a mistake that would help crack the famous Venona cables. Venona was a code word assigned only in 1961 to a series of cables that had been sent from Soviet offices in the United States to Moscow between 1940 and 1948. Because they had been encrypted using one-time pads they were very difficult to break and in fact they were only cracked beginning in 1947, owing to a duplication used in one set of messages, a mistake that the Russians should not have made. This led to Fuchs and other atomic spies but the Americans were in a bind. They did not want to use the codes in court because that would alert the Soviets to the fact that they had cracked Venona. This meant that their preferred way forward was to get a confession, after interrogation. This was obtained from Fuchs, eventually, but with highly unfortunate timing, in January 1950. This was not long after the detonation of the first Soviet bomb, and the declaration of a People's Republic in China a few weeks later.

Without Fuchs's aid, and that of the other Soviet atom spies, Russia would probably not have had a bomb until early 1951 or a year later. As we shall see, this window of time – late 1949 to early 1951 – was crucial.

Crucial because, at this uncomfortable juncture, on 25 June 1950, when events nuclear were moving rapidly, the Korean War erupted. This was the first 'hot' conflict of the Cold War and by far the most dangerous flashpoint since Hiroshima and Nagasaki. By that stage, as we have seen, and in contrast to Churchill's continual entreaties, the Americans had made strikingly little use of their atomic monopoly to extract concessions from the Soviet Union over the previous four years.

Both the American and British governments immediately condemned the invasion by North Korea of its southern neighbour, as did the newly formed UN Security Council. Within forty-eight hours the British had placed its Pacific Fleet at the disposal of the US Navy and, eventually, twenty-one members of the UN voted to support South Korea (which was not then itself a member of the UN), and the European/NATO contingent, under American leadership, comprised fifteen divisions from France, five from Germany, while Britain and America provided six divisions between them.

The Korean War had many ramifications. For a start it risked a wider war between the United States and China and that in turn threatened Britain's regional interests such as Hong Kong. Even more worrying, the Soviet Union had recently signed a mutual security treaty with China, meaning it too could be drawn into the struggle. None of this was helped by the appointment of General Douglas MacArthur as the commander of the United Nations forces in Korea. Although he had been running the occupation regime in Japan since the war, and was the undisputed star of the American Philippines campaign, he was a swaggering maverick, a man with such a 'towering ego' that he was said to have attended the same prep school as God.[46]

Despite this, or perhaps because of it, to begin with the Korean War went well from a Western point of view, before seesawing. Seoul, which had fallen to the North Koreans only three days into the conflict, was recaptured on 29 September, and MacArthur crossed the 38th parallel into North Korea and took Pyongyang, bearing down on the Yalu River, the frontier between North Korea and China. However, warnings from Beijing that these developments were a direct threat to China were downplayed or not taken on board by the swashbuckling MacArthur, with the result that, on 26 November, 250,000 Chinese troops swept south across the frontier, overwhelming the 75,000-strong UN army, who were forced into a headlong retreat. The US Eighth Army – the very core of the UN forces – was threatened with annihilation, and back home in the United States there were calls to abandon Korea and embark instead on a full-scale war with China.

This was not something Russia could allow to happen and the world found itself in a perilous place. Certainly, it was in the most perilous place since the end of the Second World War and arguably, given the arrival of nuclear weapons, the most perilous place ever. In Washington, Senator Joseph McCarthy was calling in private for nuclear weapons to be used.[47]

In a celebrated paper published in 1988, Mark Trachtenberg, a national security expert at the University of Pennsylvania, examined in detail American strategic thinking around the time it lost its crucial nuclear monopoly, and the advent of the first 'hot' conflict so soon afterwards. He notes that throughout the late 1940s, and well into the early 1950s, there were many who argued that the United States should not just sit back and allow a hostile power like the Soviet Union to acquire a nuclear arsenal. A much more 'active' or 'positive' policy had to be seriously considered.

The basic fact was that while America had nuclear bombs and Russia had far greater conventional forces, a balance of sorts existed, in Europe in particular. Even so, Trachtenberg found that aggressive ideas were taken 'very seriously' at this time, 'even at the highest levels in the administration' and there were explicit calls for 'a showdown' with the Soviets 'before it was too late'.[48]

In 1946 Churchill, as we have seen, had predicted that there would be a war with Russia in perhaps seven or eight years. 'We ought not to wait till Russia is ready,' he said in reply to a question, and argued repeatedly for a showdown, for 'bringing matters to a head' before the nuclear monopoly was broken. If this led to a war, he had told the House of Commons at the beginning of the year, having it then offered 'the best chance of coming out of it alive'. He pressed the Americans to deliver an ultimatum: the Soviets must either withdraw from East Germany or see their cities obliterated in an atomic attack.[49]

In 1948 William Laurence, the science correspondent of the *New York Times*, foresaw a war coming, once the Russians had the bomb, so argued that America should 'get in first'.[50]

As widespread as this view was – that a preventive war was the way forward – it did not prevail in the very highest echelons of the Truman administration. However, the Soviet atomic explosion in 1949, somewhat earlier than expected, as we have seen, nonetheless did provoke a major revision of American thinking. NSC 68, the famous nuclear strategy document referred to earlier, was mainly written by Paul Nitze, head of the State Department's Policy Planning staff but naturally reflected the views of Dean Acheson, secretary of state. The authors of NSC 68 believed that America's atomic monopoly was the one thing that had balanced Soviet superiority in ground forces and they were worried that, with growing Soviet atomic capabilities, America's nuclear edge was being 'neutralised more rapidly than conventional forces could be created to fill the gap'.[51]

The central point now, following the Soviet bomb, was that a nuclear balance was *not* stable. The advantages of a surprise attack were so great as to create a continuing tense stand-off and it was widely assumed in America that the Russians were so intent on world domination that a modus vivendi could never be sustained.

The essential strategy of NSC 68, therefore, was to *not* provoke the Soviet Union for the time being, but to build up all American forces – conventional and nuclear – to re-establish a commanding lead. Acheson and Nitze did not want a war, not then, in 1950. Of course, had the Russians known this, it was an invitation for them to attack. So belligerence – bluff – was needed. Meanwhile, the American medium-term policy was that, having a bigger and more efficient economy, they could outstrip the Russians in both conventional and nuclear weaponry, and if that worked out they might not have to fight at all.[52]

And it was against *this* background that the Korean War broke out.

One immediate thing for the Truman administration to consider was whether Russia was behind the outbreak of hostilities. Was the Korean War the first sign of increased Soviet aggression now that it had the bomb? The CIA took the uncompromising view that the invasion was 'undoubtedly undertaken at Soviet direction' and it was certainly taken for granted in Washington that a serious Soviet intervention in the Korean conflict would lead to a Third World War.[53] The view was also widespread that any intervention by the Chinese would mean that the war should be widened – even against Russia, which many thought was the 'heart' of the problem.[54]

Trachtenberg says that Truman himself was 'viscerally attracted' to the idea of a showdown with the Russians, warning publicly in 1950 that new acts of aggression 'might well strain to breaking point the fabric of world peace'. Several other leading figures now called all over again for a preventive war. There was a widespread feeling 'that the aggressive thrust of Soviet policy reflected in the North Korean attack was something the US could not live with for ever, and that perhaps the time was coming when it would have to be dealt with directly, before matters got completely out of hand'.[55] But America was still too weak militarily to mount an attack. Indeed, the real fear was that Russia might be the one to mount such a pre-emptive assault.

A further complication was that, without actually declaring war, Russia – whose bombers could reach Britain – could effectively neutralise her, threatening to bomb her unless she declared herself disassociated from America.

To put the matter simply, as Trachtenberg sums up: 'The North Korean attack, and then the Chinese intervention, seemed to support the idea that the Soviets were willing to accept, at the very least, an increased risk of war during this period of America's relative weakness.'[56] Should the Soviets attempt another Berlin Blockade, for example, when she tried to limit the Allies' access to their Berlin sectors in 1948, a reprise of the airlift would not be practical.[57] The army chief of staff went so far as to say that the US would not be ready to engage in global war until 1 July 1952.[58] This was the thinking behind the decision to avoid an advance beyond the 38th parallel, which MacArthur contravened.

But all this was only one side of the equation. There was also the fact – not often enough appreciated, perhaps – that from 1951 on it became clearer that Washington's worst fears about Russia had been exaggerated. There were those in the administration who thought that the Russians were bluffing in Korea and, as Trachtenberg puts it, 'East Germany did not invade the Federal Republic; there was no new Berlin crisis; Yugoslavia was not attacked; Soviet forces did not move into Iran.' Just as America didn't want to risk a nuclear war in 1951, neither did Russia. After the critical juncture – late 1949 to early 1951 – balance was being re-established.

But what interests us is the *restraint* on both sides, during the Korean 'hot' conflict, across the winter of 1950–51, and on into 1952.

Other elements in the mix included the assessment by the Joint Chiefs that there were few targets in North Korea for atomic bombs (in other words, few large population centres), and the gradual acceptance by the State Department, the military and Truman himself, that the bomb would only be used 'as a means of avoiding defeat', 'as a last resort'. Others asked whether it was worth risking a world war for the sake of Korea.[59]

What we can see here, dimly, is the uncertain emergence of a frightening nuclear standoff that would eventually come to characterise the Cold War. As events had matured, the mix of nuclear weapons, conventional forces, the confusion between civilian and military targets and the range of bombing were proving more complicated and intractable than anyone had predicted.

On the ground in the war, the difficult situation was further complicated by the deteriorating relationship between the president and MacArthur. The general had been a great fan of Roosevelt and his 'extraordinary self-control', which he contrasted tellingly with what he called Truman's 'violent temper and paroxysms of ungovernable rage'. These were swaggering words in themselves, as was the 'speech' he sent to the summer encampment of the Veterans of Foreign Wars, the largest and oldest war veterans association of the United States, dating back to before the First World War. Though he didn't attend the camp himself, MacArthur's words were designed to be read out in his absence. In these remarks he contradicted the Truman administration's policy towards Taiwan (then known as Formosa), saying: 'Nothing could be more fallacious than the threadbare argument by those who advocate appeasement and defeatism in the Pacific, that if we defend Formosa we alienate continental Asia.' 'Threadbare' wasn't exactly flattering of the administration, but it was his use of the term 'appeasement' that really challenged and infuriated Truman. This compounded other occasions when MacArthur had dismissed the possibility of Chinese or even Russian intervention in Korea.

During the Korean conflict the United States did in fact consider using atomic weapons several times. MacArthur wanted them because they would help him hit the routes that led into North Korea from Manchuria. His total requirements called for twenty-six bombs.[60] Their use came closest in early April 1951, exactly the time when MacArthur was removed, and it is now clear, says Bruce Cumings in his recent account of the war, that MacArthur was not removed simply for his acts of insubordination but also because Truman 'wanted a reliable commander on the scene should Washington decide to use nuclear weapons'. MacArthur had also asked for 'a "D" Day Atomic capability, to retain air superiority in the Korean theatre, after intelligence sources suggested the Soviets appeared ready to move air divisions to the vicinity of Korea and put Soviet bombers into air bases in Manchuria (from where they could strike not just Korea but also American bases in Japan), and after the Chinese massed huge new forces near the Korean border'.[61] On 14 March Air Force General Hoyt Vandenberg wrote: 'Finletter and Lovett alerted on atomic discussions. Believe everything is set.'*

* Thomas Finletter was secretary of the Air Force; Robert Lovett was assistant secretary of war, for Air.

By the end of the month the atomic bomb loading pens at Kadena Air Base in Okinawa were operational and the bombs delivered there, but unassembled. They were put together on the base until all they lacked were their nuclear cores or 'capsules'. On 5 April the Joint Chiefs of Staff ordered 'immediate atomic retaliation' against Manchurian bases if large numbers of new troops entered the fighting, or if bombers were launched against American assets from there. The very next day, Truman authorised the transfer of nine Mark 4 nuclear bombs to military control. On the same day the nine nuclear capsules were transferred to the 9th Bomb Group, the designated carrier. General Bradley, as chairman of the Joint Chiefs of Staff, obtained Truman's approval for the transfer 'from AEC to military custody' and at the same time the president signed an order to use them against Chinese and North Korean targets. The 9th Bomb Group deployed to Guam. Gordon Dean, chairman of the US Atomic Energy Commission, was apprehensive about passing these to MacArthur's control, and so were the Joint Chiefs of Staff, for fear he might act 'prematurely'. So instead, they decided the nuclear strike force would report to the Strategic Air Command.

The final order, though signed, was never sent. A later operation, Operation Hudson, involved flying B-29 bombers over North Korea on simulated bombing runs, dropping 'dummy' A-bombs. The operation called for 'actual functioning of all activities that would be involved in an atomic strike, including weapons assembly and testing, leading, ground control of bomb aiming'.

These 'dummy' runs never became real, one reason being technical, according to Cumings – 'the identification of large masses of enemy troops were extremely rare'.[62] But without the threat of retaliation, it now seems extremely likely that the United States would have used atomic weapons to finish the Korean War.

MacArthur's disloyalty was provocative enough, even without the threat of nuclear weapons. But the build-up had been further complicated at one point when, at a presidential press conference on 30 November 1950, in the course of a set of questions about General MacArthur, and whether he had exceeded his brief, Truman said that he was prepared to take 'whatever steps are necessary' to meet the military situation.

Did that include the atomic bomb? someone asked.

'That includes every weapon we have.'[63]

Did this mean there was 'active consideration' of use of the bomb? Truman was asked.

'There has always been active consideration of its use.'

Unable to quite believe what they were hearing, reporters asked again, 'Did we understand you clearly that the use of the bomb is under active consideration?'

'Always has been. It is one of our weapons.'

Truman then went on to say that 'the military commander in the field' (at that point MacArthur) had charge of the use of the weapons. He was later to retract this statement but within seventeen minutes of the end of the press conference, a United Press bulletin was on the wire: 'PRESIDENT TRUMAN SAID TODAY THAT THE UNITED STATES HAS UNDER CONSIDERATION THE USE OF THE ATOMIC BOMB IN CONNECTION WITH THE WAR IN KOREA.'

And the situation didn't immediately calm down. MacArthur called on the administration to recognise that a 'state of war' had been imposed by the Chinese and that the American response should be to 'drop thirty to fifty atomic bombs on Manchuria and the mainland cities of China'. The Joint Chiefs also came down in favour of the use of nuclear weapons as 'the only way to affect the situation in Korea'.[64]

Truman refused to go down that route. He later said in a speech that he had drawn a line over Korea. As David McCullough phrases it: 'If "victory" in Korea meant risking a world war – a war of atomic bombs – Truman would settle for no victory in Korea.'[65]

Nuclear deterrence, for the first time, had worked.

Even then, though, it was clear to some, and this included both Churchill and Truman, that one way to nuclear war was by accident. And this is one reason why, on 11 April 1951, five days after the nuclear weapons were transferred, the president relieved MacArthur of his command. The official reason for his dismissal was punishment 'for multiple acts of insubordination connected with the prosecution of the war in Korea'. Given MacArthur's waywardness this was no more than appropriate, but it was also courageous of Truman, for the general was hugely popular at home. 'I didn't fire him because he was dumb son of a bitch, although he was,' the president said. 'I fired him because he wouldn't respect the authority of the President.'[66]

MacArthur's maverick behaviour in the war had become megalomaniacal and reckless, so that many felt he could trigger a general Far Eastern conflict that could in turn into a global conflagration. As evidence of that, when the news of the general's sacking was received in the House of Commons, it was met with cheering on both sides.[67]

The cheering was short-lived. The Korean conflict was the flashpoint but the whole world was on edge at that time.

Truman certainly felt it. At the end of January 1952, worried by reports that the Chinese were planning a new offensive in Korea at the very moment that armistice talks were in progress at Panmunjom, just north of the de facto border between North and South Korea, he confided to his diary that the time had come,

> to target the source of communist aggression rather than to commit blood and treasure in combating its symptoms, as in Korea ... It seems to me that the proper approach now would be an ultimatum with a ten day expiration limit, informing Moscow that we intend to blockade the China coast from the Korean border to Indo-China, and that we intend to destroy every military base in Manchuria, including submarine bases ... and if there is any further interference we will eliminate any ports or cities necessary to accomplish our peaceful purposes ... We did not start this Korean affair but we intend to end it for the benefit of the Korean people, the authority of the United Nations and the peace of the world ... This means all out war. It means that Moscow, St Petersburg, Mukden, Vladivostok, Pekin [Beijing], Shanghai, Port Arthur, Dairen, Odessa, Stalingrad and every manufacturing plant in China and the Soviet Union will be eliminated. This is the final chance for the Soviet Government to decide whether it wants to survive or not.[68]

All-out war. Survive or not. Apocalyptic words. Historians have generally explained this entry as an example of Truman letting off steam in private. They do concede that it is an extremely aggressive example though they imply he would never have made such thoughts public. And he didn't act on those thoughts. But there is another explanation as to why this truly terrifying vision never took place, and one that brings to a fitting conclusion the narrative of this book.

Niels Bohr, we may say, ultimately failed in his heroic attempt to convince world leaders, Roosevelt and Churchill above all, that they should open up to the Russians *before* the Allies had advanced so far in their preparations for a bomb that any approach would look coercive rather than co-operative. In this he was proved right. As we now know, all the post-Hiroshima and post-war attempts to fashion an international control system foundered on the fairly obvious fact that the Soviets were never going to accept any arrangement where the *balance* of the agreement enshrined the Allied advantage that then existed. Again, this was exactly as Bohr had foreseen.

The one unknown in Bohr's plan, as envisaged, was and has always been the character of Stalin. Would he have agreed to a Bohr-inspired system of control in the late spring/early summer of 1944, following Bohr's meeting with Churchill?

As we saw in the last chapter, Stalin could be both friendly and duplicitous. He was able to charm and mislead Churchill and Roosevelt. We know he found atomic weapons barbarous, and that he was an unnerved by them as much as anyone else. If he had been brought in on the secret earlier, as Bohr advocated, he would most certainly have wanted a weapon of his own but, as Churchill said more than once, he kept to his word in meeting his agreed obligations in the conduct of the war. He was perfectly capable of seeing where Russia's interests lay and in seeing that the *balance* Bohr identified was a basis of trust and of peace. It was not a foregone conclusion but, as Bohr argued, and certainly in view of what happened anyway, it was worth the risk.

An agreement on atomic weapons would mark a real advance in relations between the great powers, making them allies in the post-war world just as they were in wartime. Eisenhower's point about

Britain's great fleet, America's unequalled air force and Russia's massive ground armies provided just such a basis for a nuclear league, which would have surely gone a long way to guaranteeing a durable stability.

By the time of the Second World War, many of the brutalities of the Stalin regime were known, but not fully. The show trials of former leaders had made headlines in the late 1930s but the full extent of the terror and the purges was not yet revealed in total. The Dewey Commission, the American Committee for the Defense of Leon Trotsky, had been set up following the first of the Moscow Show Trials in 1936 and had established that some, at least, of the specific charges at the trials could not be true.

Stalin was a brute, and known to be a brute. But he was a rational brute. While it is not at all clear how Stalin would have responded to an approach along the lines Bohr suggested, Bohr was right in that it was worth the risk.

Which brings us back, one last time, to Klaus Fuchs and Niels Bohr. There is no question but that the Korean crisis was the first major 'hot' conflict of the nuclear age, and a measure of the catastrophic breakdown in relations between former allies. But, as ever throughout this story, it is the timing that interests us.

President Truman, echoing his generals, such as Omar Bradley, was in a belligerent mood, as we have seen, but, not having made the most of their four-year nuclear monopoly, where Korea/China was concerned he still did not take the nuclear option. Despite his bellicose diary entry about wiping out Communist civilisation, no nuclear action took place. Trachtenberg's analysis shows us why. Months previously, the Russians had exploded their first bomb, and so any nuclear action on Truman's part risked inevitable retaliation by Russia, who had a treaty to aid its Chinese fellow communists. The interval between the first Trinity test at Alamogordo and the plutonium bombing of Nagasaki was twenty-four days. In the two years since its first test explosion, Russia had had plenty of time to produce its first deliverable atomic bombs. (We now know it had nine.)

There was unease in the Truman administration about using nuclear weapons – many felt it was the 'wrong' war and worried what the knock-on situation might be in Europe if atomic bombs were used in Korea and failed to achieve their objective. But Truman had also, quite realistically, come to the conclusion that nuclear war would be a catastrophe. 'War today . . . might dig the grave not only of our Stalinist opponents but of our own society, our world as well as theirs. Such a war is not possible for a rational man.'[69]

On these grounds, Klaus Fuchs's espionage becomes all-important. If what he (and others) did, in leaking so many Allied atomic secrets over the years, advanced the Soviet nuclear programme by as much as two years, as some have estimated and seems reasonable, then it could easily be seen that his actions, in speeding the production of Russian nuclear weapons, actually helped to *forestall* a nuclear war brought into being as an escalation of the Korean conflict. Had Russia not exploded its first bomb when it did, then by the time Truman came to write his bellicose diary entry, and if there had not been the very real threat of Soviet retaliation in the time leading up to January 1952, then the risk assessment of leading American politicians and generals would have been very different from what it was, and they might well have decided at last to enforce the advantage that they felt the nuclear monopoly – for the time being – gave them. This would surely have provoked – at the least – a Soviet invasion of Western Europe, where its forces vastly outnumbered those of its opponents. Would America have nuclear-bombed Russian forces in, say, Germany? And then what?

As it was, the time advantage that Fuchs's actions brought the Russians (more so than all the other atom spies) meant that they already had the bomb and, as Trachtenberg has shown, this tempered American behaviour, a balance of sorts eventually being achieved in 1952. As Michael Gordin put it, 'The situation in Europe had frozen in 1950 into a tableau of terror' and would remain more or less stable until the fall of the Berlin Wall in November 1989.

We shall never know if the Korean conflict would have gone nuclear if the Russians had not had the bomb by then but it is safe to assume that it was close-run thing. Truman had signed the order.

And of course, in this narrative of ironies and coincidences there is this final uncomfortable observation. Throughout this story, Niels Bohr behaved as an honest, upright, open-minded and far-sighted member of the international community but ultimately he failed in what he hoped to achieve. Fuchs, in marked contrast, schemed in the shadows, betrayed his colleagues and the people who had given him sanctuary after the Nazis frightened him out of his homeland. He grievously abused his several positions of trust and put himself beyond the bounds of decency. But, as things turned out, his crimes and betrayals, his duplicitous deceits, his treachery and his cunning in the end propelled the world into its frozen tableau of terror and in so doing saved us from disaster.

Acknowledgements

I am grateful to the staff of the following archives and libraries who have helped in the research for this book: Churchill College, Cambridge, the papers of James Chadwick (indicated by CHAD below), Lise Meitner (MTNR), John Cockcroft (CKFT), R. V. Jones (RVJO); the National Archives of the United Kingdom, Kew, London, Cabinet Papers as CAB, Prime Minister's Office as PREM, Foreign Office as FO 188/493 (espionage), FO 942/169 (technical intelligence), FO 371/33069 (Sweden), FO 188/650 (Norway), FO 188/651 (Denmark); the Royal Society Archive, London, the papers of Francis Simon (FS) and letters of Paul Rosbaud (PR); the Niels Bohr Archive, Copenhagen; the papers of General Leslie Groves at the US National Archive II, College Park, Maryland; the Samuel A. Goudsmit Papers: 1921–1979, Series IV, Alsos Mission, Digital Library for Nuclear Issues, Atomic Heritage Foundation, Washington. The Goudsmit papers are also available at the National Archive, Archive II, at College Park, Maryland. Quite a few of these are still classified, even now, and some important ones appear to have been either misreported or misfiled and could not be located, even by the archive's own research consultants (see especially reference 5 in chapter 15).

Harry Gold was debriefed in Holmesburg Prison in June 1950. These debriefs are available as Soundscriber discs, available in the Special Collections Department, Paley Library, Temple University, Philadelphia. A special report on his work as a Soviet Agent was produced in October 1950, and is available at, among other places, the Diamond Law Library, Columbia University, in New York. A Congressional subcommittee produced in 1956 a report entitled

'Scope of Soviet Activity in the United States', which covered material on Gold. The committee was set up to investigate the workings of the Internal Security Act and reported to the second session of the Eighty-Fourth Congress in April that year. An FBI report on Harry Gold was produced by Special Agent Joseph C. Walsh in New York in May 1950. This is available in the National Archives in Washington. A second report, with the same date, was prepared by SA Robert Jensen.

Harry Gold is the subject of a Notorious Offender File in the FBI Prison records in the US National Archives, 11 College Park, Maryland.

United States v. Harry Gold, 1950 is in the National Archives, Philadelphia.

The FBI file on Klaus Fuchs has now been published in 111 volumes, running to more than 50,000 pages, and is available online. This includes the statement Fuchs made to FBI agents Hugh Clegg and Robert Lamphere at Wormwood Scrubs prison in London in May 1950. The Lamphere papers are at Georgetown University Library.

The British Security Service Files on Klaus Fuchs are contained in the KV 2 series, sections 1245-1270 in particular.

KV 6/41–45 contains material on Ursula Beurton ('Sonia').

The former UK Atomic Energy Files in the AB 1 series contain correspondence between Peierls, Fuchs, Chadwick and others. AB 3 and AB 6 contain files from the Tube Alloys directorate with correspondence between Peierls and Fuchs.

The BStU in Berlin, the 'Stasi' archive, contains personal and surveillance files on Klaus Fuchs.

For Russian material, the Vassiliev Papers, the so-called Black Notebook, KGB File No. 40159, can be researched in the Manuscript Division of the Library of Congress. These are divided into eight books, and have been translated.

I would also like to thank the librarian and staff of the London Library.

Several individuals have helped in the course of my research, either with specialist information, guidance, advice, hospitality or translations: Konstantin Akinsha, Rupert Allason (Nigel West), Robert Arnold, John Bachofen, Melissa Bond, Derek Boorman, Tom

Bower, Sophie Bridges, Frank Close, Neil Cobbett, Dale Coudert, Hu Cui'e, Michael Dean, Megan Dwyre, Richard Ellis, Edward Elson, Michael Goedhuis, Peter Goss, Charles de Groot, Gaby Hahn, Eva Hajdu, Rupert Hambro, Rebecca Hart, David Henn, Charles Hill, Mark Hollingsworth, Hans Jackson, Phillip Knightley, Elisabeth Kondal, Halina Koscia, Gregorii Koslov, Constance Lowenthal, Iain MacGregor, Douglas Matthews, Kristina von Meyerfeld, Brian Moynahan, Andrew Nurnberg, Kathrine Palmer, Michele Palmer, Nicholas Pearson, Sabine Pfannenstiel, Richard Pfennig, Werner Pfennig, Michael Pochna, Alex Podell, Clive Priddle, Irina Roberts, Gerard Le Roux, Edwina Sandys, Julia Schmidt, Frank Settle, Kaiya Shang, Robin Straus, Igor Torbakov, William Veres, Ed Victor, Jeremy Wiesen, David Wilkinson, Veronique Zimmerman. Any errors, omissions or solecisms as remain are the author's responsibility alone.

Notes

Preface – Cover-Up: 'When the Righteous Sin'

1. Helge Kragh, *Quantum Generations: A History of Physics in the Twentieth Century*, Princeton and Oxford: Princeton University Press, 1999, p. 174ff.
2. See Werner Heisenberg's letter to his wife, 9 January 1946, in Werner and Elisabeth Heisenberg, *My Dear Li, Correspondence 1937–1946*, ed. Anna Maria Hirsch-Heisenberg, trans. Irene Heisenberg, New Haven and London: Yale University Press, 2016, p. 277.
3. Gar Alperovitz, *The Decision to Use the Atomic Bomb*, New York: Vintage, 1996, p. 129.
4. Alperovitz, op. cit., p. 3; see also Michael D. Gordin, *Five Days in August: How World War 2 Became a Nuclear War*, Princeton and Oxford: Princeton University Press, 2007, p. 7.
5. Alperovitz, op. cit., p. 331.
6. Ibid., p. 4. This finding has had its critics – see Barton J. Bernstein, 'A Post-war Myth: 500,000 US Lives Saved', *Bulletin of the Atomic Scientists*, vol. 42, 1986, issue 6, pp. 38–40; online 15 September 2015.
7. Alperovitz, op. cit., pp. 6, 147.
8. But see Gordin, *Five Days*, op. cit., p. 118, for a very different reaction.
9. Ibid., p. 35. Gordin has also shown that, contrary to the 'myth' that has become established orthodoxy, the insiders involved in dropping the two bombs on Hiroshima and Nagasaki did not expect Japan to cave in so quickly. Many of them did not feel that the nuclear bombs were anything 'special', that more than two would be needed, and that the war would continue, perhaps for months.
10. Alperovitz, op. cit., p. 12.
11. *New Scientist*, 20 May 2017, pp. 20–21.

Chapter 1. Zigzag

1. Otto Frisch, *What Little I Remember*, Cambridge: Cambridge University Press, 1980, p. 147.
2. *The Times* (London), 7 December 1943, p. 4.

3. Frisch, op. cit., p. 148.
4. Rudolf Peierls, *Bird of Passage: Recollections of a Physicist*, Princeton: Princeton University Press, 1985, p. 183.
5. Robert S. Norris, *Racing for the Bomb: The True Story of General Leslie R. Groves, the Man Behind the Birth of the Atomic Age*, New York: Skyhorse Publishing, 2002, pp. 528–9.
6. Peierls, op. cit., pp. 184–5.
7. Fuchs has been the subject of three biographies: Norman Moss, *Klaus Fuchs: The Man Who Stole the Atom Bomb*, London: Grafton, 1987; Robert Chadwell Williams, *Klaus Fuchs: Atom Spy*, Cambridge, MA: Harvard University Press, 1987; Mike Rossiter, *The Spy Who Changed the World*, London: Headline, 2014.
8. Peierls, op. cit., p. 186.
9. Frisch, op. cit., p. 76.
10. Peierls, op. cit., p. 188.
11. Abraham Pais, *Niels Bohr's Times: In Physics, Philosophy and Polity*, Oxford: Oxford University Press, 2007, p. 496.
12. Thomas Powers, *Heisenberg's War: The Secret History of the German Atomic Bomb*, New York: Da Capo Press, 2000, p. 229; Pais, op. cit., p. 491.
13. Norris, op. cit., p. 251.
14. Ruth Moore, *Niels Bohr: The Man, His Science, and the World They Changed*, Cambridge, MA: MIT Press, 1985, p. 324.

Chapter 2. The Taste of Fear: The Menace of Fission

1. See, for example, Kragh, op. cit., p. 230.
2. Richard Rhodes, *The Making of the Atomic Bomb*, London and New York: Penguin Books, 1986, p. 14.
3. Ibid., pp. 209–13.
4. Ibid.
5. Ibid.
6. Ibid., p. 217.
7. Ibid., pp. 218–20.
8. Fast neutrons have a speed of up to 20,000 km/sec., while slow neutrons travel at 2.2 km/sec., about the speed of a rifle bullet.
9. Ruth Sime, *Lise Meitner: A Life in Physics*, Los Angeles and Berkeley: University of California Press, 1997. p. 5.
10. Ibid., p. 10.
11. Frisch, op. cit., p. 3.
12. Sime, op. cit., pp. 144–5.
13. Ibid., p. 186.
14. Ibid., p. 195; Paul Lawrence Rose, *Heisenberg and the Nazi Atomic Bomb Project: A Study in German Culture*, Berkeley, Los Angeles and London, 1998, p. 159.
15. Sime, op. cit., p. 191.

16. Ibid., 195.
17. Ibid., p. 197.
18. Ibid., p. 199.
19. Ibid., p. 201.
20. Rose, op. cit., p. 41. Sime, op. cit., p. 204.
21. Ibid.
22. Ibid., p. 205.
23. Ibid., p. 207.
24. Ibid., p. 255.
25. Per F. Dahl, *Heavy Water and the Wartime Race for Nuclear Energy*, Bristol and Philadelphia: Institute of Physics Publishing, 1999, p. 76.
26. Frisch, op. cit., pp. 51–2.
27. Sime, op. cit., p. 78.

Chapter 3. The Beginnings of a 'Strategic Game'

1. Sime, op. cit., pp. 236–47.
2. A biography, albeit tantalisingly incomplete and largely unsourced, was published in 1987 by Arnold Kramish, himself a physicist who worked on the Manhattan Project at Los Alamos and elsewhere, who took part in the interrogation of David Greenglass (the New York-born atom spy uncovered in 1950 who implicated the Rosenbergs). Kramish later became an expert on nuclear proliferation. Arnold Kramish, *The Griffin: The Greatest Untold Espionage Story of World War II*, Boston: Houghton Mifflin, 1986, pp. 6–8. See also Paul Lawrence Rose, op. cit., p. 62.
3. Kramish, op. cit., p. 12.
4. Ibid., p. 17.
5. Sime, op. cit., pp. 148–56. NA: Samuel A. Goudsmit Papers, Box 28, Folder 23.
6. Goudsmit File, Box 28, Folder 42. Rose, op. cit., p. 26; Gerard DeGroot, *The Bomb: A Life*, Cambridge, MA: Harvard University Press, 2006, pp. 15, 51.
7. Kramish, op. cit., p. 220.
8. Moore, op. cit., p. 226.
9. Ibid., p. 246.
10. Rhodes, op. cit., pp. 246–7; Spencer Weart, 'Scientists with a Secret', *Physics Today*, February 1976, p. 23.

Chapter 4. Struggles over Secrecy

1. William Lanouette and Bela Silard, *Genius in the Shadows: A Biography of Leo Szilard, the Man Behind the Bomb*, New York: Skyhorse Publishing, 2013, p. 192.
2. Ibid.
3. Rhodes, op. cit., p. 280.
4. Weart, op. cit., p. 24. Always very conscious of priority in ideas and

patents, Szilard had devised his own method of certification. He would write the details down, date them, and send them to himself in the mail so that the envelope was suitably franked. Lanouette and Szilard, op. cit., p. 186.

5. Ibid., p. 187.
6. Weart, op. cit., p. 25.
7. Ibid., p. 28.
8. Rhodes, op. cit., 294.
9. Weart, op. cit., p. 28. NA RG 77, Box 64.
10. Lanouette, op. cit., p. 180.
11. Weart, op. cit., p. 29.
12. Spencer Weart, 'Secrecy, Simultaneous Discovery, and the Theory of Nuclear Reactors', *American Journal of Physics*, vol. 45, no. 11, November 1977, pp. 1049–60. Samuel Goudsmit, *Alsos: The Failure of German Science*, London: Sigma, 1947, p. 164.
13. Weart, op. cit., p. 28. NA RG 77, Box 64.
14. David Irving, *The Virus House*, London: Kimber, 1967, p. 34.
15. Rose, op. cit., p. 43.
16. Ibid., p. 95.
17. Ibid., p. 96.
18. Ibid., pp. 95–6.
19. Weart, op. cit., p. 29; Kramish, op. cit., pp. 52 and 54; Mark Walker, *German National Socialism and the Quest for Nuclear Power, 1939–1949*, Cambridge UK: Cambridge University Press, 1989, pp. 16, 66, 74; Samuel A. Goudsmit Papers, Box 28, Folder 42.
20. Weart, 'Secrecy, Simultaneous Discovery, and the Theory of Nuclear Reactors', pp. 1049–60.
21. Kramish, op. cit., p. 54.
22. Rose, op. cit., p. 189.
23. Thomas Powers, *Heisenberg's War: The Secret History of the German Bomb*, New York: De Capo Press, 1993, p. 67.
24. Paul Lawrence Rose, *Heisenberg and the Nazi Atomic Bomb Project, 1939–1945: A Study in German Culture*, Los Angeles and Berkeley: University of California Press, 1998, p. 95.
25. Rose, op. cit., p. 134, n. 8.
26. Kramish, op. cit., p. 53.
27. Goudsmit File, Box 28, Folder 42.
28. Jeffrey Richelson, *Spying on the Bomb: American Nuclear Intelligence from Nazi Germany to Iran and North Korea*, New York: W. W. Norton, 2007, p. 22.
29. Kramish, op. cit., p. 54.
30. Richelson, op. cit., p. 23.
31. Rose, op. cit., p. 189; Powers, op. cit., p. 67; Goudsmit File, Box 28, Folder 42.
32. Rhodes, op. cit., p. 311; Royal Society Archive: FS/7/4/5.
33. Rhodes, op. cit., p. 351; Jeremy Bernstein, *Hitler's Uranium Club*,

Woodbury, New York: American Institute of Physics, 1996, p. 127. See also Rose, op. cit., p. 134, n. 8.
34. Weart, 'Scientists with a Secret', p. 29.
35. Rose, op. cit., p. 133; Andrew Brown, *The Neutron and the Bomb: A Biography of Sir James Chadwick*, Oxford: Oxford University Press, 1997, pp. 205–6.
36. Weart, 'Scientists with a Secret,' p. 30.
37. Kramish, op. cit., p. 54.
38. Graham Farmelo, *Churchill's Bomb: A Hidden History of Britain's First Nuclear Weapons Programme*, London: Faber & Faber, 2014, p. 133.
39. Royal Society Archive: FS/7/4/5.
40. Rhodes, op. cit., pp. 351ff.; Martin Sherwin, *A World Destroyed: The Atom Bomb and the Grand Alliance*, New York: Knopf, 1975, p. 35. If anything emerged from American research, the British were told, they would be informed accordingly.
41. Weart, 'Scientists with a Secret', p. 29.
42. Rose, op. cit., p. 133; Brown, op. cit., pp. 205–6.
43. Weart, 'Scientists with a Secret', p. 30.
44. Farmelo, op. cit., p. 133.
45. Ibid., p. 134.
46. Dahl, op. cit., p. 171.
47. Sherwin, op. cit., p. 35.
48. Dahl, op. cit., p. 172.
49. CKFT 6/6/40.

Chapter 5. The Midwives

1. Kramish, op. cit., p. 81.
2. Frisch, op. cit., p. 122.
3. Peierls, op. cit., p. 145.
4. Even Einstein thought that the critical amount of uranium would 'need to be carried by warship' – but in that case how would the crew escape?
5. Rhodes, op. cit., pp. 321 and 323.
6. Ibid., pp. 324–5.
7. Ibid., pp. 330–31.
8. Farmelo, op. cit., p. 162. Note the incredible nationalism in the publication of this title, so many years after the events described. In the US it was: *Churchill's Bomb: How the US Overtook Britain in the First Nuclear Arms Race*, New York: Basic Books, 2013; in Britain it was: *Churchill's Bomb: A Hidden History of Britain's First Nuclear Weapons Programme*, London: Faber & Faber, 2014.
9. At Cambridge Halban and Kowarski, though enjoying the splendours of King's College as temporary fellows, were excluded from the inner workings of the MAUD Committee, now that heavy water wasn't needed. Their main aim was to use it to generate nuclear power rather than for use as a weapon. This was not high up on the British list of priorities just

then. Worse, from a British point of view, they had brought with them a sheaf of patent applications filed in Paris shortly before the collapse. In the circumstances, the British thought this 'trifling and time-wasting'. Thomson dismissed them later as 'the most unmitigated nuisance'.

Chapter 6. The Strategic Sabotage of Heavy Water

1. Rhodes, op. cit., pp. 330–31.
2. Farmelo, op. cit., p. 162.
3. Dahl, op. cit., p. 116; FO 942/169; FO 371/33069.
4. Dahl, op. cit., p. 116.
5. Ibid., p. 162.
6. Ibid., p. 167.
7. Ibid., pp. 231–2.

Chapter 7. The First Glimpses of Germany's Nuclear Secrets

1. Leslie Groves, *Now It Can Be Told: The Story of the Manhattan Project*, New York: Harper and Bros, 1962, p. 146.
2. Powers, op. cit., p. 66. Fermi and Dunning confessed to Laurence after the war that they were horrified by his questions, not wanting to encourage him in his story just then.
3. Rose, op. cit., p. 156.
4. Powers, op. cit., p. 22.
5. Ibid., p. 70.
6. CKFT 6/6/40.
7. Ibid.
8. Ibid.
9. CKFT 18/24.
10. CKFT 21/6. Ball, op. cit., pp. 172–3 for doubts about Debye's morality.
11. Rudolf Peierls, *Atomic Histories*, Woodburg, NY: American Institute of Physics Press, 1997, p. 112.
12. CKFT 18/24.
13. CHAD 1/30/3.
14. CHAD 1/19, letter 11 May 1942.
15. CHAD 1/19/6, 23 September 1941.
16. AB 1/356.
17. CHAD 1/19/6, 23 September 1941.
18. Dahl, op. cit., p. 189.

Chapter 8. The 'Crown Jewel' of Secrets

1. Kramish, op. cit., p. 64.
2. Ibid., pp. 66–7.
3. Ibid., p. 68. Samuel A. Goudsmit Papers, Box 28, Folder 42. One of his other contacts was Richard Kuhn at the Deutsche Chemische Gesellschaft, who was one of those who told him that he had learned from I. G. Farben that Norsk Hydro had been blown up. FBI Fuchs

Folder 42. And see R. S. Hutton, *Recollections of a Technologist*, London: Sir Isaac Pitman and Sons, 1964, p. 180.

4. RVJO B330.
5. R. V. Jones, *Reflections on Intelligence*, London: Heinemann, 1989, p. 284.
6. Kramish, op. cit., p. 126.
7. Powers, op. cit., pp. 131–2, 143–9.
8. Rose, op. cit., pp. 311–12. Even so, V.G., as he was known, always thereafter carried a capsule of hydrocyanic acid in his pocket, rather like Leo Szilard always kept two suitcases packed, 'just in case'. See also Goudsmit File, Box 28, Folder 42.
9. Kramish, op. cit., p. 131.
10. FBI Fuchs Folder 42, p. 5; Rose, op. cit., pp. 26 and 194.
11. Paul Rosbaud to Francis Simon, 28 September 1947, Royal Society Archive; FS 7/2/592/1; Goudsmit File, Box 28, Folder 42.
12. Rosbaud/Goudsmit File, Box 28, Folder 44.
13. Rosbaud/Goudsmit File, Box 28, Folder 45.
14. Ibid., Folder 42.
15. Goudsmit File, Box 1, and Box 28, Folder 44, 10.1.59; Rose, op. cit., p. 308.
16. CKFT 18/31; Walker, op. cit., p. 47.
17. Ibid., p. 48.
18. Ibid.
19. Ibid., p. 54.
20. Because of the requirements of war, the number of physics graduates had almost disappeared and, according to one SS report, German industry was hiring physicists who had *failed* their exams.
21. Walker, op. cit., p. 49. Walker adds that 'taking into account the almost universal assumption that the war would last at most only a year or two more, this judgment appears reasonable and justifiable'. The army was not being unreasonable or miserly, says Walker. They were most certainly interested in new weaponry but only if it was relevant to the war effort. After they realised that nuclear fission was too far off, they redoubled their research into rocketry, which did pay off.
22. Ibid., p. 51; NA RG 200 Box 3, Folder 1; Powers, op. cit., p. 145.
23. Ibid., p. 146.
24. Ibid.
25. Ibid., p. 148.
26. Ibid. Speer had been told – it is not known by whom – that it was possible that a nuclear chain reaction could slip out of control, set the planet on fire and turn the earth into 'a glowing star'. Could Heisenberg assure him this could be avoided? Heisenberg refused to offer such reassurance.
27. Powers, op. cit., p. 148; NA RG 200, Papers of General Groves, Box 1, entry 11.
28. Ibid., p. 152.

Chapter 9. Intelligence Blackout: The Fatal Mistake

1. F. H. Hinsley et al., *British Intelligence in the Second World War: Its Influence on Strategy and Operations*, London: HMSO, 1979, p. 472.
2. Powers, op. cit., p. 154.
3. Ibid., p. 156.
4. Ibid., p. 157.
5. He was identified in the US security files as a British agent. Box 28, Folder 44, letter dated 18 April 1950.
6. MTNR 2/18–21; Powers, op. cit., p. 158. It is even possible that Waller met Heisenberg himself on a wartime trip to Germany, since they had both studied together at Copenhagen years before.
7. Powers, op. cit., quoting an interview with Wirtz, 14 May 1989.
8. Ibid., p. 518.
9. Ibid., p. 159.
10. Bohr's son Aage, in a brief note of his own, said that the array of visits by German scientists to Denmark, Norway and Sweden during the war years reinforced the view that the Germans attached great importance to atomic energy, and that they were trying to gauge what Bohr and the others thought, or knew – in other words, the same dilemma as with Heisenberg.
11. Kramish, op. cit., p. 131. Goudsmit File, Box 28, Folder 42.
12. Powers, op. cit., p. 161. It was clear that Heisenberg's work involved the use of some heavy hydrogen (deuterium) compound and he was stated to have had half a ton of heavy water and was due to receive a further ton.
13. Ibid., p. 162.
14. Kramish, op. cit., p. 162.
15. Powers, op. cit., p. 162.
16. Kramish, op. cit., p. 163.
17. Powers, op. cit., p. 163.
18. DeGroot, op. cit., p. 32; Goudsmit, op. cit., pp. 7–8.
19. Kramish, op. cit., pp. 163–4.
20. Powers, op. cit., p. 164.
21. Ibid., pp. 282–5 for the politics.
22. Ibid., p. 304; MTNR 2/18–21; Goudsmit, op. cit., p. 105.
23. Powers, op. cit., p. 305.
24. Hinsley et al., op. cit., Appendix 19, *TA Project: Enemy Intelligence*, pp. 934ff; also from CAB 126/244; MTNR 2/18–21.
25. Hinsley et al., op. cit., p. 585.
26. Powers, op. cit., p. 283. Planck had also quoted Heisenberg as saying that a power-producing uranium machine 'might be feasible' in three to four years – perhaps 1946 or 1947, but well after the war was expected to end.
27. Werner and Elisabeth Heisenberg, op. cit., p. 183.
28. 'New Light on Hitler's Bomb', *Physics World*, 1 June 2005.
29. RVJO B323.

30. RVJO B440; Jones, op. cit., p. 472; Rose, op. cit., pp. 168–9.
31. Hinsley et al., op. cit., p. 584.
32. Margaret Gowing, *Britain and Atomic Energy, 1939–1945*, New York: St Martin's Press, 1964, p. 368; see also Christoph Laucht, *Elemental Germans: Klaus Fuchs, Rudolf Peierls and the Making of British Nuclear Culture, 1939–59*, Basingstoke: Palgrave Macmillan, 2012, p. 10.
33. Walker, op. cit., p. 174.
34. Farmelo, op. cit., p. 163; and Home Office file F. 962, KV 2/2421. NA.
35. Richard Rhodes, *Dark Sun: The Making of the Hydrogen Bomb*, London and New York: Simon & Schuster, 1995, p. 62.

Chapter 10. Fall Out

1. Dahl, op. cit., p. 183.
2. Nancy Thorndike Greenspan, *The End of a Certain World: The Life and Science of Max Born, the Nobel Physicist who Ignited the Quantum Revolution*, Chichester: John Wiley, 2005, pp. 238–9.
3. Dahl, op. cit., p. 181; CHAD 1, 28/6.
4. Dahl, op. cit., p. 182.
5. 2 August 1941.
6. Dahl, op. cit., p. 188.
7. Ibid., p. 189; Lindemann to Churchill, 27 August 1941, CAB 126/330 NA.
8. Rhodes, *Atomic Bomb*, op. cit., pp. 357ff.
9. Dahl, op. cit., p. 191.
10. Ibid., p. 198.
11. CHAD 1 19/3.
12. Dahl, op. cit., p. 173.
13. CHAD 1 12/3.
14. Dahl, op. cit., p. 369.
15. Ibid., p. 173.
16. The importance that Roosevelt attached to the note was underlined by the fact that he arranged for it to be carried to the prime minister by hand, by Frederick Hovde, head of Bush's NDRC office in London.
17. Dahl, op. cit., p. 173.
18. Ibid., p. 203.
19. Sherwin, op. cit., p. 41.
20. Ibid. p. 39.
21. Farmelo, op. cit., p. 224.
22. J. G. Hershberg, *James B. Conant*, Stanford, CA: Stanford University Press, 1993, p. 180.
23. CHAD IV 12/5.
24. Farmelo, op. cit., p. 214.
25. Gowing, op. cit., pp. 144–5 and 437–8.
26. Farmelo, op. cit., p. 217.

27. Ibid., p. 223.
28. Sherwin, op. cit., p. 80.
29. Farmelo, op. cit., p. 226.
30. Sean Malloy, *Atomic Tragedy: Henry L. Stimson and the Decision to Use the Atomic Bomb Against Japan*, Ithaca, NY: Cornell University Press, 2008, p. 45.
31. Sherwin, op. cit., p. 81.
32. Ibid., pp. 74–6.
33. Malloy, op. cit., *passim*.
34. Farmelo, op. cit., p. 227.
35. Ibid., p. 234.
36. Sherwin, op. cit., p. 83.
37. Farmelo, op. cit., p. 232.
38. Ibid.
39. Sherwin, op. cit., p. 83.
40. Farmelo, op. cit. p. 240. The agreement was a victory for Churchill, a remarkable recovery in many ways. He paid a price, though, by disclaiming any interest in the commercial and industrial aspects of nuclear power after the war, leaving the president to have the final decision.

Chapter 11. 'Hard Evidence' of Soviet Spying

1. Sherwin, op. cit., p. 47.
2. Gregg Herken, *The Brotherhood of the Bomb: The Tangled Lives and Loyalties of Robert Oppenheimer, Ernest Lawrence and Edward Teller*, New York: Henry Holt, 2002, p. 55.
3. Ibid., p. 56.
4. Ibid.
5. Ibid., p. 57.
6. Ibid., p. 71
7. Ray Monk, *Inside the Centre: The Life of J. Robert Oppenheimer*, London: Jonathan Cape, 2012, p. 335.
8. Herken, op. cit., p. 72.
9. Ibid., p. 94.
10. Ibid.
11. He amassed a personal collection of improbable disguises (wigs, voice-altering devices, fake spectacles) that he always had with him.
12. Monk, op. cit., p. 343.
13. Herken, op. cit., p. 96.
14. Ibid., p. 97, ref. 91.
15. Near the end of the seminar, discussions had also considered Teller's idea for a hydrogen bomb, that an atom bomb might release enough energy to start a thermonuclear reaction similar to those fuelling the sun and other stars.
16. John Earl Haynes, *Spies: The Rise and Fall of the KGB in America*, New Haven, and London: Yale University Press, 2009, p. 42.

17. Nelson said he had been recruited at the end of 1942 by 'a man from Moscow', and several current and prospective agents were discussed in the course of the exchange.
18. Haynes, op. cit. p. 62.
19. Herken, op. cit., p. 94.
20. Herken, op. cit., p. 102. See also: Allen Weinstein and Alexander Vassiliev, *The Haunted Wood: Soviet Espionage in America – the Stalin Era*, New York, Modern Library, 2000, p. 185.
21. Monk, op. cit., p. 366.
22. Herken, op. cit., p. 106.
23. Ibid., p. 107.
24. Monk, op. cit., p. 376.
25. The 'acid-tongued' director had almost certainly deceived various investigators. Barton J. Bernstein, 'Reconsidering the "Atomic General": Leslie H. Groves', *Journal of Military History*, vol. 67, July 2003, p. 897. And so the director continued to be questioned from time to time.
26. Herken, op. cit., p. 114.
27. Ibid., p. 115; see also Bernstein, 'Reconsidering the "Atomic General"', p. 898.
28. Two more episodes of note took place in early 1944. In one, Martin Kamen, a chemist who commuted between Berkeley and Oak Ridge, was followed to Bernstein's Fish Grotto in San Francisco. There he was seen to meet with Kheifets and Grigori Kasparov and their overheard conversation included the words 'Lawrence', 'radiation' and 'military boys'. Told of this, Groves ordered Kamen to be dismissed immediately. In the other episode, one of Pash's men, James Murray, was transferred to Chicago, where the FBI and Groves had uncovered an espionage ring passing secrets of the Met Lab to Moscow, using the Russian New York consulate. Herken, op. cit., p. 124.
29. Sherwin, op. cit., p. 255, n. 28. Joseph Albright and Marcia Kunstel, *Bombshell: The Secret Story of America's Unknown Atomic Spy Conspiracy*, New York, Times Books, 1997, pp. 103–6; RG 200, Box 15, Folder 3.

Chapter 12. The Secret Agenda of General Groves

1. Haynes, op. cit., pp., 61–2.
2. Ronald W. Clark, *The Greatest Power on Earth: The Story of Nuclear Fission*, London: Sidgwick & Jackson, 1980, p. 133.
3. Groves, op. cit., pp. 11 and 49.
4. NA, RG 200, Papers of General Leslie R. Groves, Box 3.
5. Groves, op. cit., p. 187.
6. Malloy, op. cit., p. 83. We shall return to this but it is also worth pointing out here that Groves's comment was made at exactly the time the meeting between Nelson and Lehman was being bugged. So that may have had something to do with it.

7. Groves, op. cit., p. 45; NA RG 200 Papers of General Leslie R. Groves, Box 1.
8. Groves, op. cit., p. 48.
9. Sherwin, op. cit., p. 56.
10. Ibid., p. 50.
11. Ibid., p. 58.
12. Philip Ball, *Serving the Reich: The Struggle for the Soul of Physics under Hitler*, London: Bodley Head, 2013, pp. 164, 166.
13. Ibid., p. 165.
14. Jurrie Reiding, 'Peter Debye: Nazi Collaborator or Secret Opponent?', *Ambix*, vol. 57, no. 3, 2010, pp. 275–300.
15. Ball, op. cit., p. 176. Rockefeller Foundation Archives, RF Officer Diaries, disk 16 (Warren Weaver), memo of 6 February 1940, pp. 19–20.
16. Groves, op. cit., p. 199.
17. Jeffrey Richelson, *Spying on the Bomb: American Nuclear Intelligence from Nazi Germany to Iran and North Korea*, New York: W. W. Norton, 2007, p. 27.
18. Ibid.
19. Ibid., p. 30.
20. CHAD IV 3/1.
21. Ibid.
22. CHAD IV 11/6.
23. Haynes, op. cit., p. 35.
24. Hinsley et al., op. cit., pp. 585–6. Finally, on the intelligence front, there were the reports received from the Belgian underground to the effect that 700 tons of sodium uranate were still at the Union Minière refinery at Oolen in Belgium. 'If true they were in the Directorate of Tube Alloys's view "the strongest possible proof" that Germany was not doing any large-scale T-A work.'
25. NA Samuel A. Goudsmit Papers, Box 28, Folder 42. See also Groves, op. cit., p. 217.
26. Irving, op. cit., p. 203.
27. Ibid., p. 44. Furman was not troubled by the fact that no large-scale industrial plant had been found in connection with a bomb because no large-scale facilities had been found in connection with Germany's rocket and pilotless aircraft either, and yet they were known to exist.
28. Harold Urey, in a cover note attached to Cohen's report, expressed a certain scepticism that the Germans had made as much progress as the young man thought.
29. Richelson, op. cit., p. 50.
30. Norris, op. cit., p. 295.
31. Ibid., p. 640.
32. Groves, op. cit., p. 222. NA Samuel A. Goudsmit Papers, Box 64.
33. Ibid., p. 336.
34. Ibid.
35. Ibid., pp. 194 and 214.

36. Ibid., pp. 230, 240 and 245.
37. Gowing, op. cit., p. 368.
38. RVJO B 354.
39. Groves, op. cit., p. 185.
40. Hinsley et al., op. cit., p. 584.
41. Ibid., p. 937. The same report, confirming intelligence reported earlier, said that Dr Kurt Diebner, 'a very active Nazi, though an indifferent physicist', had written an annual survey of work on artificial radioactivity but it had not been continued beyond 1942. See also p. 940.
42. Ibid., p. 941.
43. Ibid. The report also cautioned that intelligence work inside Germany, when the time came, should be circumspect since the very nature of inquiries might show what the Allies did – and didn't – know, and thereby alert the enemy.

    This can't be quite right for it doesn't square with information and conclusions in other documents and because of the very real differences in the understanding and appreciation of the German threat as between the two Allies. The most alarmist US reports came after this date.
44. This is one reason why they took so seriously a report in May 1943 from Sam Wood, one of their diplomats in Bern, Switzerland, who had previously been in Berlin, where he formed a good contact with Erwin Respondek, a former professor of economics who had become a high-ranking civil servant. Respondek had excellent contacts and had been one of the first to warn the United States that Germany was planning to invade Russia, a warning that was hardly heeded at first. In the May 1943 report, Respondek said that Germany was setting aside five million Reichsmarks, which had been placed at the disposal of 'leading professors and scientific institutions', to test the principles involved in the creation of an atomic bomb. Another thirty million marks, he said, were set aside for 'technical tests'. (John V. H. Dippel, *Two Against Hitler, Stealing the Nazis' Best-kept Secrets*, New York: Praeger, 1992, p. 93.)

    Although the report added that the work was 'probably connected with Otto Hahn's splitting of the atom', it did not say at first that the 'inventor' of the bomb was a certain Walter Dallenbach, a Swiss under contract to AEG, whose agreement was to develop various devices 'for atomic energy processes' and 'make them available for commercial production as soon as possible'.

    The problem with this account is that it contains a number of improbables. Dallenbach was Swiss and therefore would never have been allowed to work on weapons projects. (Recall what happened to the Dutchman Peter Debye, a Nobel Prize-winning physicist who had to give up his directorship of the Kaiser Wilhelm Institute for Physics because he wouldn't take up German nationality.) Dallenbach had been working on commercial applications of electromagnetic waves until 1942, for AEG, where he had indeed already been banned from working on weapons. He had studied under Einstein and his ideas for a cyclotron had been recommended to Speer by Heisenberg. If a cyclotron *was* being built, beginning in 1943, by a Swiss, it could not have been used for a bomb, or even have military implications, and in any case useful results were at least two years away.

Dallenbach himself always claimed, after the war, that he was engaged only in basic research, 'unrelated to the war'. Rosbaud didn't like him, thinking he was able but agreed too much with the Nazis. See Goudsmit File, Box 28, Folder 42.

Here too the failure of the Allies to share atomic intelligence helped to create confusion and wasted effort. The quality of the American atomic intelligence was, as Kramish said, much inferior to that of the British. The Dallenbach story, like the Houtermans-Dessauer-Szilard-Conant story (p. 111), was garbled, unsatisfactory and wrong. Rosbaud appears to have liked and approved of Houtermans during the war but felt that he had rather lost his way afterwards. See also Rose, op. cit., p. 144.

45. Groves, op. cit., pp. 243–4.
46. Peierls, op. cit., p. 113.
47. Richelson, op. cit., p. 51.
48. RVJO B354.
49. Powers, op. cit., p. 379.
50. This is made more unsatisfactory and bizarre by the fact that the Americans did have similar thoughts where Samuel Goudsmit was concerned. The man chosen to run the Alsos mission was an accomplished physicist, well able to judge how far the Germans had got in their nuclear project (and a good friend of Heisenberg, who had stayed with him in 1939), but he was not part of the Manhattan Project itself and was not privy to its most important secrets. It was a sensible security measure. But though he had thought of it himself, it seems it was not a mode of tactical thinking that Groves felt the Germans would employ.
51. Bernstein, 'Reconsidering the "Atomic General"', p. 895.
52. Norris, op. cit., pp. 316–17.
53. CHAD IV 3/1.
54. Groves, op. cit., pp. 407–8.
55. Laucht, op. cit., p. 79. Albright et al., op. cit., p. 185.
56. CHAD IV 3/1; Bernstein, 'Reconsidering the "Atomic General"', p. 900.
57. RVJO B 354.
58. Groves, op. cit., pp. 197–8.
59. Bernstein, 'Reconsidering the "Atomic General"', p. 887.
60. Norris, op. cit., p. 311.
61. General George Marshall, chief of staff of the army, was sceptical on the grounds that such a committee was premature and susceptible to leaks. Malloy, op. cit., p. 56.
62. M1109 microfilm, Cabinet 48, draw 1, roll 3. Malloy, op. cit., p. 56. Einstein later remarked that he 'would never have lifted a finger' had he known that the Nazis were not going to produce a bomb during the course of the war. Even after the American nuclear programme began in

earnest, some Manhattan Project insiders worried that the Allies might have to 'stand the first punishing blows' of Nazi A-bombs before they could respond in kind.

63. Malloy, op. cit., p. 56. Bush to Conant, 18 January 1943. Bush-Conant file 18; RG 200, Box 1, entry 11.
64. Groves, op. cit., p. 187; Malloy, op. cit., p. 57.
65. Moreover, the Japanese had some accomplished (if not very numerous) nuclear scientists of their own, and Allied physicists knew it. One was Dr Yoshio Nishina, who had established the Rikken Institute's Laboratory for Chemistry and Physics near Tokyo, in 1931, and had built several cyclotrons (and bought one from the University of California) between 1934 and 1938. He was a good friend of both Niels Bohr and Albert Einstein and had studied at Bohr's institute in Copenhagen. The second man was Bunsaku Arakatsu at the Imperial University in Kyoto. He had worked at the Cavendish Laboratory at Cambridge under Ernest Rutherford and at Berlin University under Albert Einstein.
66. Stimson in particular was aware of this and wondered whether it would affect cooperation after the war was over, especially in respect to atomic matters. Malloy, op. cit., p. 74.
67. David Holloway, *Stalin and the Bomb: The Soviet Union and Atomic Energy, 1939–1956*, New Haven and London: Yale University Press, 1994, p. 90. RG 200, Box 1, entry 11.
68. Malloy, op. cit., p. 58.
69. Ibid., p. 60.
70. Groves, op. cit., p. 199.
71. Hinsley et al., op. cit., p. 586.
72. The British did accept, however, that fission products could probably be produced more quickly than a bomb and that Germany therefore might turn to them as a last resort in an attempt to avert final defeat. Because of this the Tube Alloys Consultative Council had authorised work to start on a simple radioactivity detector.
73. Hinsley et al., op cit., p. 587.
74. Groves, op. cit., p. 200.
75. Richelson, op. cit., p. 50; Hinsley et al., op cit., p. 587.
76. Michael Goodman, *The Official History of the Joint Intelligence Committee, volume 1: From the Approach of the Second World War to the Suez Crisis*, London: Routledge, 2014, pp. 136–7.
77. Malloy, op. cit., p. 106.

Chapter 13. The Bohr Scare

1. FO 188/651; FO 371/33069.
2. Farmelo, op. cit., p. 247.
3. Ibid., p. 248.
4. Ibid., p. 257.
5. Moore, op. cit., p. 315.

6. Powers, op. cit., p. 199.
7. Ibid., p. 200.

Chapter 14. A Vital Clue Vanishes

1. Moore, op. cit., p. 325.
2. DeGroot, op. cit., p. 43.
3. Moore, op. cit., p. 326.
4. In the summer of 1942, at the meeting referred to in the previous chapter, which discussed thermonuclear weapons – H-bombs – Edward Teller had wondered if such a bomb could ignite the earth's atmosphere. Although this was thought unlikely, 'in view of the importance of the consequences', Hans Bethe was asked to take a look at the calculations. While he was doing this, Oppenheimer took the trouble to travel from Los Angeles to Chicago to discuss with Arthur Compton 'the possibility of a global catastrophe, which could not be passed over lightly ... Was there really any chance that an atomic bomb would trigger the explosion of the nitrogen in the atmosphere or of the hydrogen in the oceans? ... Better to accept the slavery of the Nazis than to run a chance of drawing the final curtain on mankind!' By the time Oppenheimer returned to Los Alamos, Bethe had discovered 'some unjustified assumptions' in Teller's arguments, and the possibilities of catastrophe were discounted. Monk, op. cit., p. 320.
5. Rose, op. cit., p. 156.
6. Ibid., pp. 160 and 161, n. 49.
7. Bethe and Teller appear to have been chosen because, a few months earlier, they were the ones who had written to Oppenheimer about their concern at newspaper stories which, in contrast to the one in *Stockholms Tidningen*, suggested that a German bomb might soon be ready.
8. More specifically, they added that the proposed 'pile' must have been made of uranium sheets immersed in heavy water. And, they concluded: 'In the course of the explosion the major part of the energy is liberated in the uranium plates in the form of heat ... The multiplication is expected to stop when the pile has expanded to roughly twice its initial linear dimensions ... [This] means an energy liberation of about one quarter of that due to the same mass of TNT.' See, for example, the exchange of letters between Bernstein, Thomas Powers and Michael Frayn, in the *New York Review of Books*, 25 May 2000, 15 October 2000 and 8 February 2001.
9. Rose, op. cit., p. 163.
10. Aage, who was nineteen in 1941, later wrote a letter to Powers: 'Heisenberg certainly drew no sketch of a reactor during his visit in 1941. The operation of a reactor was not discussed at all.' Bernstein, Powers et al., op. cit., p. 3.
11. For the train ride, Powers, op. cit., pp. 247–8.
12. Bernstein, Powers et al., op. cit., p. 3.

13. Ibid.
14. Ibid. For Rosbaud's role, see also Rose, op. cit., p. 159.
15. Powers, op. cit., pp. 247–8. Most crucially, Bohr, being in occupied Denmark and therefore isolated from other developments, had not thought through the role of heavy water, which both the Allies and the Germans (and the Russians, come to that) knew could be used to manufacture plutonium, which was just as fissionable as U-235. Bernstein says Bohr had some 'midway idea' that the heavy water would be 'blown off' by an explosion, which would then cause the explosion to stop. Such a bomb might be 100 times more deadly than TNT but it would be phenomenally expensive to produce even one such bomb. Even so, Bohr was worried that the Germans were making headway in this direction.
16. Ibid. For Rose's disagreement, see Rose, op. cit., p. 160, note 47.
17. Rose, op. cit., p. 146. Charles Frank, *Operation Epsilon: The Farm Hall Transcripts*, Los Angeles and Berkeley, University of California Press, 1993, op. cit., p. 94.
18. These included the 144,000 German patents and patent applications held in the Reichspatentamt in Berlin, which were microfilmed by the Office of Technical Services, and subsequently transferred to the Library of Congress. A second Library of Congress source is a collection of over 700 files confiscated by the ALSOS Mission, which were returned to Germany after being microfilmed. Rose, op. cit., p. 149.
19. Rose also found some correspondence in the Weapons Research Office that appeared to relate to this patent application, in which Basche, the relevant administrator, sent a note to Paul Harteck, a main member of the Uranverein, who, as we have seen, had done important work on isotope separation and the use of heavy water as a moderator. Basche's note requested information 'as to what persons of your Institute collaborated on the "Uranium Pile" and who accordingly are to be listed with the Patent office as inventors. Prof. Bothe suggests that the patent be looked upon as [the result of] common efforts by all institutes.' On 20 August 1942 Basche also wrote to Otto Hahn, enclosing the *Patentmeldung 'Uranmaschine'*, which he described as an important product of teamwork. Hahn replied asking for sixteen co-workers to share in the invention. Rose, op. cit., p. 148.
20. Rainer Karlsch, *Hitlers Bombe*, Berlin: Deutsche Verlag-Anstalt, GmBH, 2005. See the summary in English in *Forum*, June 2005.
21. Ibid.
22. Ibid.
23. nbl.ku.dk. See 'Release of Documents Relating to the 1941 Bohr-Heisenberg Meeting'. Documents released 6 February 2002. See also Werner Heisenberg, *Physics and Beyond*, London: Allen & Unwin, 1971, chapter 15 for his own account (post-war).
24. Powers, op. cit., pp. 113–15.
25. Rose, op. cit., pp. 280–82.
26. Monk, op. cit., p. 298.

27. Rose, op. cit., p. 161; RG 200, Box 78.
28. Brown, op. cit., p. 253.

Chapter 15. 'Don't Bother Me With Your Scruples': The Loss of Innocence

1. When he had moved to Britain, Rotblat had left his wife in Poland, the plan being for her to join him later. He never saw her again.
2. J. Rotblat, 'Leaving the Bomb Project', *Bulletin of the Atomic Scientists*, vol. 41, no. 7, 1985, pp. 16–19. See also Bernstein's article calling Rotblat's memory into question: Bernstein, 'Reconsidering the "Atomic General"', p. 903.
3. DeGroot, op. cit., p. 37; FBI Fuchs file 65-58805, Boston FBI report BS 65-3319, of 15 February 1950.
4. Powers, op. cit., pp. 256–7.
5. Groves, op. cit., p. 194. The references that Paul Rose gives in his book on Heisenberg, describing what the British gave Groves, and Groves's reaction, RG 77, MED, Foreign Intelligence Unit, entry 22, box 170, folder 32.69-1, and box 168, folder 202.3-1, no longer exist or at least could not be traced by the in-house research consultants at College Park.
6. DeGroot, op. cit., p. 69. For Rotblat's intelligence dossier, see Albright et al., op. cit., p. 101. Gowing, op. cit., p. 368 for the Oppenheimer quote.
7. Groves, op. cit., p. 297.
8. DeGroot, op. cit., p. 125.
9. Monk, op. cit., p. 403.
10. Richard G. Hewlett and Oscar F. Anderson, *The New World 1939/1946: Volume 1 of a History of the United States Atomic Energy Commission*, Philadelphia: Pennsylvania State University Press, 1962, p. 253.
11. Laucht, op. cit., p. 127.
12. CHAD IV 3/3
13. Farmelo, op. cit., p. 179; Brown, op. cit., p. 123.
14. Groves, op. cit., p. 265.
15. DeGroot, op. cit., p. 28; Sherwin, op. cit., p. 180.
16. And this in a book that itself ranks the discovery of atomic energy with the discovery of America in historical importance. 'It is a big, fundamental discovery that is sharply changing the direction of history.'

Chapter 16. The Claws of the Bear

1. Holloway, op. cit., p. 68.
2. Ibid., p. 11.
3. Ibid., p. 12.
4. Ibid., p. 29.
5. Ibid., p. 35.
6. Ibid., p. 39.
7. Ibid., p. 42.
8. Ibid., pp. 46 and 48.

9. Ibid., p. 51.
10. Ibid., p. 55. Brown, op. cit., pp. 202–3.
11. Holloway, op. cit., p. 51.
12. Ibid., p. 58.
13. Ibid., pp. 75–6.
14. David Burke, *The Spy Who Came in from the Coop: Melita Norwood and the Ending of Cold War Espionage*, Woodbridge: Boydell Press, 2008; Kindle edition: 52 per cent.
15. Holloway, op. cit., p. 76.
16. He also provided a drawing of an experimental bomb, divided into two hemispheres, with conventional weapons used to propel one hemisphere very rapidly into the other to form a critical mass. He couldn't know it but it was a mechanism not dissimilar to that suggested by Frisch and Peierls. Ibid., p. 77.
17. G. N. Flerov, 'U semu my moshem pouchit'sia u Kurchatova', in p. Aleksandrov, ed., *Vospominaniia ob Igore Vasil'eviche Kurchatov*, Moscow: Nauka, 1988, pp. 72 and 81. Quoted in Holloway, op. cit., p. 78.
18. Holloway, op. cit., p. 83.
19. Fuchs's contacts were with GRU, the Chief Intelligence Directorate of the General Staff, not the NKVD. There was bitter rivalry between these two outfits, so it is improbable that the NKVD transmitted Fuchs's information to Moscow.
20. Holloway, op. cit., p. 84.
21. Ibid. For the Taganrog material, see Nigel West and Oleg Tsarev, *The Crown Jewels: The British Secrets at the Heart of the KGB Archives*, London: HarperCollins, 1998, p. 229.
22. Ibid., p. 85. For the discussion of 'K' and 'Moor', see West and Tsarev, op. cit., pp. 231–6.
23. Holloway, op. cit., p. 86.
24. Ibid., p. 88.
25. Ibid., p. 90; West and Tsarev, op. cit.
26. Ibid., p. 91.
27. Ibid. and ref.
28. Ibid., p. 93.
29. Ibid., p. 90.
30. Ibid., p. 93.
31. Rhodes, *Dark Sun*, pp. 74 and 77.
32. Holloway, op. cit., pp. 94–5. Holloway adds this: 'The tone of Kurchatov's memoranda is revealing. There is no gloating about obtaining information that Western governments were trying to keep secret; nor is there any bitterness at the fact that the war had speeded up research in Britain and the United States, but slowed it down in the Soviet Union. Kurchatov does not attempt to belittle the achievements of British and American scientists, or to magnify the work of his own colleagues. His excitement at learning of what was being done abroad comes through, along with

his admiration for the quality of the research.'
33. Rhodes, *Dark Sun*, p. 146.
34. Holloway, op. cit., p. 101.
35. Rhodes, *Dark Sun*, p. 100.
36. Holloway, op. cit., p. 104. In fact, we now know that a special Department S was set up in February 1944, separate from the NKGB and GRU, to deal exclusively with atomic espionage, with six translators and technical help.
37. Michael D. Gordin, *Red Cloud at Dawn: Truman, Stalin and the End of the Atomic Monopoly*, New York: Picador, 2010, p. 109. Rhodes, *Dark Sun*, p. 121, details on p. 80.
38. RVJO B323.

Chapter 17. The Little Fox

1. Norman Moss, *Klaus Fuchs: The Man Who Stole the Atom Bomb*, London: Grafton, 1987, p. 4. For other biographies of Fuchs, see reference 7, chapter 1.
2. Moss, op. cit., p. 5.
3. Ibid., p. 5.
4. Ibid., p. 6.
5. Ibid.
6. Laucht, op. cit., p. 84; Moss, op. cit., p. 7.
7. Ibid., p. 8.
8. Ibid., p. 9.
9. Ibid., p. 10.
10. Ibid., p. 11.
11. Ibid.
12. Ibid., p. 12.
13. Ibid., p. 10.
14. Ibid., p. 13. The 174-page document by General Groves is at: RG 200, Box 3, Folder 1.
15. Moss, op. cit., p. 14.
16. Ibid.
17. Ibid., p. 16.
18. 'They joined at a time when it was much easier than it is now to believe in the [Russian] revolution, and to believe that it was for the betterment of humanity, and to believe also that it was best served by serving the interests of the Soviet state.' Ibid., pp. 8ff.
19. Ibid., p. 20.
20. Greenspan, op. cit.; Moss, op. cit., p. 20.
21. Weinstein and Vassiliev, op. cit., p. 316.
22. Moss, op. cit., p. 22.
23. Ibid., p. 24.
24. Ibid., p. 26.
25. Ibid., p. 31.

26. Rossiter, op. cit., p. 55.
27. Greenspan, op. cit., p. 239.
28. Moss, op. cit., p. 34.
29. Ibid., p. 37.
30. Ibid., p. 38.
31. Ibid., p. 36.
32. Ibid., p. 37.
33. Ibid., p. 43.
34. Ibid.
35. Ibid., p. 44. In Birmingham, Fuchs was becoming fond of Rudolf and Genia Peierls, a feeling that was reciprocated. At one point, Fuchs travelled to Edinburgh for a short break with Max Born. While there he became ill, developing a dry cough that was to last for several years. The local doctor who treated him took a shine to him and invited him home for dinner. Later, the doctor himself fell ill with leukaemia. Fuchs, who was in America by then, sent him a food parcel.
36. Ibid., p. 45.
37. Farmelo, op. cit., p. 239.
38. Holloway, op. cit., p. 90.
39. CHAD IV 11/5, and 11/6.
40. Groves, op. cit., p. 143; Gowing, op. cit., vol. 2, *Policy Execution*, p. 147.
41. Fuchs FBI archive, section 16, p. 34; section 18, p. 12; David Drake, *French Intellectuals and Politics from the Dreyfus Affair to the Occupation*, Basingstoke: Palgrave Macmillan, 2005, p. 111.
42. Robert J. Lamphere and Tom Shachtman, *The FBI-KGB War: A Special Agent's Story*, Macon, GA: Mercer University Press, 1985, p. 154.
43. Klaus Fuchs FBI file, part 27, pp. 55ff. According to Weinstein and Vassiliev in *The Haunted Wood*, p. 312, a 1945 Moscow memorandum claimed that Kuczynski worked for US military Intelligence in 1944–5.
44. Fuchs FBI File, part 27, p. 63.
45. Rhodes, *Dark Sun*, p. 105 and ref.
46. Weinstein and Vassiliev, op. cit., p. 187.
47. Haynes, op. cit., p. 43; Fuchs FBI archive, section 34, p. 1; West and Tsarev, op. cit., p. 238.

Chapter 18. Lunch in the Supreme Court

1. Sherwin, op. cit., p. 99.
2. DeGroot, op. cit., p. 69.
3. Sherwin, op. cit., p. 91.
4. Ibid., p. 93.
5. Ibid., p. 94.
6. Ibid., p. 95.
7. Ibid., p. 97.
8. Ibid., p. 100.

Chapter 19. 'The Mission He Had Been Waiting For'

1. Rhodes, *Dark Sun*, p. 107.
2. Allen M. Hornblum, *The Invisible Harry Gold: The Man Who Gave the Soviets the Atom Bomb*, New Haven and London: Yale University Press, 2010, p. 7.
3. Hornblum, op. cit., p. 9. Despite Sam's enviable worth ethic, he never made much money and Harry grew up without too much regard for material things.
4. Hornblum, op. cit., p. 15.
5. Ibid., p. 18.
6. Ibid., p. 24.
7. Ibid., p. 29.
8. Ibid., p. 34.
9. Ibid., p. 68. See also Weinstein and Vassiliev, op. cit., pp. 176–7.
10. Hornblum, op. cit., p. 89.
11. Ibid., p. 111.
12. Klaus Fuchs FBI file, Part 43/8.
13. Rhodes, *Dark Sun*, p. 104.
14. Ibid., p. 108.

Chapter 20. A President 'Eager for Help'

1. Sherwin, op. cit., p. 100.
2. Nor was this the first time Roosevelt had been forced to give thought to whether he should inform the Russians about the possible advent of an atomic bomb. On 26 December 1942, a little over a year earlier, Secretary of War Stimson had been told about an Anglo-Soviet agreement proposing the exchange of scientific information. This was during the time when the Allies were at loggerheads over their own co-operation on the bomb and Stimson told the president that in his view, 'this agreement seemed to put us in a very serious situation in regard to S-1 [the American code for the bomb]'. The president agreed, telling Stimson that it would be a mistake for the US 'to enter into any similar exchange pact with the Soviets'. Sherwin, op. cit., pp. 100–101.

   In the circumstances, the president's view at that time is understandable. Although it was clear that, if the bomb should come off (not certain in late 1942, but likely after Fermi's pile had gone critical in Chicago), it would play a significant role in post-war diplomacy. But it was by no means certain *how* it would feature, or exactly how the Allies would or could use it. There was at that time already in place a committee of the Department of State exploring post-war planning, but the members of that committee were not privy to the great secret, and so stringent were security concerns that there was not even the suggestion that the committee be told. Roosevelt, like Churchill, was jealous of his standing and believed that diplomacy, like politics, needed the personal touch. Given the fact that he had not yet met Stalin face to face, that was another reason for delay. That encounter would not take place until the Teheran Conference, in November 1943, eleven months away, and in the meantime the Battle of Stalingrad was still raging, the outcome still uncertain. So there seemed no point to any agreement with Russia in December 1942. Better to wait, the more so as, after fighting alongside the Russians for only twelve months, their combined actions had singularly failed to remove a distrust that had existed for a generation.

Nevertheless, the exchange between Stimson and Roosevelt that December in 1942 showed the sort of problems – and opportunities – that were over the horizon, but not too far over. Scientific arguments for the bomb's overwhelming dimensions, the fact that the Russians could be so difficult to deal with at times and their emerging power as Stalingrad proceeded, meant that the sharing of information about even the existence of the atomic bomb was viewed with great suspicion, the more so as at that stage the outcome was so uncertain. For the moment, it seemed safer to do nothing. See also Weinstein and Vassiliev, op. cit., p. 188.

3. Sherwin, op. cit., p. 102.
4. Farmelo, op. cit., p. 260.
5. Sherwin, op. cit., p, 100.
6. Ibid.
7. Fuchs FBI File, 38/15, p. 88.
8. And how far behind the Russians were – Holloway, op. cit., p. 104.
9. Fuchs FBI, part 15, p. 88.
10. Fuchs never knew the names of other spies in the Project but he could infer that he was not alone from some of the questions he was asked by Moscow.

Chapter 21. The Letter from Moscow

1. Farmelo, op. cit., p. 261.
2. Holloway, op. cit., p. 113. Pais, op. cit., p. 500, quoting *Moscow News*, 8 October 1989.
3. Holloway, op. cit., p. 113.
4. 'One of the principal weapons of modern warfare are explosives,' Kapitsa had said, pointedly. Saying that 'explosives' are useful in war is hardly profound unless some sort of code, or hidden agenda, was being used. Was this yet another attempt to make the same appeal?
5. Pais, op. cit., p. 500.
6. NA Microfilm M1109, cabinet 48, drawer 1, roll 3. Holloway, op. cit., pp. 112–13.
7. Holloway, op. cit., p. 113.
8. Ibid., p. 138.
9. Rhodes, *Dark Sun*, pp. 110–11.
10. Rossiter, op. cit., p. 116.
11. Ibid., p. 114.
12. Fuchs may have mentioned at this meeting, or the one immediately

following, that the British scientists in the Manhattan Project were just then rather dissatisfied with the way that their work was being regarded. Much later, in 1949, when the FBI had had their breakthrough in deciphering the Russian wartime cables, there was one in the file, in J. Edgar Hoover's hand, to the effect that, on 8 May 1944, Fuchs had advised the Russians that the British mission was meeting with little success in the US, and that the Soviets wondered whether it was in their interests for him to be sent back to Britain.

The significance of this is twofold. First, that it alerted the Russians to yet more disagreements between the American and British Allies. And second, that Fuchs and Gold discussed quite a lot more than either was willing to admit later.

Chapter 22. The Prime Minister's Mistake

1. Farmelo, op. cit., p. 262.
2. Lindemann to Churchill, n.d., May 1944. PREM 3/139/2. Kevin Ruane, *Churchill and the Bomb in War and Cold War*, London: Bloomsbury Academic, 2016, p. 80.
3. Not enough has been made of this point.
4. R. V. Jones, *Most Secret War*, London: Hamish Hamilton, 1978, p. 475.
5. Ibid., pp. 475–6.
6. Ibid., p. 476.
7. Ibid. Here is a typical sample of Bohr's English syntax, which will give some idea of Jones's reservations, and show why others were equally apprehensive of what the Downing Street meeting might – or might not – achieve. It comes from a document he wrote later in the year for President Roosevelt. In the course of it he said that his connections with German scientists had allowed him 'rather closely to follow the work on such lines that from the very beginning of the war was organized by the German government. Although thorough preparations were made by a most energetic scientific effort, disposing of expert knowledge and considerable material resources, it appears from all information available to us, that at any rate in the initial for Germany so favourable stages of the war it was never by the government deemed worthwhile to attempt the immense and hazardous technical enterprise that an accomplishment of the project would require.' Powers, op. cit., pp. 244–5.
8. At this time there was another development across the Atlantic that might have had momentous consequences had it been more widely known.

   On 28 April 1944, James Sterling Murray, the security chief at the Metallurgical Laboratory in Chicago, had become so 'highly suspicious' of the chemist, Clarence Hiskey, that he had him drafted to the 'rather undesirable assignment' of the Yukon Territory to serve as 'property survey master' (unofficially 'to count underwear'). Murray's suspicions were well founded. Hiskey had worked at two atomic research sites, the

Substitute Alloy Material Lab (SAM) at Columbia and the University of Chicago's DSM Laboratory. Born Clarence Szczechowski in Milwaukee, Wisconsin, Hiskey earned a PhD in chemistry from the University of Wisconsin in 1939 after which he became director of the Rhenium Research Project at the University of Tennessee (rhenium is a very rare element, important strategically for use in jet engines and rockets). At Tennessee, Hiskey 'was remembered as an outspoken supporter of communism'. He also mixed at party gatherings in the San Francisco area.

Not long after he had become an instructor at Columbia in September 1941, Hiskey had met a veteran Soviet spy, Arthur Adams, at a left-wing music store. A year later he had joined Columbia's SAM Lab, which was where the Manhattan Project was developing its gaseous diffusion process for separating U-235. At around this time he had admitted to an undercover NKVD agent, Franklin Zelman (a 'plant' posing as a post-doctoral student), that he was working on a 'radio-active bomb' that could destroy the city of New York. Hiskey did not know that Zelman was an agent at the time, and tried to withdraw his remarks, though he was plainly conflicted, expressing also his 'hopes that the Soviets knew all about his project'. Zelman maintained contact with Hiskey, passing reports to Zarubin that contained some of the earliest atomic information the Russians received.

In October 1943 Hiskey moved to Chicago's metallurgical lab, which as we have seen was working on the large-scale production of plutonium for a bomb. In Chicago he met up again with Adams. Adams was an accomplished Soviet spy, who had several cover jobs.

Hiskey's premises were raided by the FBI – this would have been in May 1944, after he had been sent to the Yukon – where seven pages of notes on DSM were discovered and a notebook on the SAM lab. Adams's home was also broken into later, on 25 September 1944, and was found to contain copious notes on nuclear matters, including mention of the Oak Ridge plant, material on isotope separation, Norway's production of heavy water and 'speculations' about uranium salt and sources of radium and uranium in Czechoslovakia, Germany and Sweden. His notes, the FBI concluded 'reflect an intimate knowledge concerning highly secret phases' of the DSM project. Adams was judged 'the most dangerous espionage agent yet discovered'.

But it is the timing that interests us. The FBI kept Adams under surveillance and had done since 1941 and it was his meetings with Hiskey in Chicago, from October 1943 onwards, that attracted the suspicions of James Sterling Murray. This was yet another instance of intelligence being collected but not shared, and again it was in the crucial period, autumn 1943–summer 1944. We cannot expect the FBI to have informed Bohr but R. V. Jones was the chief of British scientific intelligence and if he had known about Hiskey, and his relationship with Adams, and since his relationship with Churchill was so strong, and since he felt as Bohr did on nuclear matters, then here was yet more information to lay before the prime minister. See Katherine A. S. Sibley, *Red Spies in America, Stolen Secrets and the Dawn of the Cold War*, Kansas City: University Press of Kansas, 2004, pp. 157–60.

9. Then there was the worrisome state of the prime minister's health. At sixty-nine, he had suffered a serious illness at the end of 1943 which had left him 'tired, yawning in meetings, bereft of his usual drive, feuding acrimoniously with his ministers and Chiefs of Staff'. His House of Commons performances and broadcasts were not what they had been and Alexander Cadogan, the permanent under-secretary for foreign affairs, confessed to his diary that spring that he doubted whether Churchill could continue. Farmelo, op. cit., p. 263.
10. Farmelo, op. cit., p. 262.
11. Pais, op. cit., pp. 500–501.
12. Farmelo, op. cit., p. 266.
13. Jones, *Most Secret War*, p. 477.
14. Farmelo, op. cit., p. 267.
15. Jones, *Most Secret War*, p. 417.
16. Rhodes, *Dark Sun*, p. 111.
17. Klaus Fuchs FBI File, part 16, pp. 17ff.

Chapter 23. The Bomb in Trouble: Mistakes at Los Alamos

1. Farmelo, op. cit., p. 261; Jones, op. cit., p. 476.
2. Rhodes, *Dark Sun*, p. 112.
3. Moore, op. cit., p. 347.
4. Moss, op. cit., p. 55.
5. Klaus Fuchs FBI File, Part 28, p. 44.
6. Rhodes, *Dark Sun*, p. 116; Weinstein and Vassiliev, op. cit., p. 200. Albright and Kunstel, op. cit., p. 70.
7. Sherwin, op. cit., p. 109; CHAD IV 12/5.
8. Sherwin, op. cit., p. 109.
9. Ibid., p. 105.
10. Rhodes, *Dark Sun*, p. 113.
11. Moss, op. cit., p. 60.

Chapter 24. The President's Mistake

1. Farmelo, op. cit., p. 270.
2. Sherwin, op. cit., p. 111.
3. Ibid., p. 110; Brown, op. cit., p. 269, fn.
4. Farmelo, op. cit., p. 270ff.
5. And he added, pointedly, 'I do not know whether you realize that the possibilities of a super weapon ... have been publicly discussed for at least six or seven years.' In other words, secrecy might buy time, though the basic knowledge about fission was 'out there'. Farmelo, op. cit., p. 273.
6. Pais, op. cit., p. 503.

7. Farmelo, op. cit., p. 273. Bush, who, Cherwell noted, was silent at this remark, agreed to 'check up' on Bohr. Nothing came of it.
8. Martin Sherwin, however, argues that the Hyde Park agreement has been overshadowed by the more dramatic events of the Yalta Conference (of February 1945, which determined the military occupation zones of post-war Germany and the shape of post-war liberated Europe). The Hyde Park agreement confirmed the 'special relationship' that Churchill had worked so hard for, and the specific and original meaning of which is too often oversimplified or not thought through. Ten days after the meeting, Roosevelt wrote to Cordell Hull, his secretary of state: 'The real nub of the situation is to keep Britain from going into complete bankruptcy at the end of the war. I just cannot go along with the idea of seeing the British empire collapse financially, and Germany at the same time building up a potential re-armament machine to make another war possible in twenty years.' Sherwin, op. cit., p. 113.
9. Ibid.
10. Ibid., p. 114.

Chapter 25. Bohr and Stalin

1. Jones, op. cit., pp. 476–7.
2. Hewlett and Anderson, op. cit., p. 322.
3. Sherwin, op. cit., p. 120.
4. Alperovitz, op. cit., p. 187.
5. Ibid., pp. 188–90.
6. Ibid., pp. 188–9; DeGroot, op. cit., p. 72 and Barton Bernstein, 'The Atomic Bomb Reconsidered', *Foreign Affairs*, January/February 1995.
7. It appears that some scientists were unsure of the difference between a 'military demonstration' and 'full' use, which was another option they were presented with.

   Szilard also devised a petition of his own, which sixty-nine scientists at Chicago signed, calling for the surrender terms to be announced publicly in advance, and the Japanese given the chance to respond, before any weapon was used. Szilard tried to circulate this petition at other Manhattan Project sites, but was prevented from doing so by Groves. Szilard gave the petition as it was to Compton, who passed it on to Groves, who held on to it for six days before passing it to Stimson's office, just as Truman was about to leave Potsdam for the voyage home (during which time Hiroshima was bombed).
8. NA RG 200, Box 3, Folder 1, p. 130; DeGroot, op. cit., p. 73.
9. Sherwin, op. cit., p. 124.
10. Hewlett and Anderson, op. cit., p. 329.
11. Ibid.
12. This particularly applied to a hydrogen bomb since the supplies of heavy hydrogen were 'essentially unlimited'. Hewlett and Anderson, op. cit., p. 329.

13. Malloy, op. cit., pp. 150–51 and 153.
14. Hewlett and Anderson, op. cit., p. 342.
15. Sherwin, op. cit., p. 201.
16. Ruane, op. cit., p. 98.
17. Laucht, op. cit., p. 75. He was so impressed by Fuchs that he requested he stay on. See also: Gregg Herken, *The Winning Weapon: The Atomic Bomb in the Cold War 1945–1950*, New York: Vintage, 1982.
18. See Sherwin, op. cit., p. 139, for the difficulties.
19. Ruane, op. cit., p. 101.
20. Ibid.
21. Ibid.
22. Ibid., p.104. As Richard Rhodes sums it up: 'If Stalin needed evidence that the nations that called themselves his allies were colluding against him to deny him nuclear weapons while they built up an arsenal, Donald Maclean could supply it. Someone did; a discussion of the "question of the existence and reserves of uranium deposits" and who controlled them turned up in a general NKVD review of Anglo-American bomb development that went to Beria on February 28, 1945.' By then Groves had reviewed more than 67,000 reports, more than half in foreign languages, which described the whereabouts of uranium deposits around the world. Rhodes, *Dark Sun*, p. 130.
23. DeGroot, op. cit., p. 55.
24. Hewlett and Anderson, op. cit., p. 344.
25. Sherwin, op. cit., p. 151.
26. Ibid., pp. 161 and 170; Brown, op. cit., p. 294.
27. Sherwin, op. cit., p. 178; Malloy, op. cit., p. 34.
28. Hewlett and Anderson, op. cit., p. 344.
29. Ibid., p. 346.
30. Sherwin, op. cit., p. 215.
31. Gordin, op. cit., pp. 5 and 44.
32. Ibid., pp. 50–52.
33. Bernstein, 'Reconsidering the "Atomic General"', p. 907.
34. Gordin, op. cit. pp. 53–4.
35. Bernstein, 'Reconsidering the "Atomic General"', p. 907.
36. Gordin, op. cit., pp. 55–6. Before the surrender came, General Marshall even considered using up to nine atomic bombs to support the invasion then set for November. This also confirms the 'nonchalant' attitude to radioactivity (not to say naivety). That too would have been suicidal on a mass scale. Gordin, op. cit., pp. 5 and 101. The circumstances surrounding the run-up to the use of the bomb against Japan were introduced in the Preface. From our point of view, equally as important as the actual dropping of the bombs were the twin issues of an Anglo-American monopoly and the arithmetic of international control. Bohr's central point was that peaceful coexistence could only be achieved if there was a *balance* between the powers. But how was that to be achieved if there were *two* Western Allies and one Soviet nation?

As we shall see, this arithmetic would come to matter very much, to Churchill and the British more than to the Americans, but the monopoly and balance arguments in themselves were not helped by the prime minister's actions in the immediate wake of the end of the war in Europe. Inside a week, he had issued an instruction to the Chiefs of Staff to plan for a surprise Anglo-American military assault *on* the USSR. This became known as Operation Unthinkable, where the planning called for ten German divisions forced to support forty-seven American and British divisions, in a vast advance with a provisional launch date of 1 July 1945.

The whole idea was indeed unthinkable – fantastic, unrealisable and even unhinged – and the Chiefs of Staff told Churchill as much in no uncertain terms. Moreover, Stalin got wind of what Churchill was up to and, together with Truman's 'nuclear-fuelled cockiness' at Potsdam, Operation Unthinkable created suspicion and bitterness among the former Allies when the war against Japan still had some weeks to run. See: 'Operation Unthinkable, Churchill's plan to start WWIII', *Russia and India Report*, 13 June 2013.

37. Ruane, op. cit., p. 142.
38. Alperovitz, op. cit., p. 663.
39. CHAD IV, 3/1.
40. Ruane, op. cit., pp. 133–4.
41. Laucht, op. cit., pp. 40–41.
42. Alperovitz, op. cit., p. 386.
43. Michael D. Gordin, *Five Days in August: How World War II Became a Nuclear War*, Princeton and Oxford: Princeton University Press, 2015, pp. 117–18.
44. Rhodes, *Dark Sun*, p. 176.
45. Ibid.
46. Ibid., p. 177.
47. Ibid., p. 179.
48. Simon Sebag Montefiore, *Stalin: The Court of the Red Tsar*, London: Weidenfeld & Nicolson, 2007, p. 443.
49. Gordin, *Five Days*, p. 27.
50. Ruane, op. cit., p. 146.
51. Sherwin, op. cit., pp. 195–7.
52. Alperovitz, p. 434.
53. Ibid., p. 435.
54. Ibid., p. 353.
55. Ibid., p. 654.
56. Ibid., p. 654, fn.
57. Gordin, *Red Cloud*, p. 234.
58. Bernstein, 'Reconsidering the "Atomic General"', pp. 683–920.
59. Alperovitz, op. cit., p. 591.
60. Ibid., p. 480.
61. Farmelo, op. cit., p. 312.
62. Groves, op. cit., pp. 350–51.

63. Powers, op. cit., p. 118.
64. Francis Simon Papers, Royal Society Archive: Folder FS/7/4/6/39.
65. Powers, op. cit., p. 508.
66. 24 May 1991, p. 18.
67. Holloway, op. cit., p. 113.
68. Ibid., pp. 118–19.
69. Abraham Pais, *Niels Bohr's Times: In Physics, Philosophy and Polity*, Oxford: Clarendon Press, 1991, pp. 499–500.
70. Holloway, op. cit., p. 370.
71. Sebag Montefiore, op. cit., pp. 3–4.
72. Ibid., p. 42.
73. Ibid., p. 373.
74. Ibid., p. 414.
75. Ibid., p. 429.
76. Ibid., p. 443.

Chapter 26. Fuchs: Shining in the Shadows

1. Herken, *Winning Weapon*, pp. 296–8.
2. The Russian response is discussed in ibid., pp. 82–94.
3. Bernstein, 'Reconsidering the "Atomic General"', p. 914.
4. Herken, *Winning Weapon*, see for example pp. 97–100.
5. Ibid., p. 110.
6. Ibid., p. 112, fn. See also Bernstein, 'Reconsidering the "Atomic General"', p. 915.
7. Herken, *Winning Weapon*, pp. 123–8. Strangely though, as Herken also points out, Groves never did much to stop Soviet spying, and when, in February 1946, the *New York Times* published an article on the widespread availability of uranium 'for those who are willing to pay a high price', Groves paid little heed. He himself was the source of a number of leaks about nuclear spies, all of which were aimed at keeping the 'atomic secret' from being shared with the Russians.
8. Herken, *Winning Weapon*, p. 227.
9. Ibid., pp. 230 and 157.
10. Ibid., pp. 168 and 273.
11. Ibid., pp. 232–3.
12. Ruane, op. cit., p. 173.
13. Ibid.
14. James Reston, *Deadline: A Memoir*, New York: Random House, 1991, p. 185.
15. Gordin, *Red Cloud*, p. 66.
16. Groves, op. cit., p. 409. Even managers of the companies that had been involved in manufacturing the Manhattan Project sites (Tennessee Eastman, Du Pont) opted for four or five years, but only if they had German help (which of course they did). DeGroot, op. cit., p. 133.
17. Weinstein and Vassiliev, op. cit., p. 316. For what Fuchs told the Russians

about Britain's bomb, see West and Tsarev, op. cit., p. 242. He also told them that America had exhausted its supplies of plutonium after Nagasaki but in 1946 was planning a stockpile of 125 units and that Britain's medium-term plan was a stockpile of 200 units by 1957. West and Tsarev, op. cit., pp. 243–5.

18. Gordin, *Red Cloud*, p. 161.
19. Ibid., p. 18.
20. Burke, op. cit., 7 per cent.
21. Herken, *Winning Weapon*, pp. 229–34. See also, Gordin, *Red Cloud*, p. 70.
22. Gordin, *Red Cloud*, op. cit., pp. 71–3.
23. Ibid., p. 75.
24. Herken, *Winning Weapon*, pp. 186 and 224.
25. Ibid., p. 126.
26. Gordin, *Red Cloud*, p. 78. In October 1946 the Office of Reports and Estimates chose ten years 'but only on a projection from known facts in the light of past experience and conjecture'. Ibid., p. 83. On the German view, see *Operation Epsilon*, op. cit., p. 92.
27. Burke, op. cit., 7 per cent.
28. Ibid., 9 per cent.
29. Gordin, *Red Cloud*, p. 238.
30. Ibid., p. 87.
31. Ibid., p. 240.
32. Ibid., pp. 248–9.
33. Ibid., p. 175.
34. Ibid., p. 179.
35. Ibid., p. 220. Groves's comments are in: NA RG 200, Box 1, NN-366-108, Folder 7.
36. Russian claims that he saved them six years are generally discounted. Laucht, op. cit., p. 92.
37. Gordin, *Red Cloud*, p. 132.
38. Rhodes, *Dark Sun*, p. 138.
39. Ibid., pp. 152–4. Albright et al., op. cit., p. 127.
40. Laucht, op. cit., p. 91; Albright et al., op. cit., p. 125; NA RG 200, Box 3, folder 7.
41. Rhodes, *Dark Sun*, pp. 158–9.
42. Ibid., p. 168; West and Tsarev, op. cit., p. 250.
43. Gordin, *Red Cloud*, pp. 164 and 168.
44. Ibid., p. 117.
45. During this time he attended some meetings on nuclear subjects where a fellow committee member was Donald Maclean, one of the notorious 'Cambridge five' spy ring, though neither knew of the other's espionage activities just then. See also Weinstein and Vassiliev, op. cit., p. 209.
46. Gordin, *Red Cloud*, p. 180.
47. Reston, op. cit., p. 216. In fact, we now know that Bradley's worry was overcooked. The CIA predicted ten to twenty Soviet bombs by mid-1950

and up to 135 by mid-1953. In fact, Russia had nine bombs in 1950 and, by May 1953, when Stalin died, no more than fifty. Again, no one knew that at the time. Gordin, *Red Cloud*, p. 259.

48. Trachtenberg, op. cit., p. 7.
49. Lord Moran's diary, *Winston Churchill: Struggle for Survival 1940–65*, London: Constable, 1966, p. 315 but also pp. 505 and 545. Churchill was not entirely alone in his belligerence: Christoph Laucht points out that it was the British (rather than the American) government that sought confrontation with the Russians between 1945 and 1947. Laucht, op. cit., p. 99.
50. 'How Soon will Russia have the A-bomb?', *Saturday Evening Post*, 6 November 1948, p. 182. This idea of a 'preventive war' was a live issue at think tanks like RAND and even in the State Department, where Charles Bohlen and George Kennan, senior diplomats who both served as ambassador to Moscow, worried about what would happen if matters 'were allowed to drift'. It might even be preferable, they suggested, to provoke Russia into opening a conflict, when the US could 'retaliate' with nuclear weapons.
51. Trachtenberg, op. cit., p. 11.
52. This was a coherent strategy, up to a point, depending as it did on Russia being deterred by the *number* of bombs the US possessed. But even this was a more complex matter than it may appear in retrospect. This is because, in 1949 and 1950, as Trachtenberg makes clear, it was not taken for granted, as it is in a post H-bomb world, that all-out nuclear war would mean the destruction of entire societies. Two official studies at the time, for example, made it clear that the initial 'atomic blitz' could not be counted on (as Churchill for one appeared to believe) to destroy the war-making power of the Soviet Union. In like manner, a Soviet atomic attack on the United States in the early 1950s would have had only a 'limited effect' on the American war economy – a Soviet assault would not prevent the US from carrying out a major retaliation. The reason was that fission bombs in 1949–50 were, relatively speaking, of limited power. In 1947, in a chilling calculation, Edward Teller had pointed out that even if 'say a thousand or ten thousand' bombs were launched against America, 'many millions' would die 'but if certain elementary precautions were taken, the country as a whole could survive heavy atomic bombardments and go on to win the war'. (This certainly shows how thinking has changed over the years.) Trachtenberg, op. cit., p. 22.
53. Herken, *Winning Weapon*, p. 330.
54. Ibid. p. 19.
55. Ibid., p. 21.
56. Ibid., p. 24.
57. Ibid., p. 320.
58. Ibid., p. 26.
59. Ibid., pp. 334–6.
60. DeGroot, op. cit., p. 187.

61. Bruce Cumings, *The Korean War: A History*, New York: Modern Library, 2011, pp. 156 and 162.
62. Ibid., p. 158.
63. The Indian Treaty Room in the White House went very quiet. 'The topic that had never been considered appropriate for a press conference had suddenly become the focal point.' David McCullough, *Truman*, New York: Simon & Schuster, 1992, p. 812.
64. Ibid., p. 832.
65. Ibid., p. 856.
66. Ruane, op. cit., p. 190.
67. Ibid., p. 191.
68. Ibid., p. 205 and ref.
69. DeGroot, op. cit., p. 188. In 1956 a curious book was published, originally in German, by A. M. Biew, which argued that in fact Kapitsa was given 'absolute control' over the development of nuclear weapons and that he 'invented' the Russian hydrogen bomb. But the status of this book is open to question. It included direct dialogue between Kapitsa and Stalin (across a glass table), conversations between the author and Ioffe, direct quotes of Beria and many others. Many colleagues have contradicted this and have drawn attention to Kapitsa's absence as director of the Institute for Physical Problems between 1946 and 1955 as due to his being out of favour for his independent attitudes and for having pursued some unacceptable lines of work during the war. As these critics also point out, if Kapitsa was removed after helping to start the Soviet project, Kurchatov's programme seemed not to have been delayed as a result. Kapitsa himself denied working on nuclear physics. See Lawrence Badash, *Kapitsa, Rutherford and the Kremlin*, New Haven and London, Yale University Press, 1985, p. 112. For Kapitsa's relations with Beria, see J. W. Bong (compiler) et al. (eds), *Kapitsa and Cambridge and Moscow: The Life and Letters of a Russian Physicist*, Amsterdam: Elsevier Science, 1990.

Index

Credit: Bert Hulsemans

Peter Watson is a journalist, television presenter and historian of intellectual movements. He has written for the *Observer*, the *Daily Telegraph*, *The Times*, *The Sunday Times*, the *New York Times* and the *Spectator*. His books include *The Modern Mind; Ideas: A History from Fire to Freud;* and *The German Genius; The Age of Atheists;* and *Convergence*—and he has been published in twenty-six countries.

PublicAffairs is a publishing house founded in 1997. It is a tribute to the standards, values, and flair of three persons who have served as mentors to countless reporters, writers, editors, and book people of all kinds, including me.

I. F. STONE, proprietor of *I. F. Stone's Weekly*, combined a commitment to the First Amendment with entrepreneurial zeal and reporting skill and became one of the great independent journalists in American history. At the age of eighty, Izzy published *The Trial of Socrates*, which was a national bestseller. He wrote the book after he taught himself ancient Greek.

BENJAMIN C. BRADLEE was for nearly thirty years the charismatic editorial leader of *The Washington Post*. It was Ben who gave the *Post* the range and courage to pursue such historic issues as Watergate. He supported his reporters with a tenacity that made them fearless and it is no accident that so many became authors of influential, best-selling books.

ROBERT L. BERNSTEIN, the chief executive of Random House for more than a quarter century, guided one of the nation's premier publishing houses. Bob was personally responsible for many books of political dissent and argument that challenged tyranny around the globe. He is also the founder and longtime chair of Human Rights Watch, one of the most respected human rights organizations in the world.

• • •

For fifty years, the banner of Public Affairs Press was carried by its owner Morris B. Schnapper, who published Gandhi, Nasser, Toynbee, Truman, and about 1,500 other authors. In 1983, Schnapper was described by *The Washington Post* as "a redoubtable gadfly." His legacy will endure in the books to come.

Peter Osnos, *Founder*